Traditional Chinese Architecture Surveying and Mapping Series:
Religious Architecture

BA-SHU BUDDHIST ARCHITECTURE

Compiled by School of Architecture and Urban Planning, Chongqing University
Edited by ZHANG Xingguo, FENG Di, GUO Xuan, WANG Zhiyang

China Architecture & Building Press

国家出版基金项目
NATIONAL PUBLICATION FOUNDATION

『十二五』国家重点图书出版规划项目

中国古建筑测绘大系·宗教建筑

巴蜀佛寺

重庆大学建筑城规学院 编写
张兴国 冯 棣 郭 璇 汪智洋 主编

中国建筑工业出版社

Editorial Board of the Traditional Chinese Architecture Surveying and Mapping Series

『中国古建筑测绘大系』编委会

Contents

目录

Introduction

The School of Architecture and Urban Planning of Chongqing University was the former Department of Architecture in Chongqing Construction Engineering College. Since the 1950s, the faculty and students of the Department carried out a series of survey and research of the ancient buildings in the Ba-Shu region and southwestern China along with their teaching programs. For decades, over 200 ancient buildings of different types have been surveyed, which have become an important documentation for cultural heritage conservation. This book is an edited publication of the Ba-Shu Buddhist architecture.

Buddhist culture was introduced into the Ba-Shu region in the Eastern Han Dynasty (AD25-200), when depiction of Buddha appeared on carved stones and bricks in the Ba-Shu region. In Tang Dynasty (AD618-907), large numbers of Buddhist architectural carvings were seen on Ba-Shu cliff carvings. Because of the special historic and climatic environment of the Ba-Shu region, very few timber structure of the Ming Dynasty (AD1386-1644) has survived, and those from Yuan Dynasty (AD1271-1368) are very rare. On the other hand, reconstructed and new built Buddhist buildings since the Qing Dynasty (AD1636-1912) are found all over the Ba-Shu region. Since the Ming and Qing dynasties, Ba-Shu Buddhist architecture has developed distinctive regional and diversified characteristics, and several representative surveys of Buddhist temples and monasteries of different styles have been selected here.

导言

重庆大学建筑城规学院建筑系原重庆建筑工程学院建筑系，从20世纪50年代以来，建筑系的师生结合教学实践，对巴蜀地区以及西南地区的古建筑进行了系列的测绘研究，几十年来先后测绘不同类型的古建筑200余处，成为文化遗产保护的重要文献档案资料。本书编辑出版的是巴蜀佛教建筑类型。

佛教在东汉时期已传入巴蜀地区。东汉时期的巴蜀画像、石画像砖已有佛像的描绘，唐代的巴蜀摩崖石刻更有大量的佛教建筑雕刻形象。由于巴蜀地区特殊的历史环境和气候环境，明代的木构建筑遗存极少，元代的木构建筑遗存已是罕见，清代以来重建和新建的佛教建筑却遍布巴蜀大地。明清以来的巴蜀佛教建筑已有明显的地域化和多元化特色，这里选择了几种不同风格特色的代表性佛教寺院测绘资料。

平武报恩寺兴建于明代，是典型的大木大式建筑，具有浓郁的北方官式建筑风格，应是遗存的孤例。梁平双桂堂的空间布局有明代以来南方禅宗寺院的遗风，但建筑构架、建筑风格形态已有明显的地域特色，应是巴蜀禅宗寺院布局的典型例证。峨眉山是我国四大佛教名山之一，历史文化悠久，佛教寺院结合山地环境，创造出气势宏大而有灵气的佛教建筑群落，可谓达到地域文化特色的最高境界。潼南大佛寺则是伴随着大佛造像出现的摩崖式楼阁殿堂，这种建筑类型在巴蜀大地遗存较多，大足圣寿寺的大悲殿等都属于这种类型，成为巴蜀摩崖佛寺建筑的一大特色。

The Bao'en Temple in Pingwu County, built in the Ming Dynasty, is a typical timber structure with profound characteristics of the northern official architectural style, probably the sole existing example. The spatial layout of Shuangguitang Temple in Chongqing Liangping District shows features of southern Zen monasteries in the Ming Dynasty, but the architectural structure and style have apparent local characteristics, which should be considered a typical example of Ba-Shu Zen temple layout. Mount Emei is one of the four famous Buddhist mountains in China with a long history and rich culture. The Buddhist monastery formed a magnificent and spiritually enlightening architectural complex in harmony with the mountainous environment, which has arguably reached a supreme level of regional cultural characteristics. The Great Buddha Temple in Tongnan County is a cliff building developed with the emergence of huge Buddha statue creation. This type of building is frequently seen in the Ba-Shu region, with notable examples like Shengshou Temple in Chongqing Dazu District, which is celebrated as a unique representative of the Ba-Shu cliff temples.

图版

Drawings

峨眉山佛寺建筑群

峨眉山为我国著名的四大佛教名山之一，佛教传入的历史悠久，相传东晋时由智持大师最早在峨眉山建普贤寺，到晚唐时期佛教文化在峨眉山达到高潮，宋代是峨眉山佛教发展最盛的时期。由于历史的变迁，唐宋时期的佛寺建筑已不复存在，峨眉山现存最早的佛寺建筑有元代的飞来寺和明代的万年寺无量殿等，而更多的佛教寺庙是清代重修、复原重建或是新建。清代以来的巴蜀佛寺建筑从空间布局到建筑风貌都有浓郁的地方特色，建筑多采用大木小式而不用斗栱，空间组合多采用巴蜀地域特色的四合院。峨眉山现存的佛教建筑，更是适应险峻而优美的山地环境，利用筑台、出挑、吊脚等山地建筑手法，创建出独具魅力与气势的佛教寺庙建筑群。峨眉山的佛寺建筑多采用抬梁构架与穿斗构架组合的构筑方式，屋面材料多采用素筒瓦或小青瓦，建筑风格朴实大方，与山水环境和谐统一；适应多雨雾的气候环境，一些寺庙建筑还大胆地采用锡和铅等金属材料作为屋面覆盖构件，构成了峨眉山又一独特的建筑风格。

Buddhist Temple Complex of Mount Emei

Mount Emei is one of the four famous Buddhist mountains in China, well known for its long history. It is said that, in the Eastern Jin Dynasty, Master Zhichi first built a Samantabhadra Temple on Mount Emei. By the late Tang Dynasty, Buddhist culture reached its climax on Mount Emei, and the Song Dynasty saw the most prosperous period of Buddhist development on Mount Emei. Yet with the change of time, the Buddhist temples of the Tang and Song dynasties no longer exist. The earliest existing Buddhist temples on Mount Emei include the Feilai Temple from Yuan Dynasty and the Wuliang Hall in Wannian Temple from Ming Dynasty among others, with many more restored, reconstructed, or new built in Qing Dynasty. Since the Qing Dynasty, Ba-Shu Buddhist temples show rich local characteristics in from the spatial layout to the architectural style. Buildings are built with the *Xiaoshi* style without bracket sets, and the spatial combination is more often that of the four-courtyard with Ba-Shu regional characteristics. The existing Buddhist buildings on Mount Emei are well adapted to the steep and beautiful mountainous environment, with mountain architectural features like terrace, overhang and stilt, which create a unique charming and magnificent Buddhist temple complex. Most Buddhist temples on Mount Emei are built with a combined timber structure of overlaying beams and tying braces. The roof is often covered with plain imbrex tiles or flat grey tiles. The architectural style is simple and unadorned, and in harmony with the landscape. For adaptation to the rainy and foggy climate, metals such as tin and lead are boldly used on the roof in some temples, forming another unique architectural style of Mount Emei.

Baoguo Temple

Baoguo Temple is located on Fenghuangping of Mount Emei, at an altitude of 533m. First built in the yi-mao year of Wanli period in Ming Dynasty (AD1615), the temple was originally situated beside Fuhu Temple. In the 9th years of Shunzhi's reign of Qing Dynasty (AD1652), the building was moved to Fenghuangping, reconstructed during Jiaqing reign, and expanded in the 5th year of Tongzhi's reign (AD1866). There are five main buildings on the central axis successively: the entrance gate, the Maitreya Hall, the Great Buddha Hall, the Hall of Seven Buddhas, and the Samantabhadra Hall. The structures, well adapted to the terrain, stand on gradually rising terraces with spatial layouts that are both regular and flexible. The entrance gate is on the central axis, with a slanted orientation following the *feng shui* principle. The Maitreya Hall and the Great Buddha Hall create a huge courtyard with the two sides halls, generating a solemn sense of symmetry and expansive space. The Hall of Seven Buddhas and the Samantabhadra Hall are located on terraces, in a transverse alignment parallel to the contour. Climbing the mountain stairs and looking up will evoke a feeling of astounding awe, thanks to the unique spatial treatment of Baoguo Temple in harmony with the mountainous environment.

报国寺

报国寺位于峨眉山的凤凰坪，海拔533米。始建于明万历四十三年（1615年），原址位于伏虎寺一侧。清顺治九年（1652年）迁建于凤凰坪，清嘉庆年间重建，清同治五年（1866年）进行了扩建。主要建筑有殿堂五重，在中轴线上依次是山门、弥勒殿、大雄宝殿、七佛宝殿、普贤殿。建筑结合地形逐渐升高分台布置，空间布局既严谨又灵活多变。山门在中轴线上而朝向适应风水环境而转折；弥勒殿和大雄宝殿与两侧配殿围合成巨型的四合院空间，具有庄严肃穆的对称感，又有宏阔的空间尺度感；七佛宝殿和普贤殿分别布置在高台上，建筑平行等高线呈一字形横向展开，通过爬山梯道登临仰望极具震撼力，这是报国寺建筑群利用山地环境的独特空间处理手法。

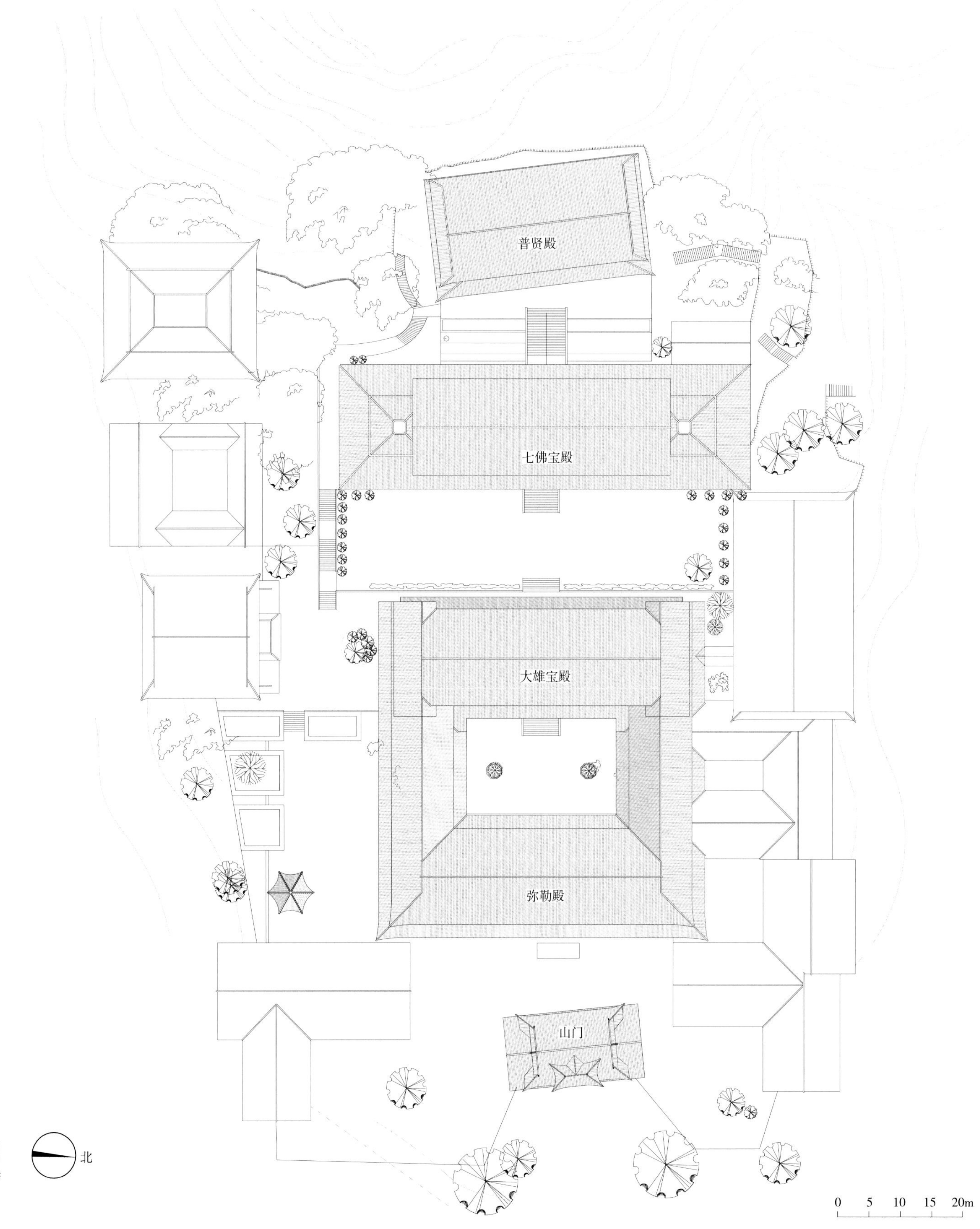
普贤殿
七佛宝殿
大雄宝殿
弥勒殿
山门
北
0 5 10 15 20m

报国寺总平面图
Site Plan of Baoguo Temple

大雄寶殿

图二　报国寺大雄宝殿

图一　报国寺鸟瞰图

報國寺

图三　报国寺山门

Fig.1 Bird's Eye View of Baoguo Temple
Fig.2 Mahavira Hall, Baoguo Temple
Fig.3 Main Gate (Shan Men), Baoguo Temple

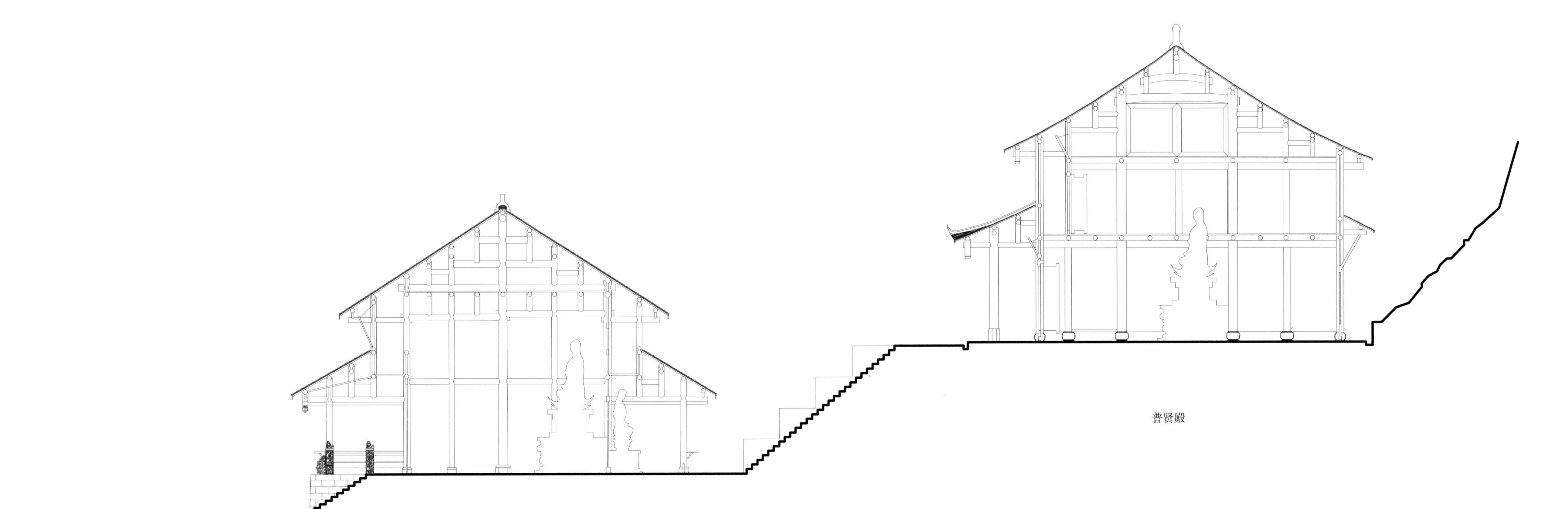
普贤殿
七佛宝殿

报国寺中轴剖面图
Cross Section on Axis of Baoguo Temple

0 5 10m

山门
弥勒殿
大雄宝殿

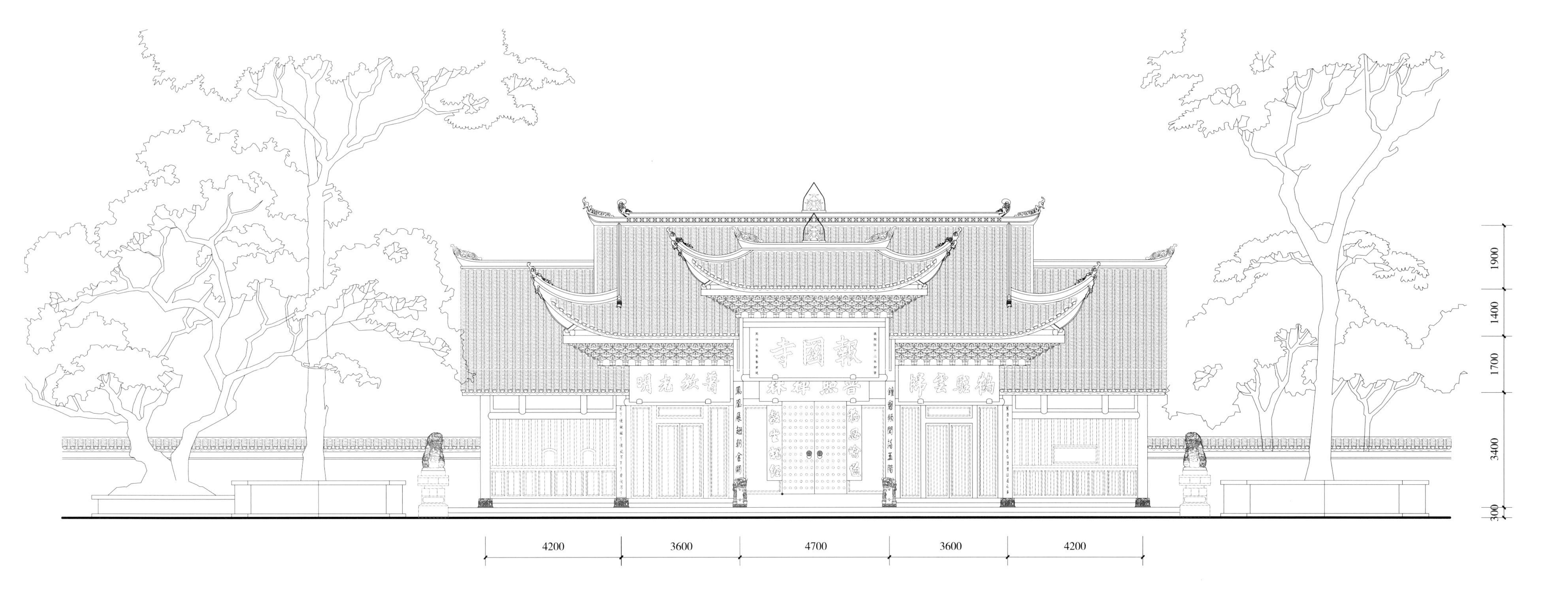
報國寺
1900
1400
1700
3400
300
4200
3600
4700
3600
4200

报国寺山门正立面图

Front Elevation of Main Gate (Shan Men), Baoguo Temple

0 1 2 3 4 5m

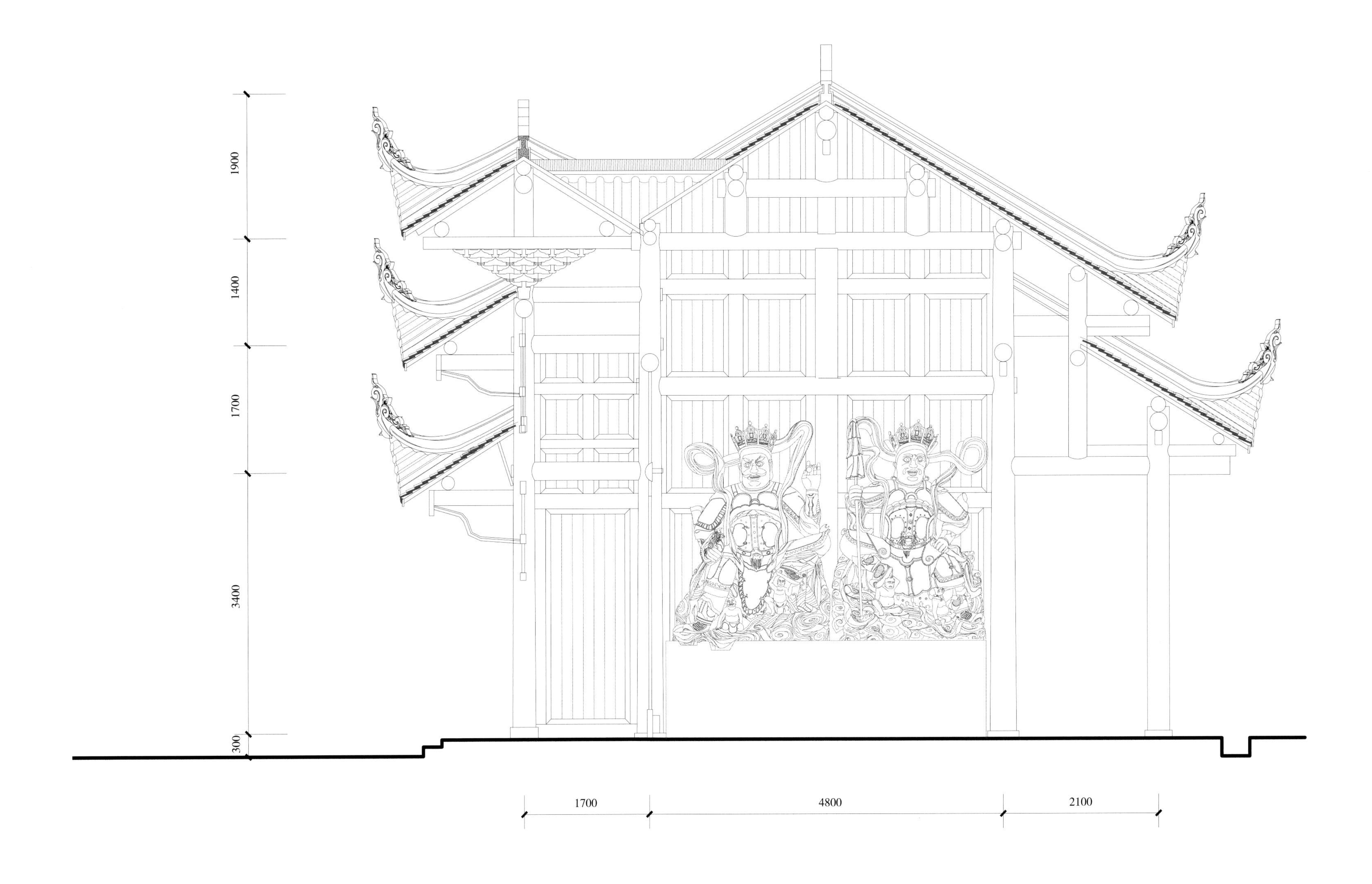
1900
1400
1700
3400
300
1700
4800
2100
0
1
2
3m

报国寺山门剖面图

Cross Section of Main Gate (Shan Men), Baoguo Temple

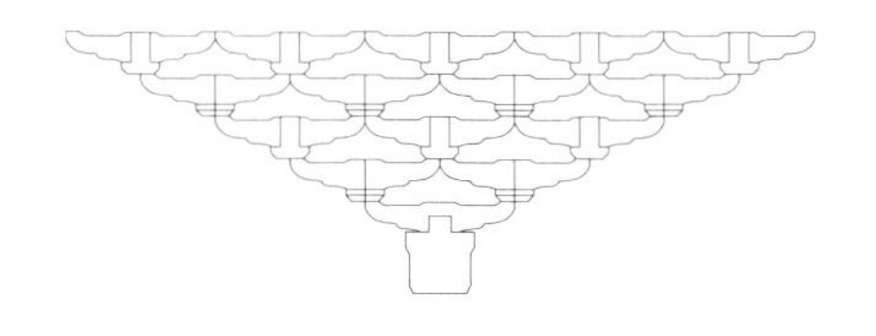

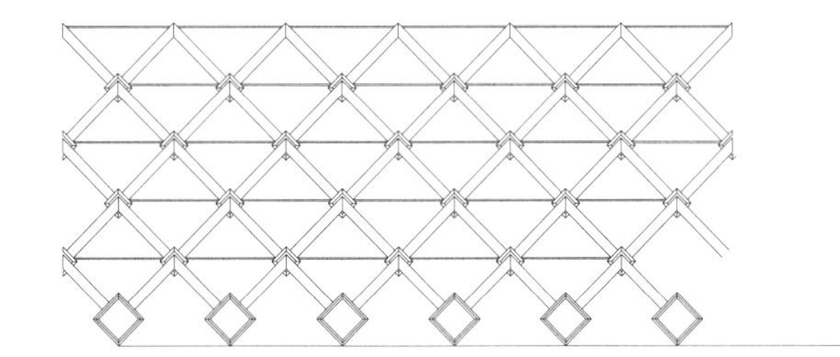

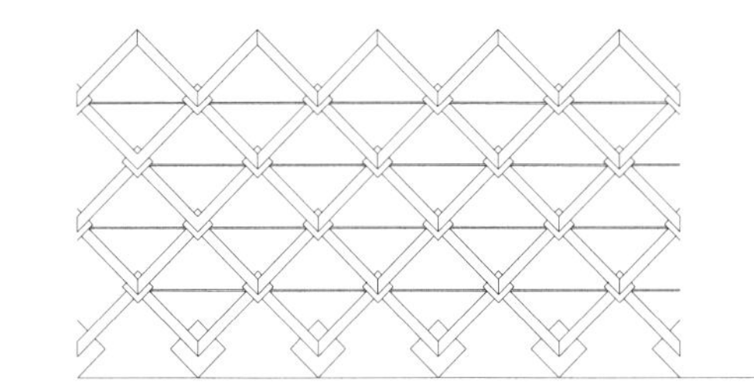

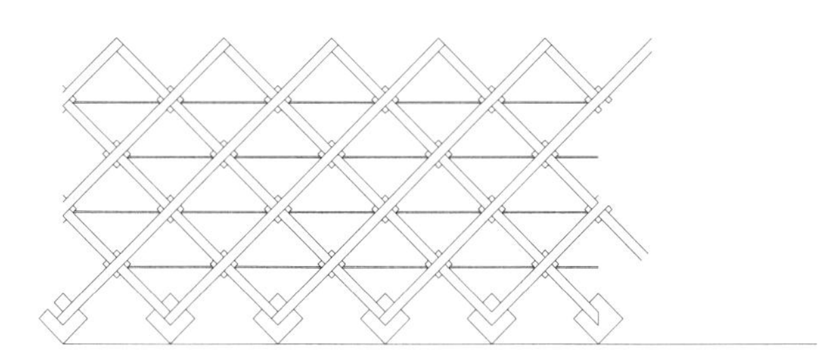

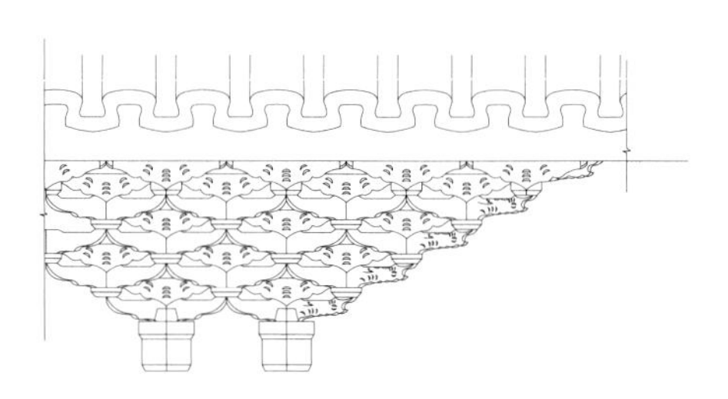

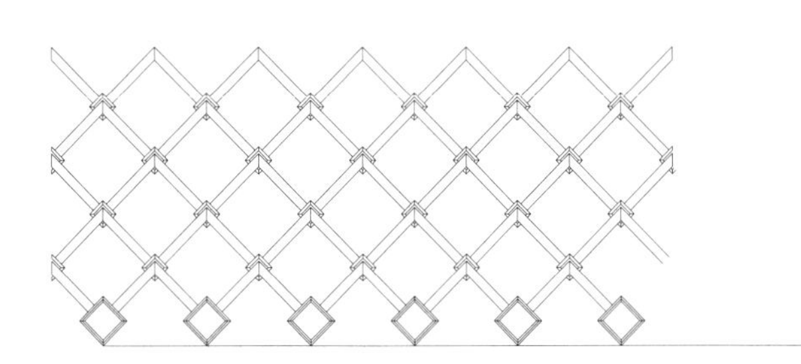

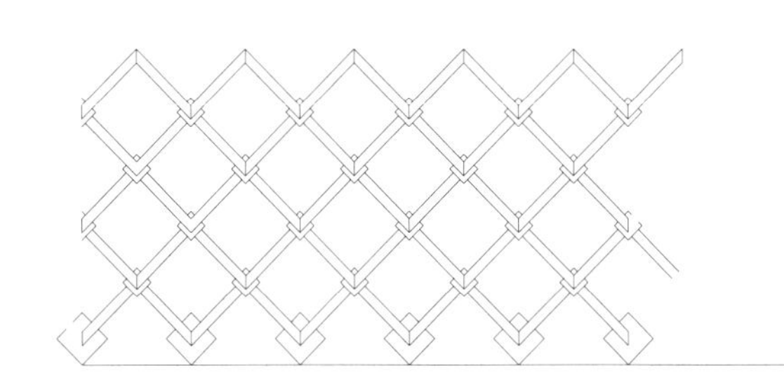

报国寺山门如意斗栱大样

Detail, Ruyi Dougong Sets, Main Gate (Shan Men), Baoguo Temple

0 1m

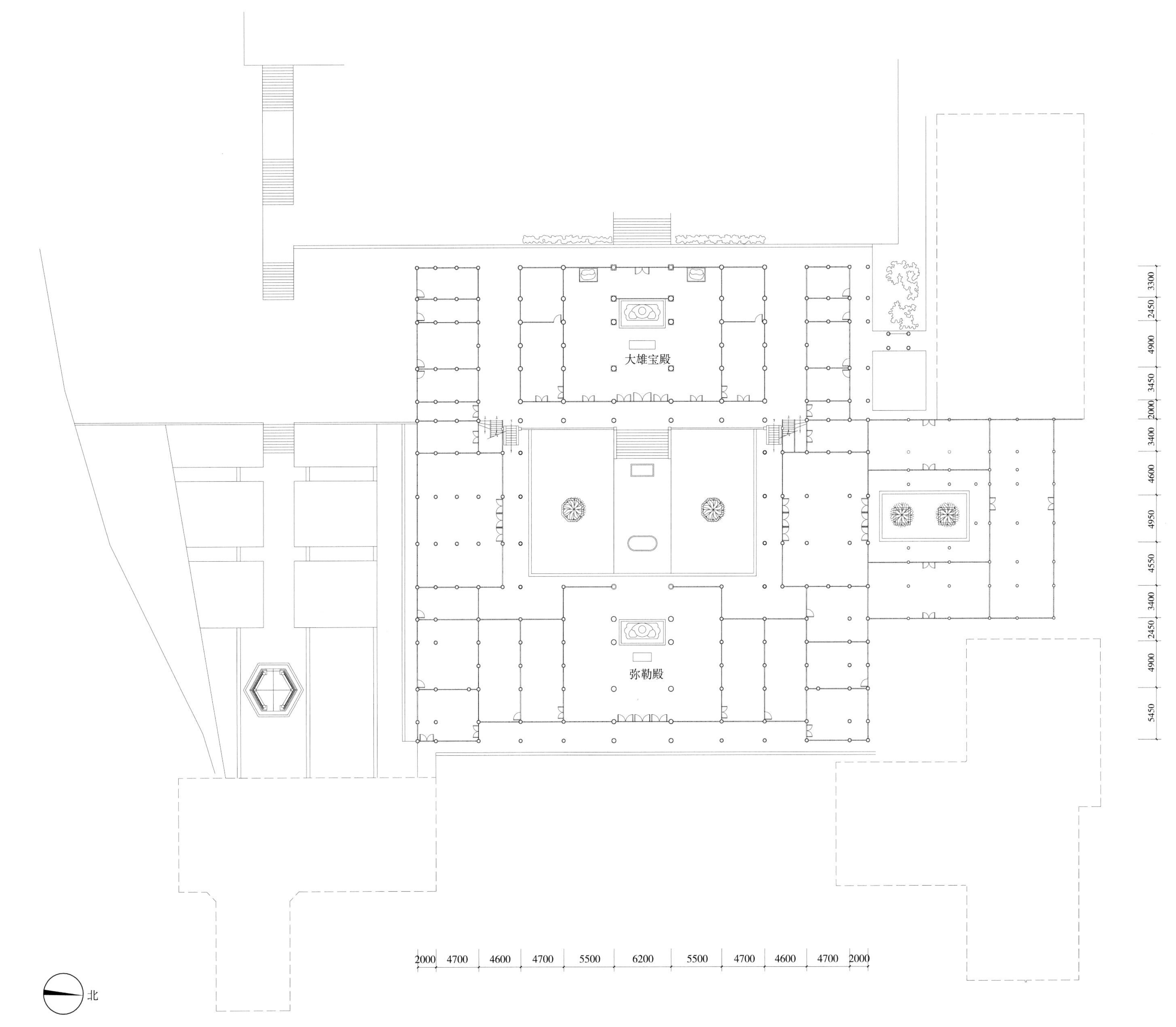
大雄宝殿
弥勒殿
3300
2450
4900
3450
2000
3400
4600
4950
4550
3400
2450
4900
5450
2000
4700
4600
4700
5500
6200
5500
4700
4600
4700
2000
北
0
5
10m

报国寺弥勒殿和大雄宝殿一层平面图
Ground Floor Plan of Maitreya Hall and Mahavira Hall, Baoguo Temple

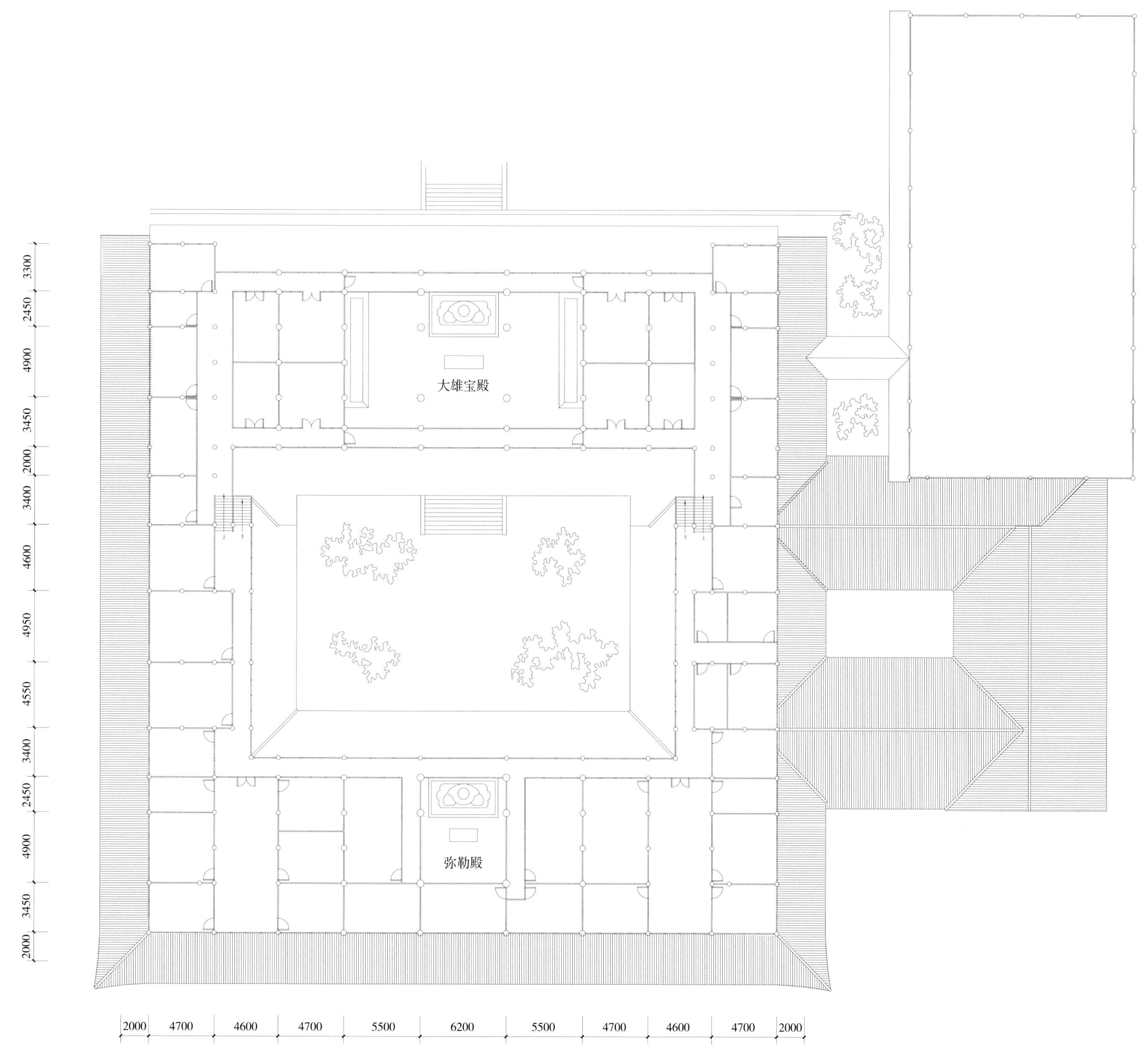

3300
2450
4900
3450
2000
3400
4600
4950
4550
3400
2450
4900
3450
2000
大雄宝殿
弥勒殿
2000
4700
4600
4700
5500
6200
5500
4700
4600
4700
2000

报国寺弥勒殿和大雄宝殿二层平面图
First Floor Plan of Maitreya Hall and Mahavira Hall, Baoguo Temple

0
5
10m

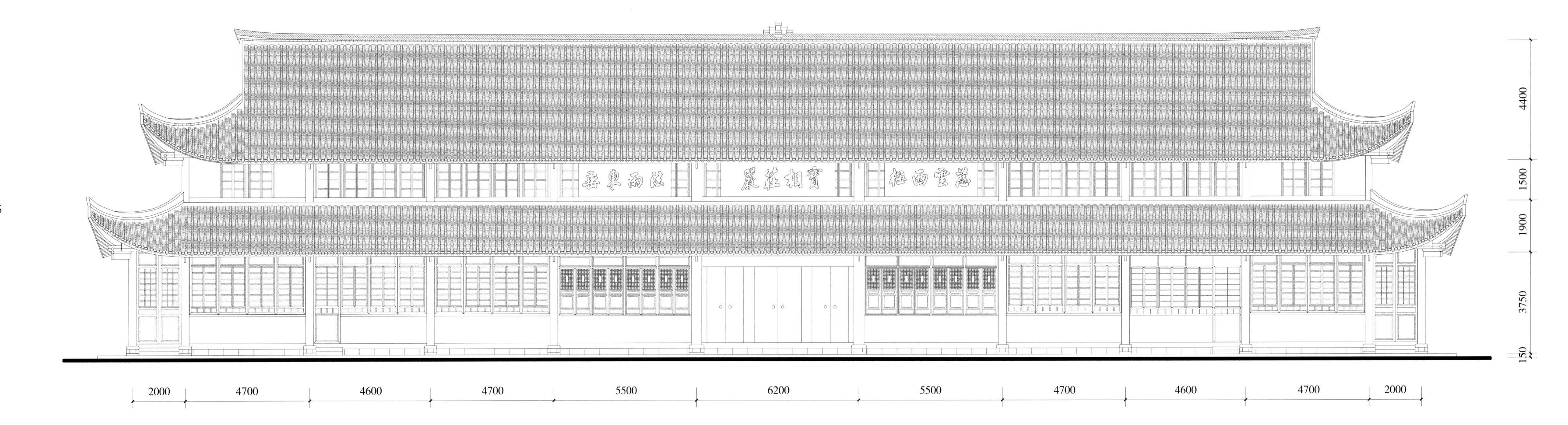
法雨東垂
寶相莊嚴
慈雲西極
4400
1500
1900
3750
150
2000
4700
4600
4700
5500
6200
5500
4700
4600
4700
2000

报国寺弥勒殿正立面图
Front Elevation of Maitreya Hall, Baoguo Temple

0 1 2 3 4 5m

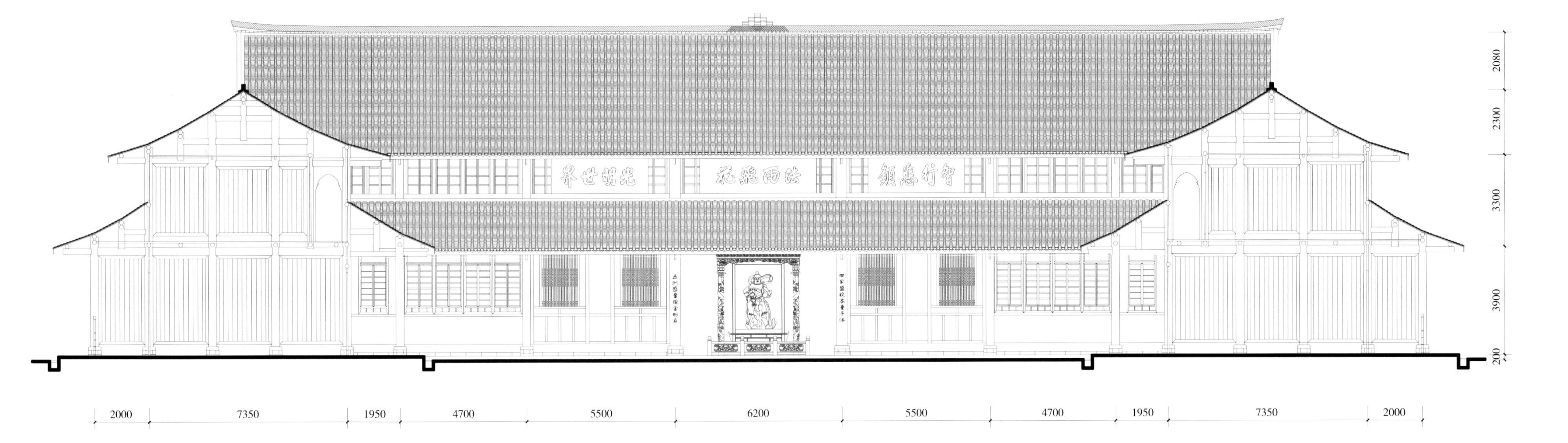
光明世界
法雨飛花
智行悲願
2080
2300
3300
3900
200
2000
7350
1950
4700
5500
6200
5500
4700
1950
7350
2000

报国寺弥勒殿背立面图
Back Elevation of Maitreya Hall, Baoguo Temple

0 1 2 3 4 5m

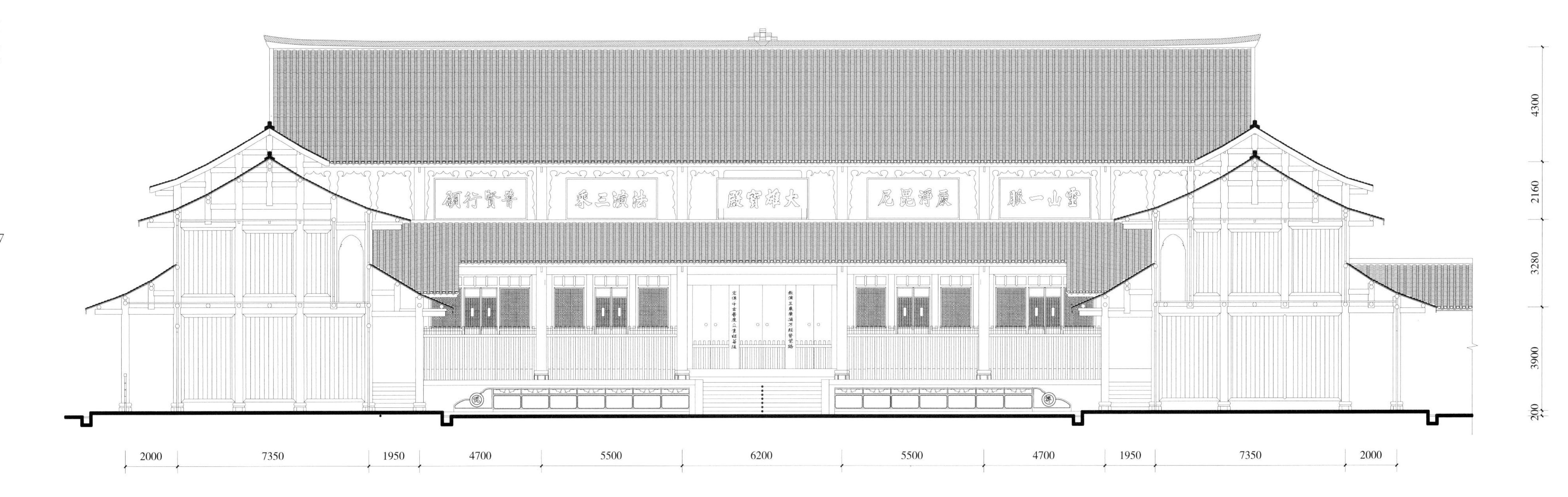
普賢行願
法演三乘
大雄寶殿
嚴淨毘尼
靈山一脈
4300
2160
3280
3900
200
2000
7350
1950
4700
5500
6200
5500
4700
1950
7350
2000

报国寺大雄宝殿正立面图
Front Elevation of Mahavira Hall, Baoguo Temple

0 1 2 3 4 5m

3000
2250
2150
700
4400
350
4700
4600
4700
5500
6200
5500
4700
4600
4700

报国寺大雄宝殿背立面图

Back Elevation of Mahavira Hall, Baoguo Temple

0 1 2 3 4 5m

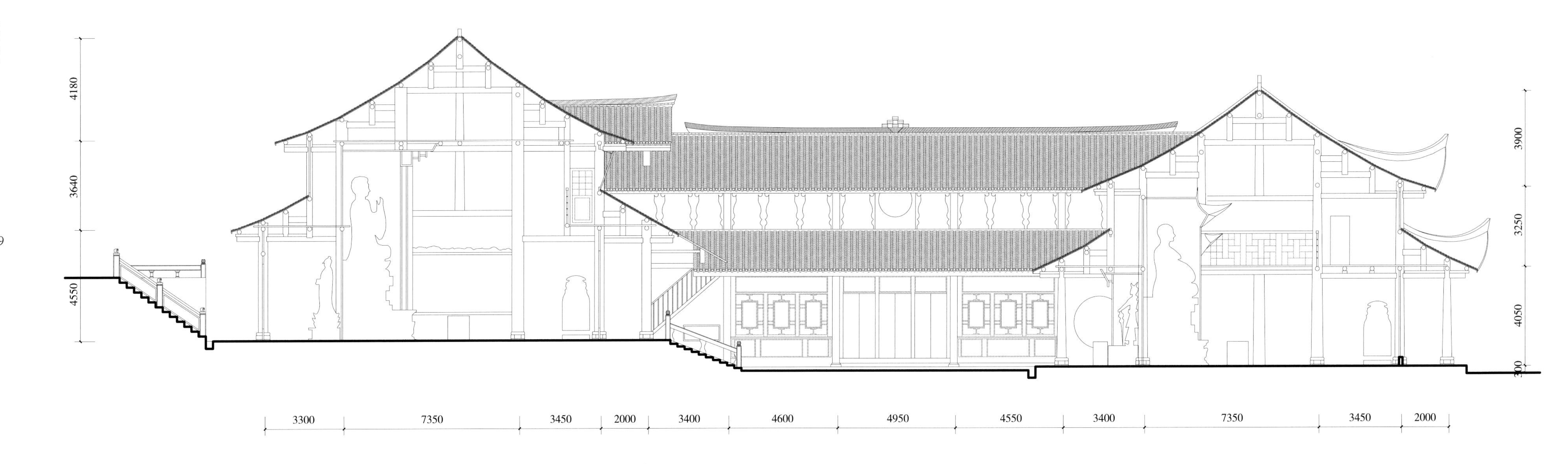
4180
3640
4550
3900
3250
4050
300
3300
7350
3450
2000
3400
4600
4950
4550
3400
7350
3450
2000

报国寺弥勒殿和大雄宝殿横剖面图

Cross Section of Maitreya Hall and Mahavira Hall, Baoguo Temple

0 1 2 3 4 5m

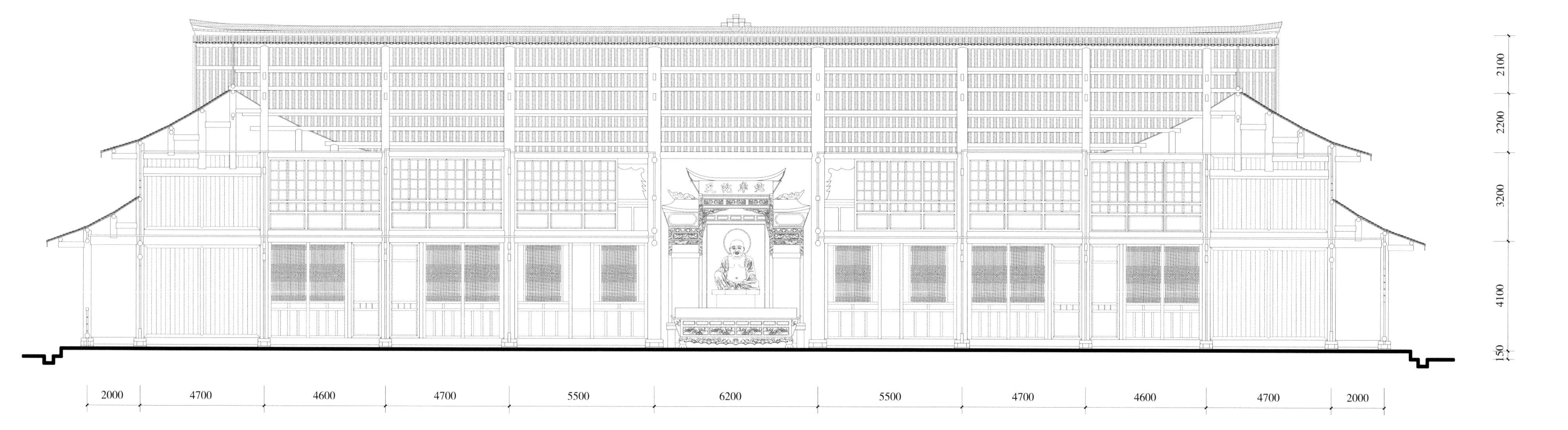
2100
2200
3200
4100
150
2000
4700
4600
4700
5500
6200
5500
4700
4600
4700
2000

报国寺弥勒殿纵剖面图
Longitudinal Section of Maitreya Hall, Baoguo Temple

0 1 2 3 4 5m

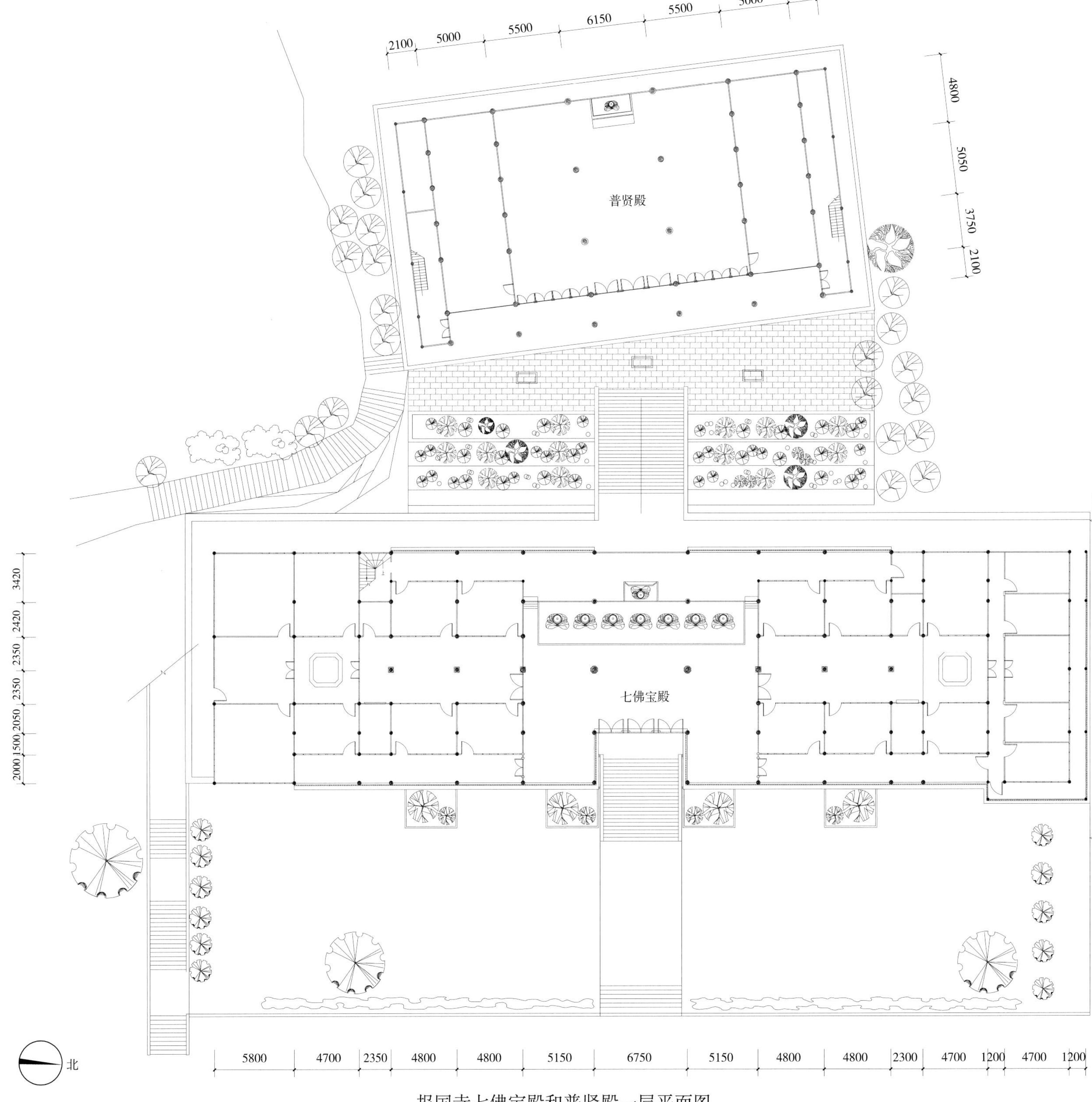
2100
5000
5500
6150
5500
5000
2100
4800
5050
3750
2100
普贤殿
七佛宝殿
3420
2420
2350
2350
2050
1500
2000
5800
4700
2350
4800
4800
5150
6750
5150
4800
4800
2300
4700
1200
4700
1200
北
0
5
10m

报国寺七佛宝殿和普贤殿一层平面图

Ground Floor Plan of Seven Buddhas Hall and Samantabhadra Hall

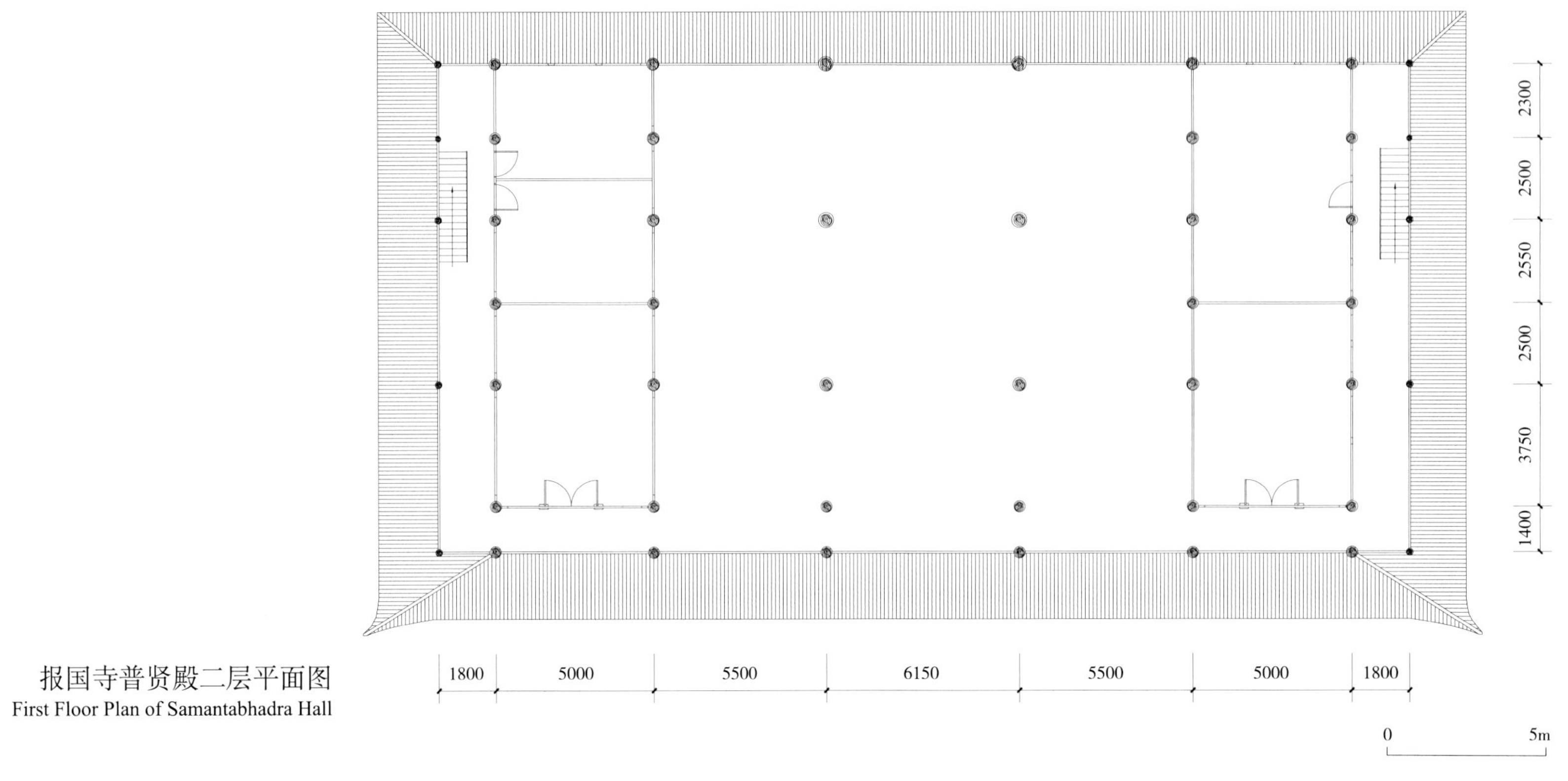
2300
2500
2550
2500
3750
1400
1800
5000
5500
6150
5500
5000
1800
0
5m

报国寺普贤殿二层平面图
First Floor Plan of Samantabhadra Hall

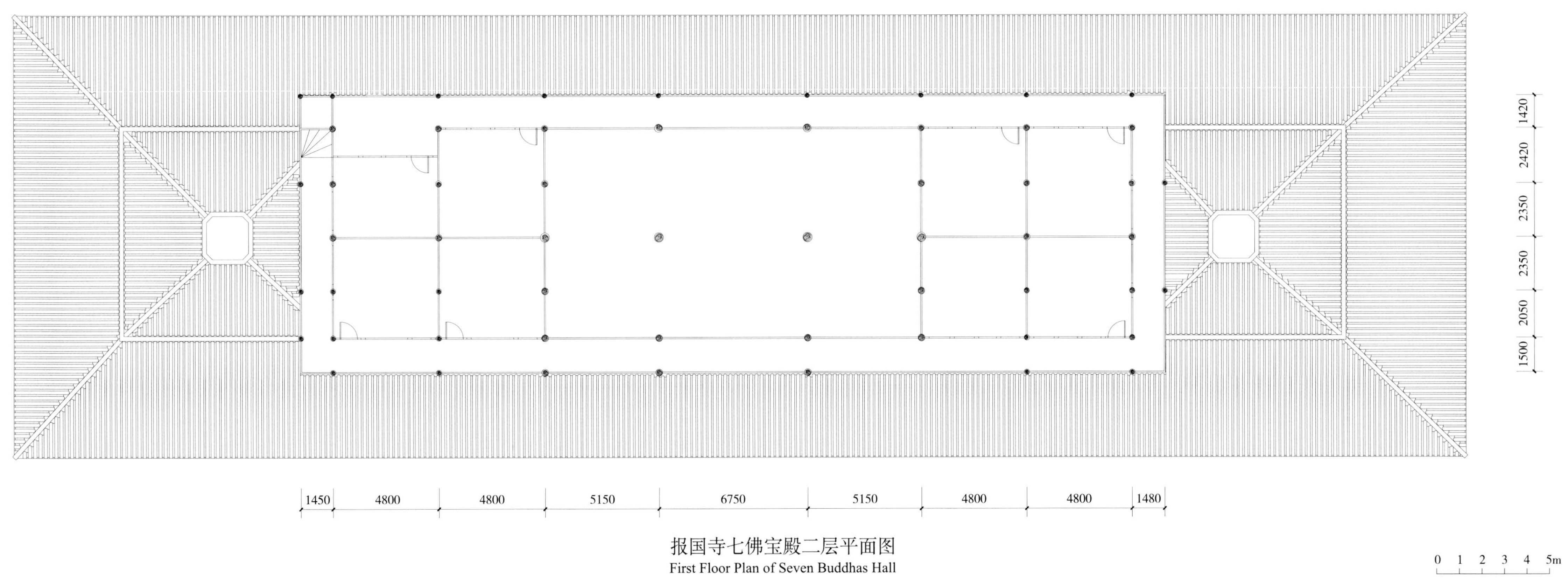
1420
2420
2350
2350
2050
1500
1450
4800
4800
5150
6750
5150
4800
4800
1480

报国寺七佛宝殿二层平面图
First Floor Plan of Seven Buddhas Hall

0 1 2 3 4 5m

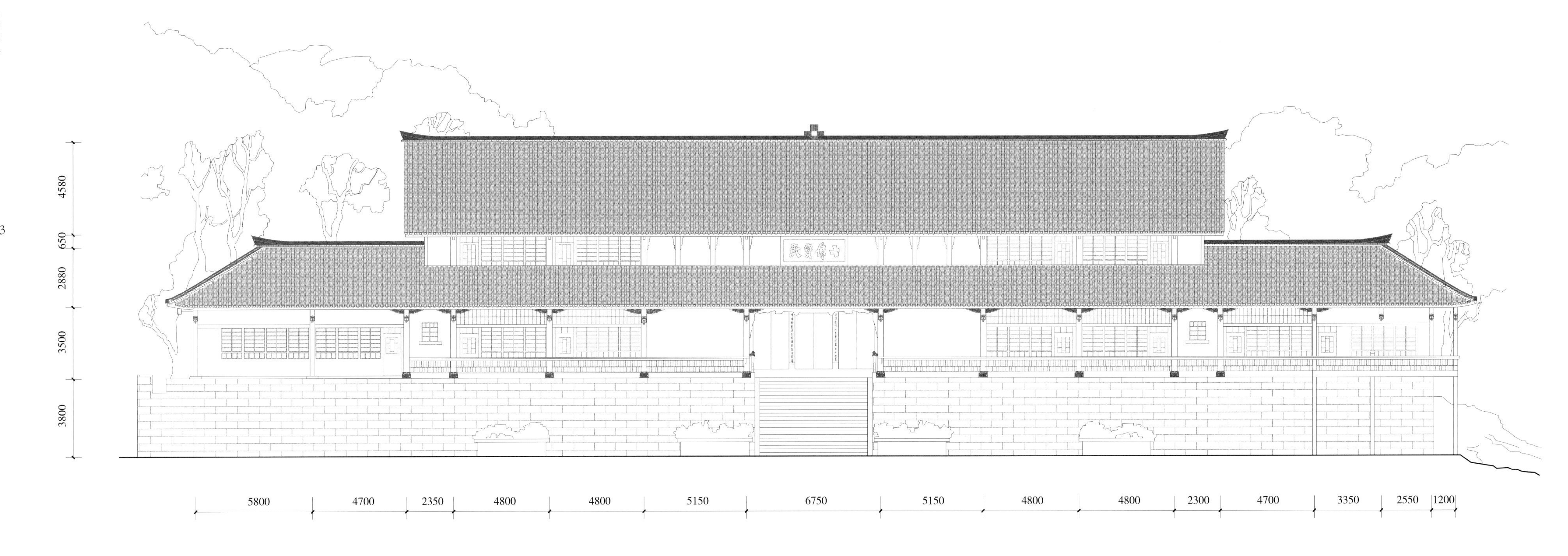
4580
650
2880
3500
3800
5800
4700
2350
4800
4800
5150
6750
5150
4800
4800
2300
4700
3350
2550
1200
0 1 2 3 4 5m

报国寺七佛宝殿正立面图
Front Elevation of Seven Buddhas Hall

4670
2230
1600
4350
5900
2100
5000
5500
6150
5500
5000
2100
0 1 2 3 4 5m

报国寺普贤殿正立面图
Front Elevation of Samantabhadra Hall

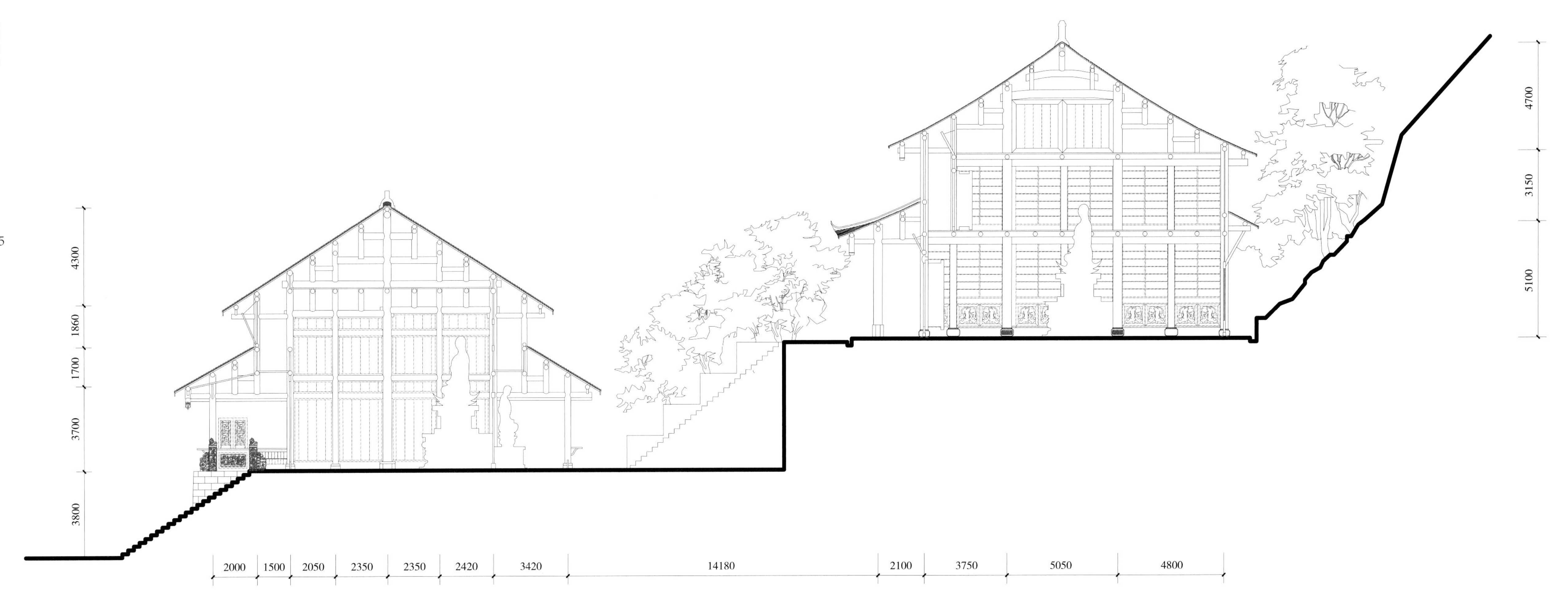

4300
1860
1700
3700
3800
2000
1500
2050
2350
2350
2420
3420
14180
2100
3750
5050
4800
4700
3150
5100

报国寺七佛宝殿和普贤殿横剖面图
Cross Section of Seven Buddhas Hall and Samantabhadra Hall

0 1 2 3 4 5m

伏虎寺

伏虎寺位于峨眉山麓伏虎岭下，海拔630米。唐代末叶已有伏虎寺的相关文献记载，又载南宋年间因镇虎患而得名伏虎寺，明崇祯年间毁于兵火，现存建筑为清顺治八年（1651年）重修。现存清代遗构主要由虎溪精舍和离垢园两进院落空间构成，建筑坐南朝北，分别布置在前后高差较大的台地上。虎溪精舍为三合院式建筑，离垢园为四合院式建筑，两进院落有明显的轴线对称秩序，中轴线上从低到高分别是弥勒殿、普贤殿、大雄宝殿，院落两侧是配殿和客堂等。大殿两侧有华严宝殿亭和御书楼。虎溪精舍是一组反向的三合院，正面为二层楼阁置于石砌台地上，呈一字形横向展开，显得宏伟壮观；从三合院拾级而上，是乾隆帝颁题的离垢园，这是一组规模宏大的四合院，院落入口正面和两侧厢房均为二层楼阁，而北面轴线尽端的大雄宝殿正面看似三层，实际上内部空间为一层。大雄宝殿通过筑台、退台和出挑吊脚等山地建筑手法，达到外部空间尺度与四合院整体和谐，同时殿内供奉释迦牟尼佛像，从而创造出宏阔的空间环境氛围。寺院周边古木参天，景色格外雄伟秀丽。

Fuhu Temple

Fuhu Temple is located at the foot Fuhu (literally, taming of tigers) Ridge on Mount Emei, at an altitude of 630m. Early records related to Fuhu Temple can be found in late Tang Dynasty literature, which was given its name due to the taming of tigers in Southern Song Dynasty (AD1127-1279). The temple was destroyed by warfare during Chongzhen periods in Ming Dynasty. The existing structure is a reconstruction in the 8th year of Shunzhi's reign in Qing Dynasty (AD1651), which is mainly composed of two courtyards: the Huxi Buddhist house and the Ligou Garden. These buildings, facing north, are arranged on terraces with notable height difference between the front and rear. The Huxi Buddhist house is surrounded by buildings on three sides, while the Ligou Garden is on four. The two courtyards have a distinctive symmetrical axial order, and on the central axis, from the lower level to the higher, are the Maitreya Hall, Samantabhadra Hall, and the Great Buddha Hall. On either side of the courtyard are side buildings and guest halls. The Pavilion of Huayan Hall and Imperial Library flank the Great Hall. The Huxi Buddhist house is a courtyard with a reverse orientation. In the front is a two-story pavilion erected on a stone platform, of which a transverse alignment gives it a magnificent appearance. Through stairs from this courtyard up is found the Ligou Garden with Emperor Qianlong's inscription. In this large courtyard, there are two-story buildings in the front and on the sides. The Great Buddha Hall at the north end of the axis appears to be three-story, which, in fact, has a single story interior. By means of terrace, setback, overhang, tilts and other mountain architectural features, the external spatial scale and the whole courtyard are well harmonized. The hall that accommodates the Sakyamuni statue creates an expansive spatial atmosphere. The ancient trees dotting the temple make up a magnificent and beautiful landscape.

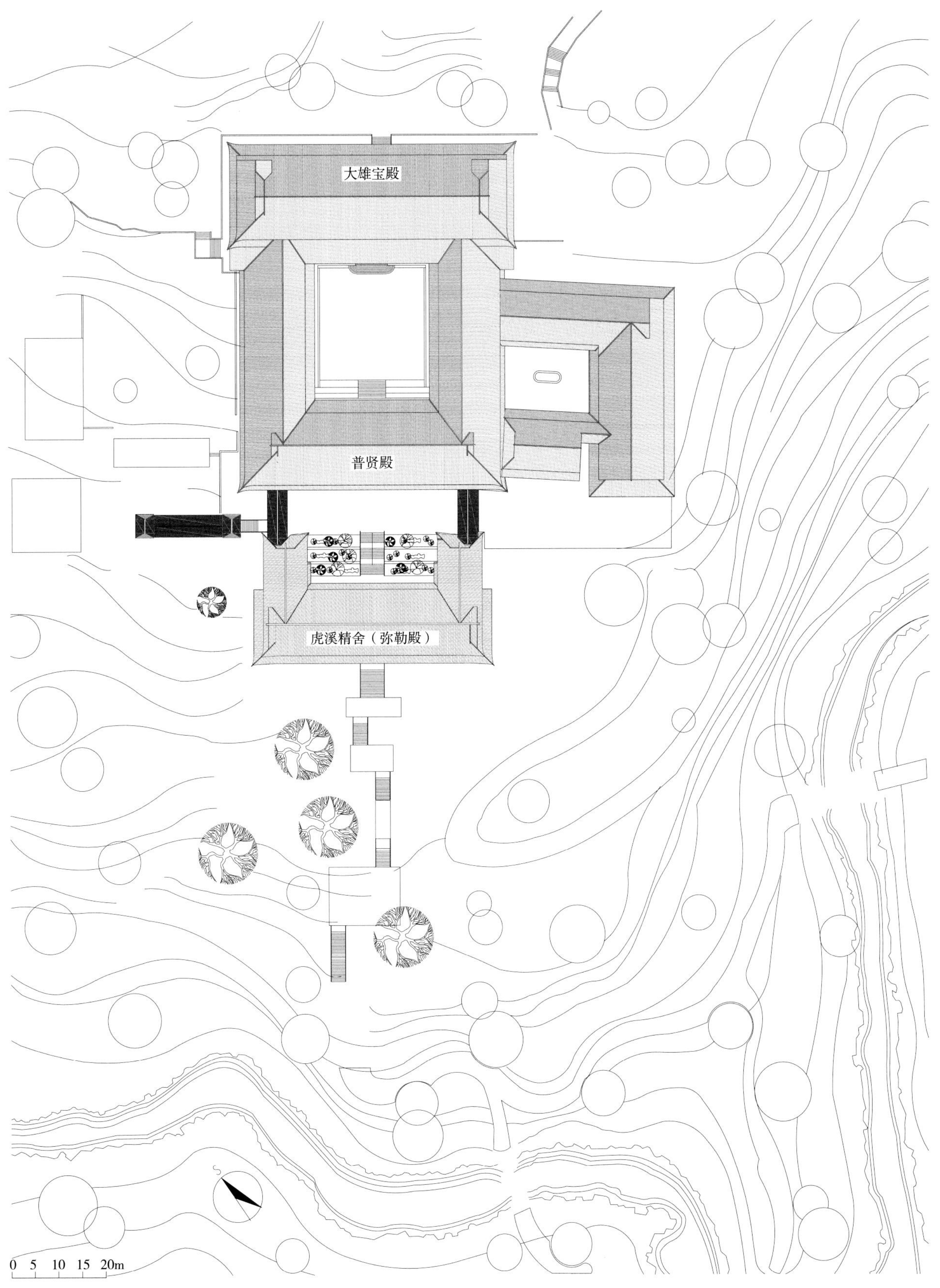
大雄宝殿
普贤殿
虎溪精舍（弥勒殿）
0 5 10 15 20m

伏虎寺总平面图
Site Plan of Fuhu Temple

图二 伏虎寺鸟瞰图

图一 伏虎寺鸟瞰图

Fig.1 Bird's Eye View of Fuhu Temple
Fig.2 Bird's Eye View of Fuhu Temple

图五　伏虎寺虎溪精舍（弥勒殿）

大雄寶殿

图三　伏虎寺大雄宝殿

图四　伏虎寺普贤殿

Fig.3 Mahavira Hall, Fuhu Temple
Fig.4 Samantabhadra Hall, Fuhu Temple
Fig.5 Huxi Jingshe (Maitreya Hall), Fuhu Temple

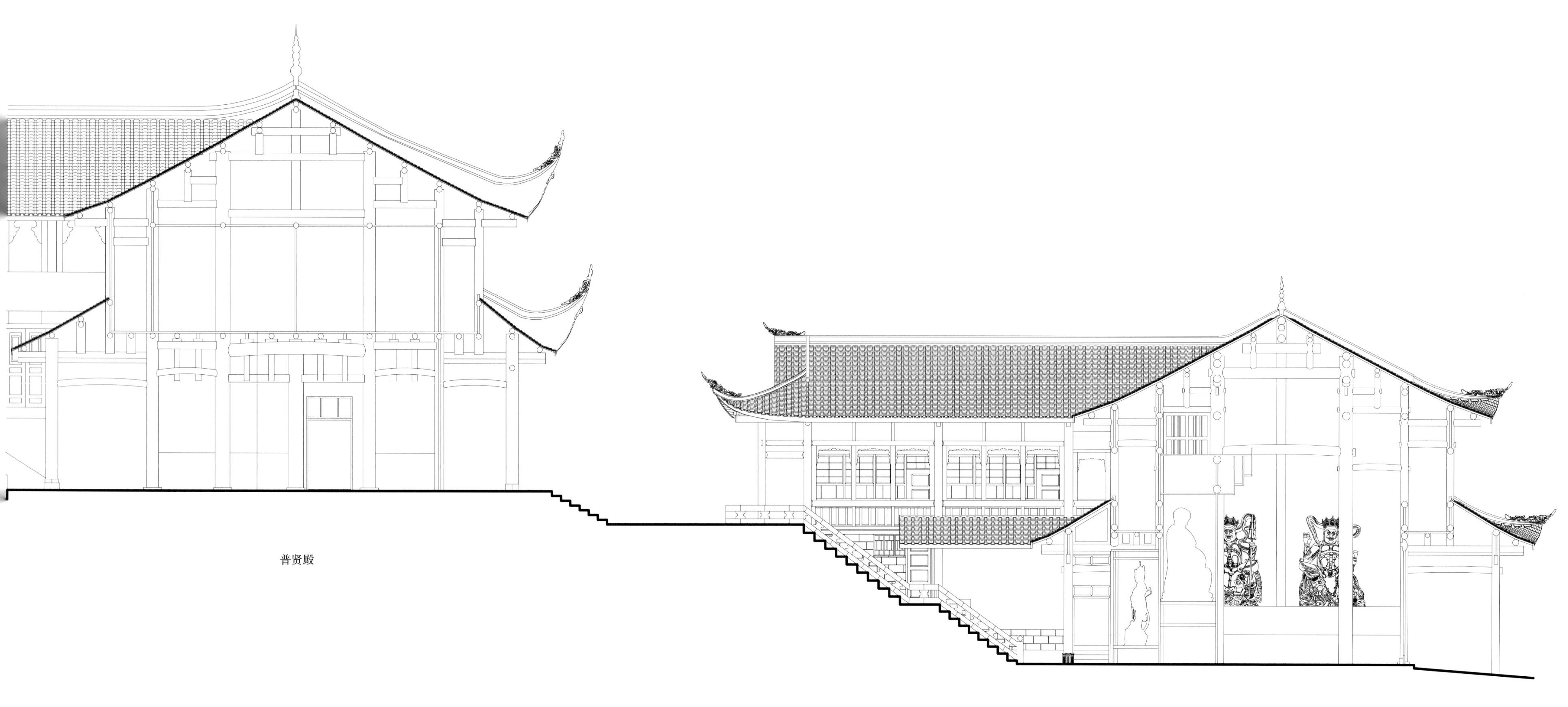
普贤殿
虎溪精舍（弥勒殿）
0
5
10m

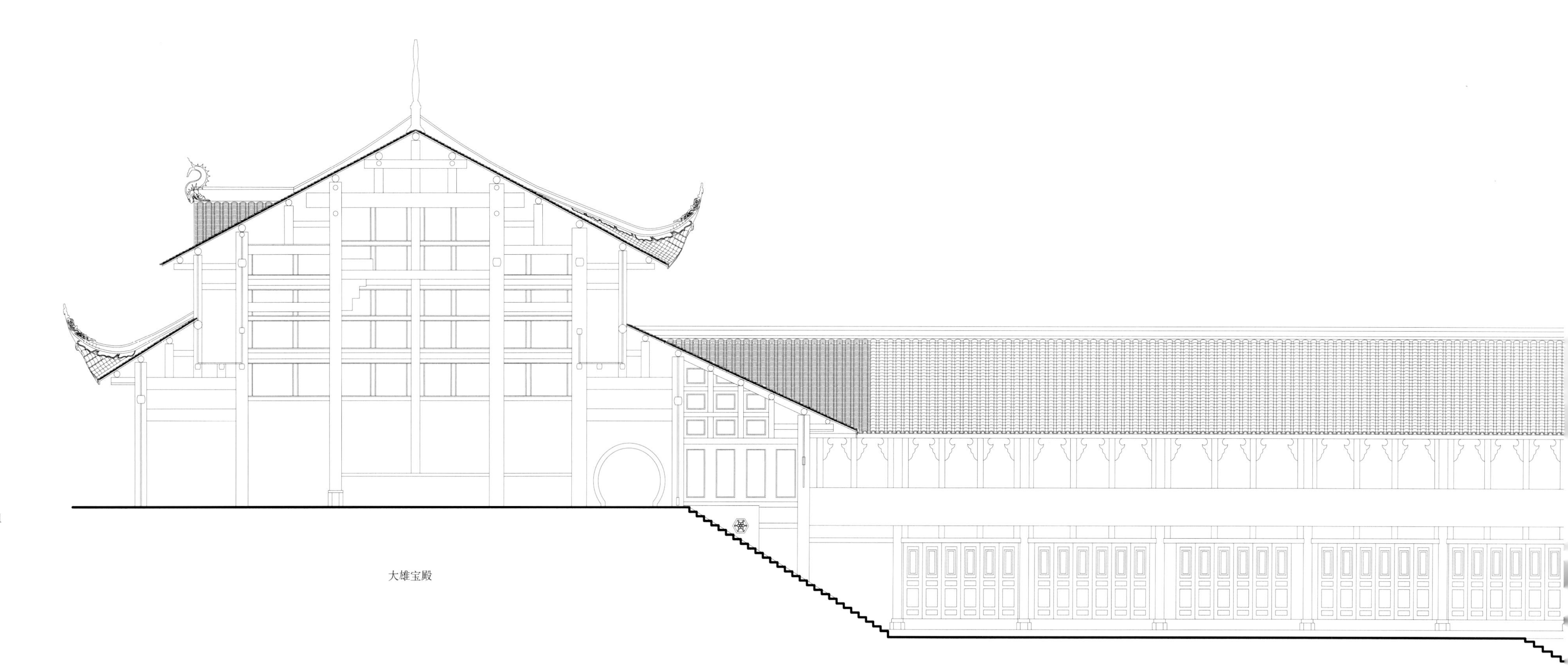
大雄宝殿

伏虎寺中轴剖面图
Cross Section on Axis of Fuhu Temple

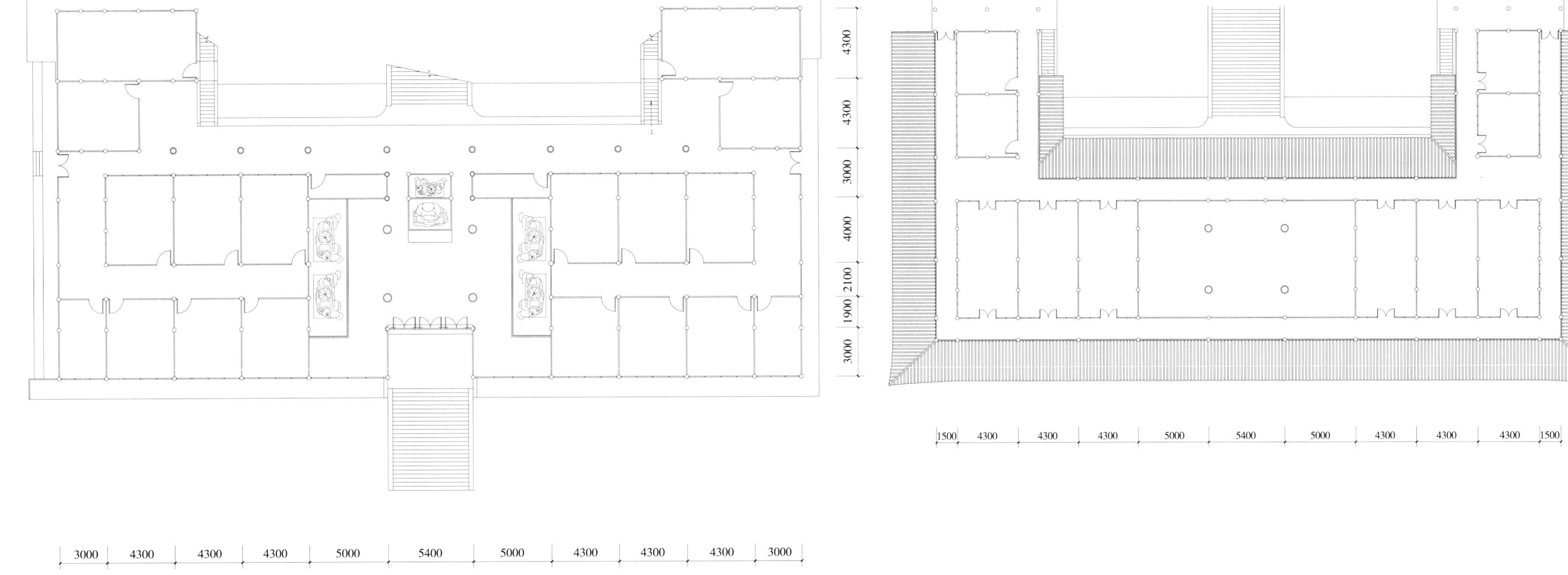

4300
4300
3000
4000
2100
1900
3000
3000 4300 4300 4300 5000 5400 5000 4300 4300 4300 3000
1500
4300
4300
3000
1900
2100
2100
1900
1500
1500 4300 4300 4300 5000 5400 5000 4300 4300 4300 1500

伏虎寺虎溪精舍（弥勒殿）一层平面图

Ground Floor Plan of Huxi Jingshe (Maitreya Hall), Fuhu Temple

0 1 2 3 4 5m

伏虎寺虎溪精舍（弥勒殿）二层平面图

First Floor Plan of Huxi Jingshe (Maitreya Hall), Fuhu Temple

0 1 2 3 4 5m

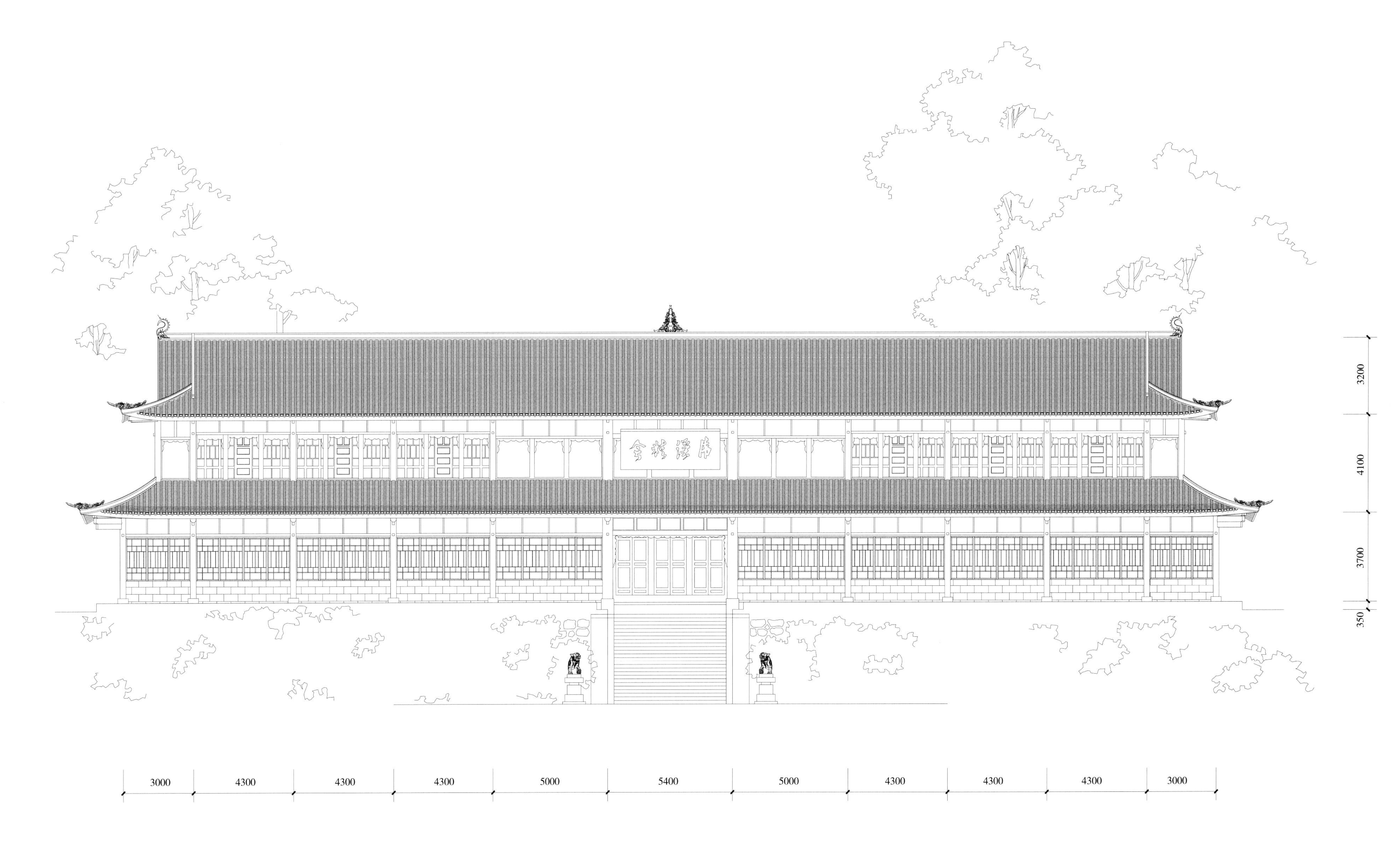

3200
4100
3700
350
3000
4300
4300
4300
5000
5400
5000
4300
4300
4300
3000
0 1 2 3 4 5m

伏虎寺虎溪精舍（弥勒殿）正立面图
Front Elevation of Huxi Jingshe (Maitreya Hall), Fuhu Temple

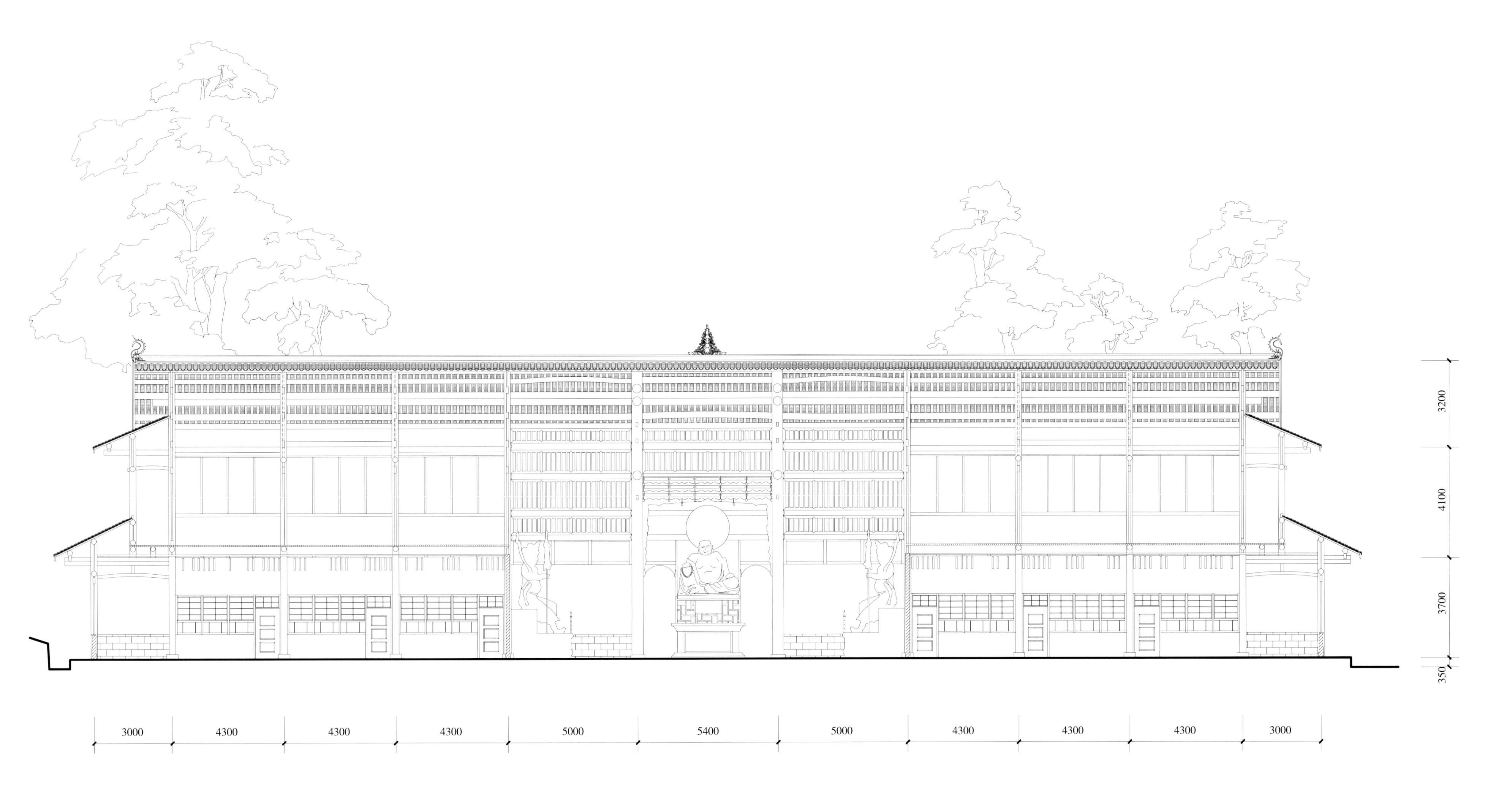
3200
4100
3700
350
3000
4300
4300
4300
5000
5400
5000
4300
4300
4300
3000

0 1 2 3 4 5m

伏虎寺虎溪精舍（弥勒殿）纵剖面图
Longitudinal Section of Huxi Jingshe (Maitreya Hall), Fuhu Temple

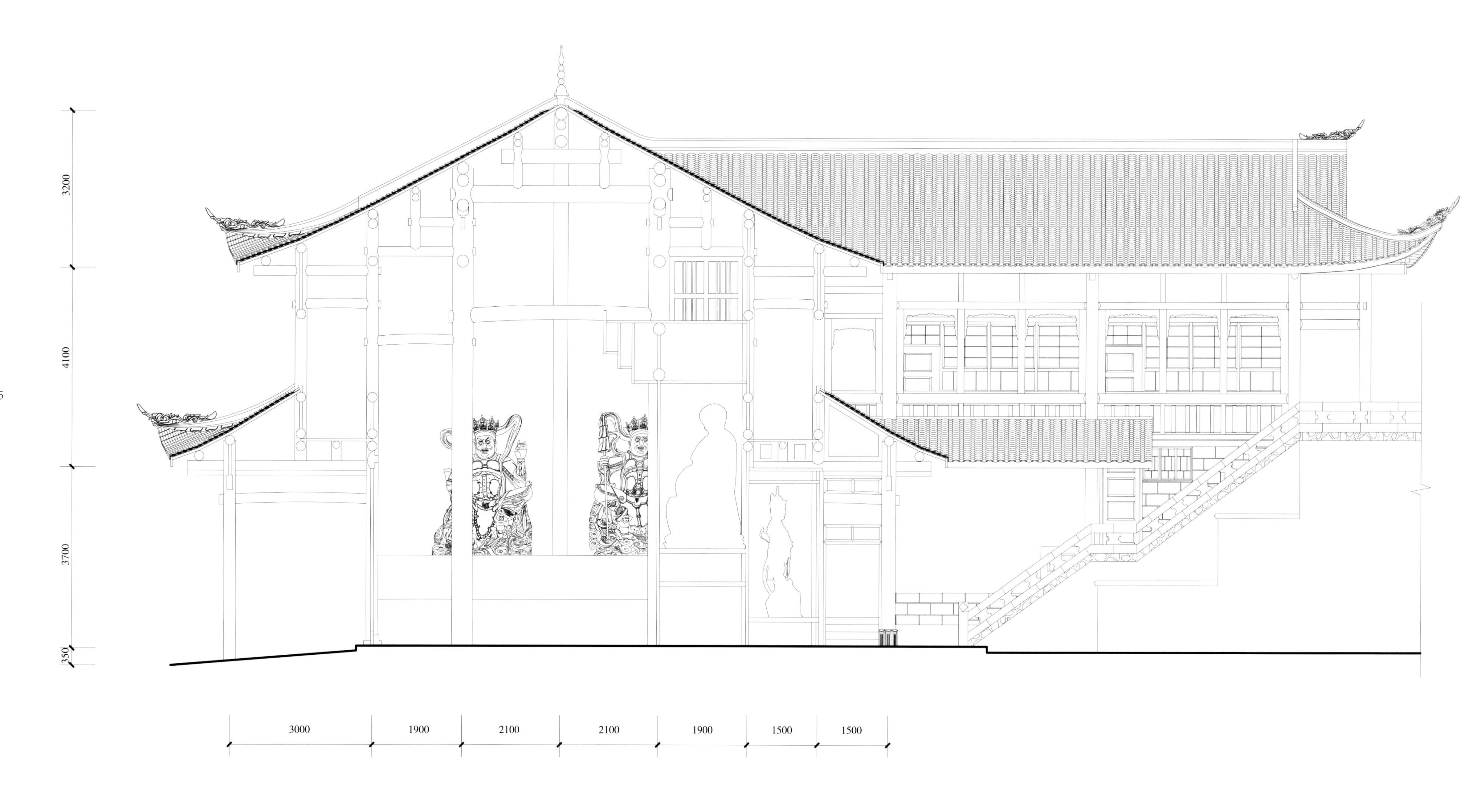
3200
4100
3700
350
3000
1900
2100
2100
1900
1500
1500

伏虎寺虎溪精舍（弥勒殿）横剖面图
Cross Section of Huxi Jingshe (Maitreya Hall), Fuhu Temple

0 1 2 3 4 5m

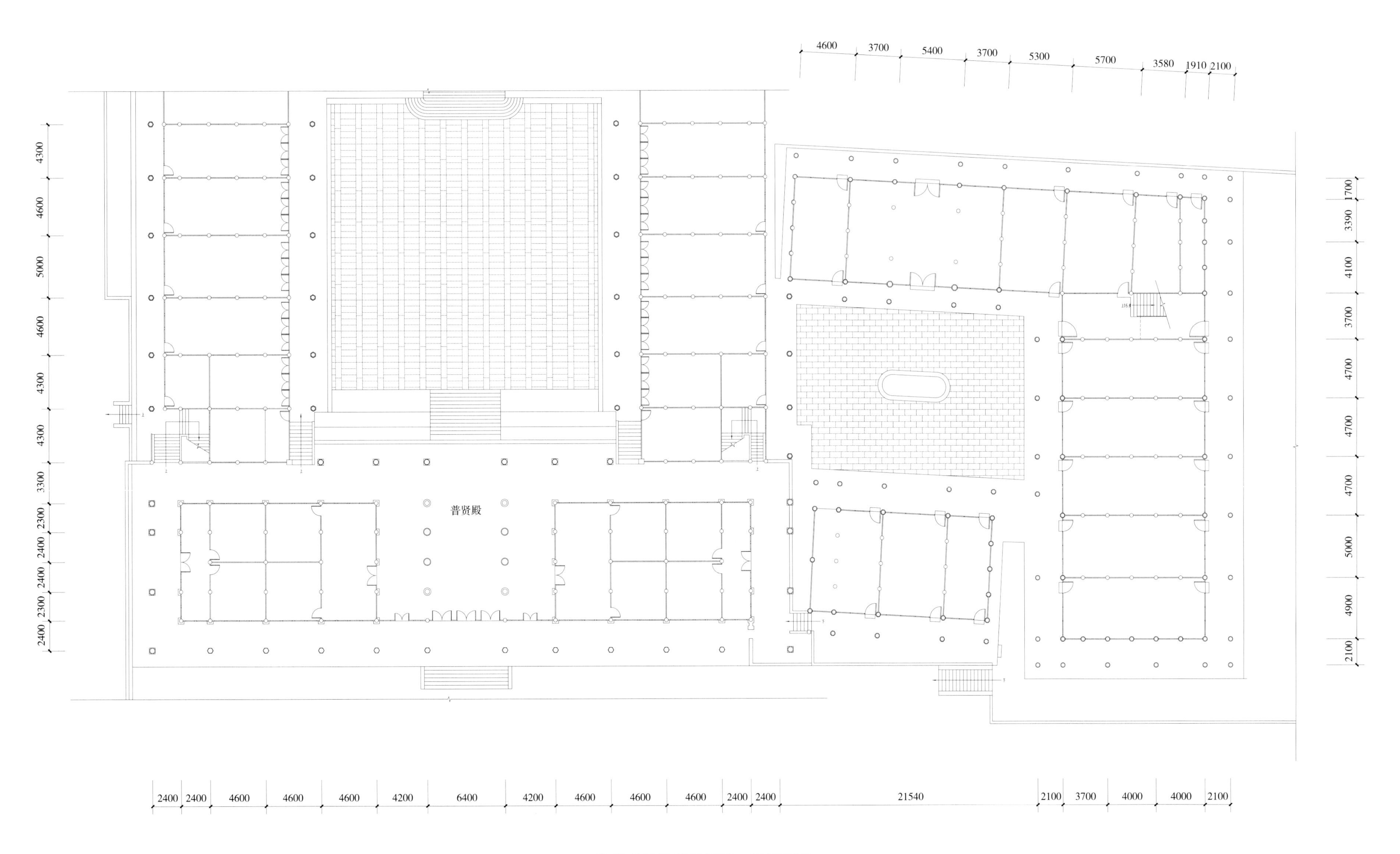
普贤殿
4600
3700
5400
3700
5300
5700
3580
1910
2100
4300
4600
5000
4600
4300
4300
3300
2300
2400
2400
2300
2400
1700
3390
4100
3700
4700
4700
4700
5000
4900
2100
2400
2400
4600
4600
4600
4200
6400
4200
4600
4600
4600
2400
2400
21540
2100
3700
4000
4000
2100
0
5
10m

伏虎寺普贤殿和侧院一层平面图
Ground Floor Plan of Samantabhadra Hall and Side Yard, Fuhu Temple

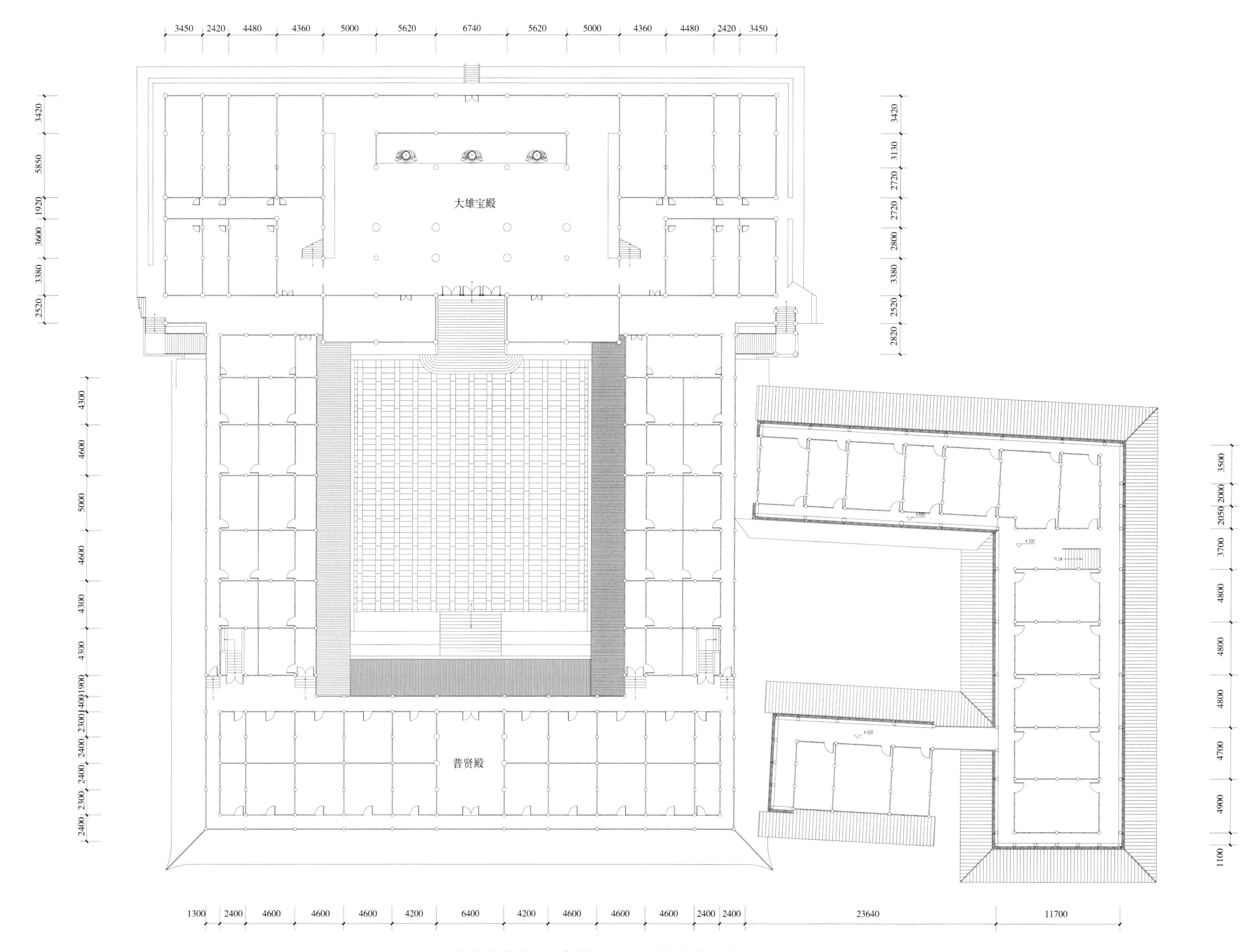
大雄宝殿
普贤殿

伏虎寺普贤殿和侧院二层平面与大雄宝殿一层平面图

First Floor Plan of Samantabhadra Hall and Side Yard and Ground Floor Plan of Mahavira Hall, Fuhu Temple

3880
3800
3400
150
10300
2180
4600
3700
5400
3700
3200
2100
11700
2100
0 1 2 3 4 5m

4640
2190
1820
4100
4250

1100 10300 2000 3430 5620 6740 5620 3432 2000

大雄寶殿

伏虎寺大雄宝殿正立面图
Front Elevation of Mahavira Hall, Fuhu Temple

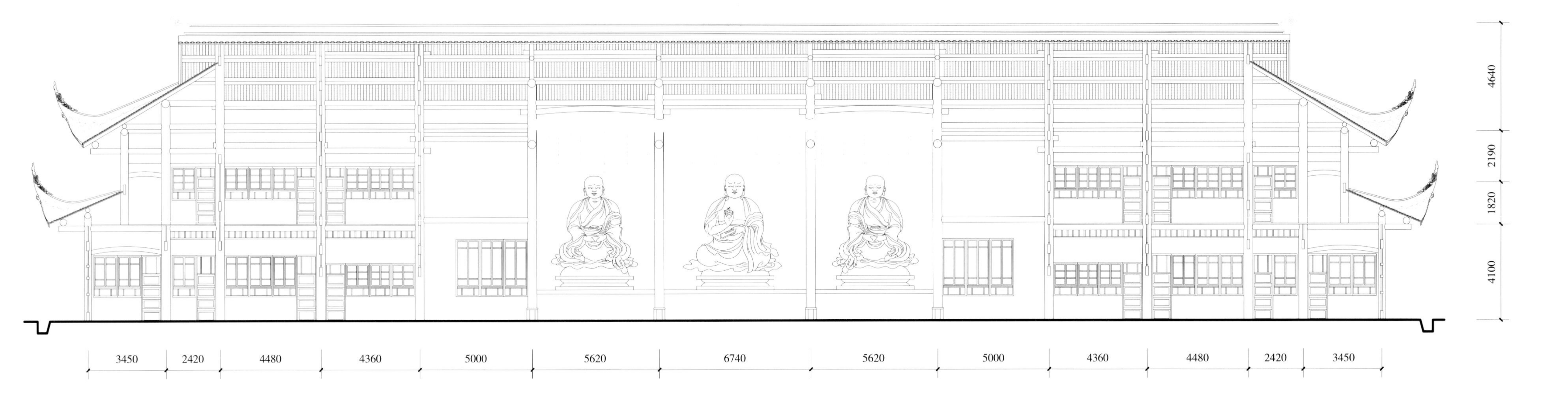
4640
2190
1820
4100
3450
2420
4480
4360
5000
5620
6740
5620
5000
4360
4480
2420
3450

伏虎寺大雄宝殿纵剖面图
Longitudinal Section of Mahavira Hall, Fuhu Temple

0 1 2 3 4 5m

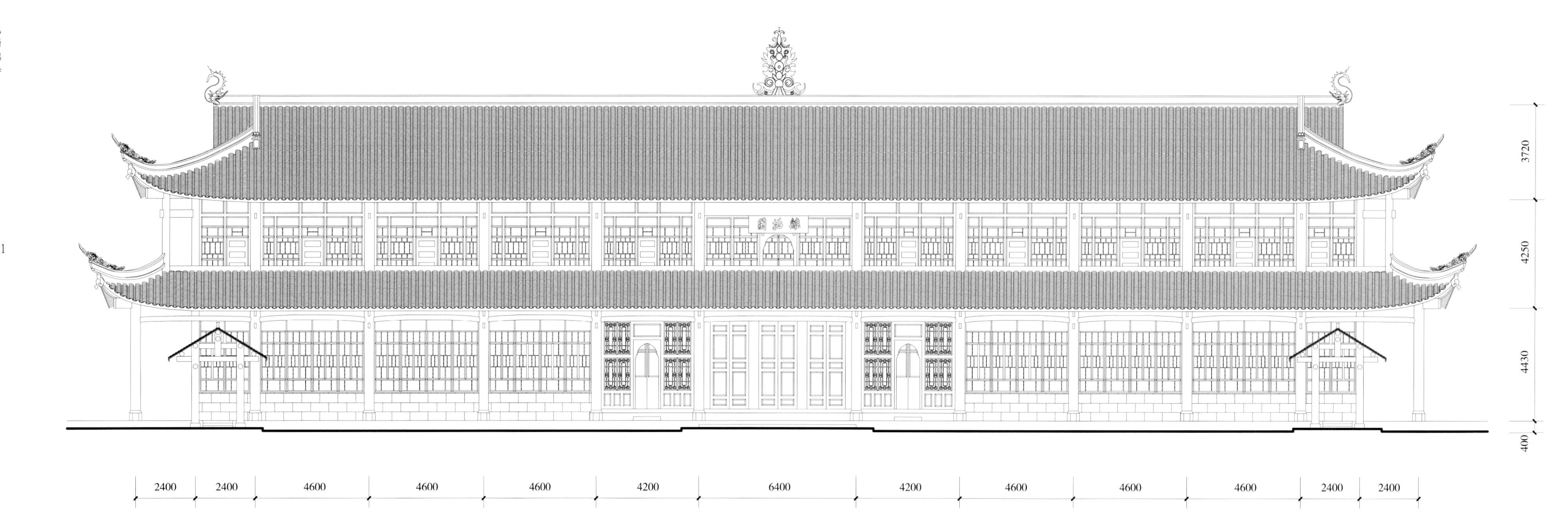
3720
4250
4430
400
2400
2400
4600
4600
4600
4200
6400
4200
4600
4600
4600
2400
2400
0 1 2 3 4 5m

伏虎寺普贤殿正立面图
Front Elevation of Samantabhadra Hall, Fuhu Temple

清音阁

清音阁位于峨眉山的牛心岭下，海拔710米。始建于唐，因其位于牛心岭附近曾得名牛心寺，明代更名清音阁，清康熙四十一年（1702年）重建，现存建筑为1917年重修。现存建筑主要有大雄宝殿和双飞亭。大雄宝殿建造在崖壁台地上，陡峭的踏步蜿蜒而上，建筑利用地形出挑吊脚与台地环境融为一体，显得宏伟壮观而秀丽多姿。清音阁环境优美，古木参天，双飞亭前黑龙江和白龙江交汇处，有孤石形如牛心，黑白二水牛心为独特景点『双桥清音』，是峨眉十景之一。

Qingyin Building

Qingyin Building is located at the foot of Niuxin (literally, cow heart) Ridge on Mount Emei, at an altitude of 710m. Originally built in Tang Dynasty, it was known by the name of Niuxin Temple because of its location near Niuxin Ridge. In Ming Dynasty it was renamed Qingyin Building, and reconstructed in the 41st year of Kangxi's reign in Qing Dynasty (AD1702). The existing structure was the result of a repair in 1917. The existing buildings include the Great Buddha Hall and the Shuangfei Pavilion. The Great Buddha Hall is built on a cliff terrace, where steep steps wind up. Thanks to ingenious adaptation to the topography with overhang, the building is integrated with the terrace, creating a magnificent and beautiful look. Qingyin Building is noted for its graceful environment and ancient trees. In front of the Shuangfei Pavilion, where Heilong and Bailong Rivers intersect, there is an isolated stone in the shape of a cow heart. The two rivers and the stone make up a unique picture of “Shuangqiao Qingyin”, one of the ten famous Emei Scenes.

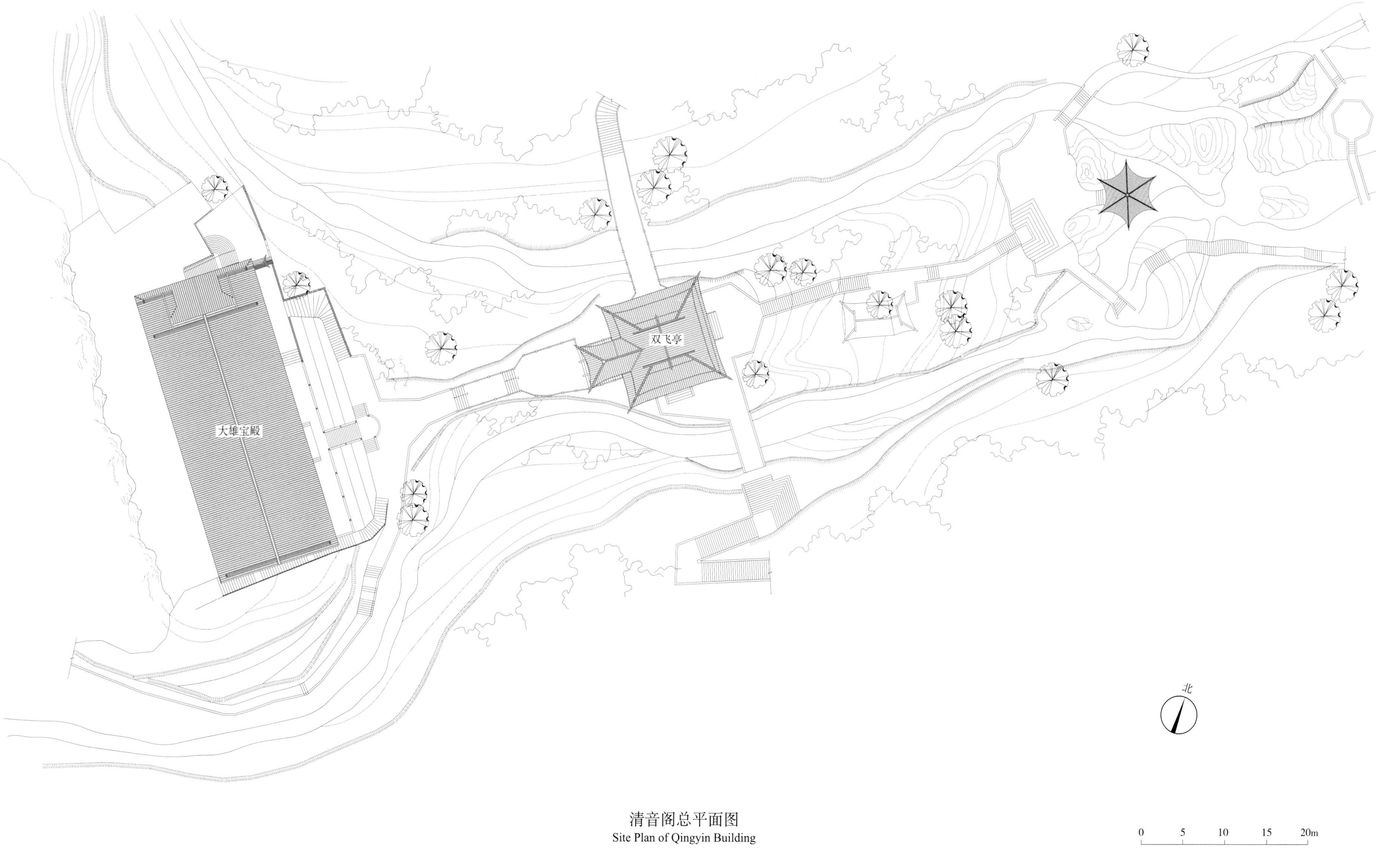

大雄宝殿

双飞亭

北

0 5 10 15 20m

清音阁总平面图

Site Plan of Qingyin Building

图一　清音阁鸟瞰图

图二　清音阁双飞亭

图三　清音阁大雄宝殿

Fig.1 Bird's Eye View of Qingyin Building
Fig.2 Shuangfei Pavilion, Qingyin Building
Fig.3 Mahavira Hall, Qingyin Building

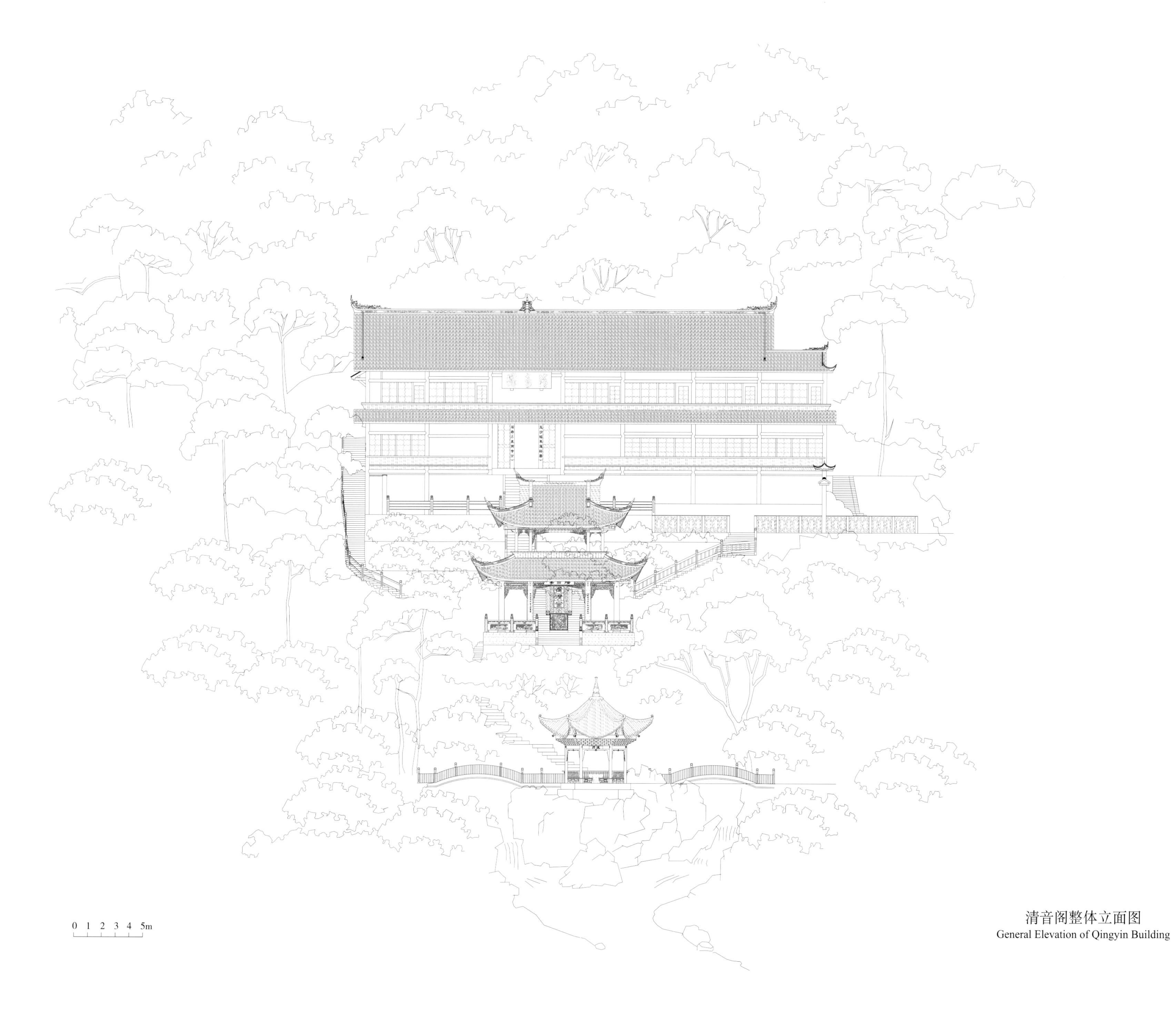
0 1 2 3 4 5m

清音阁整体立面图
General Elevation of Qingyin Building

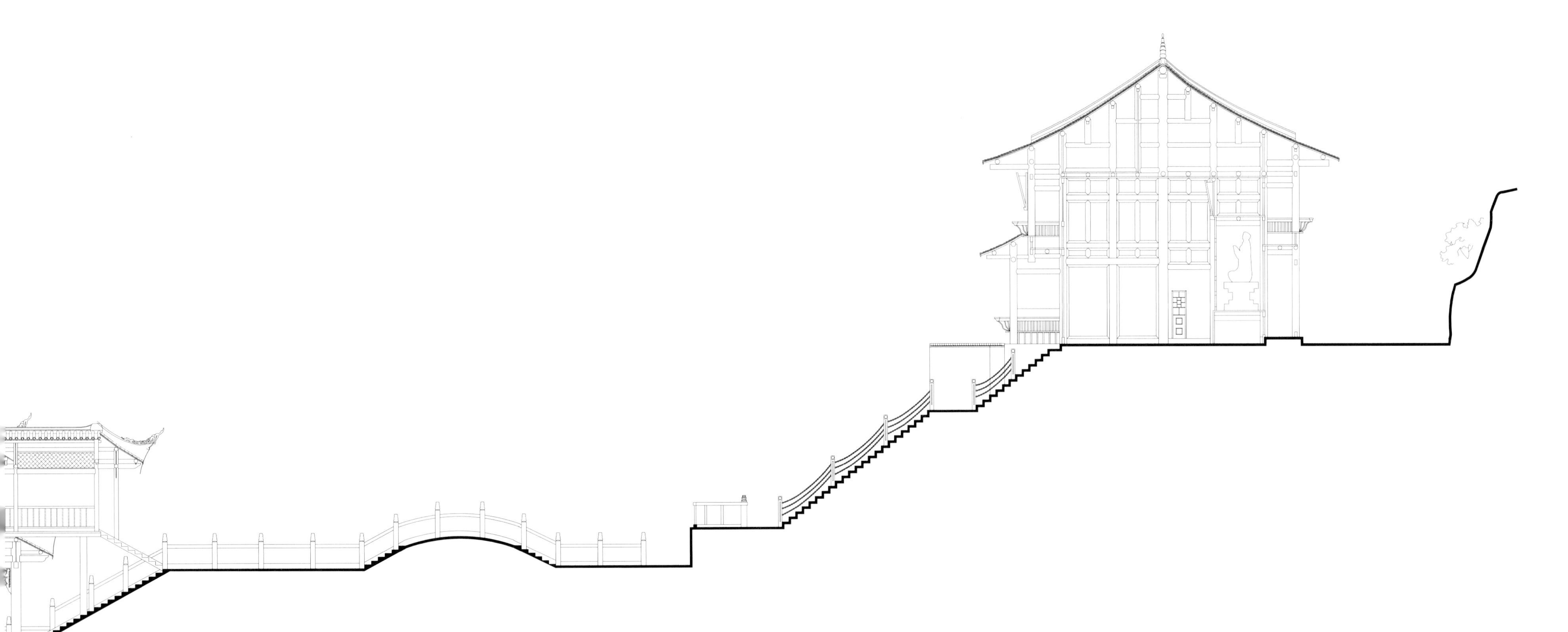

清音阁中轴剖面图
Cross Section on Axis of Qingyin Building

0 1 2 3 4 5m

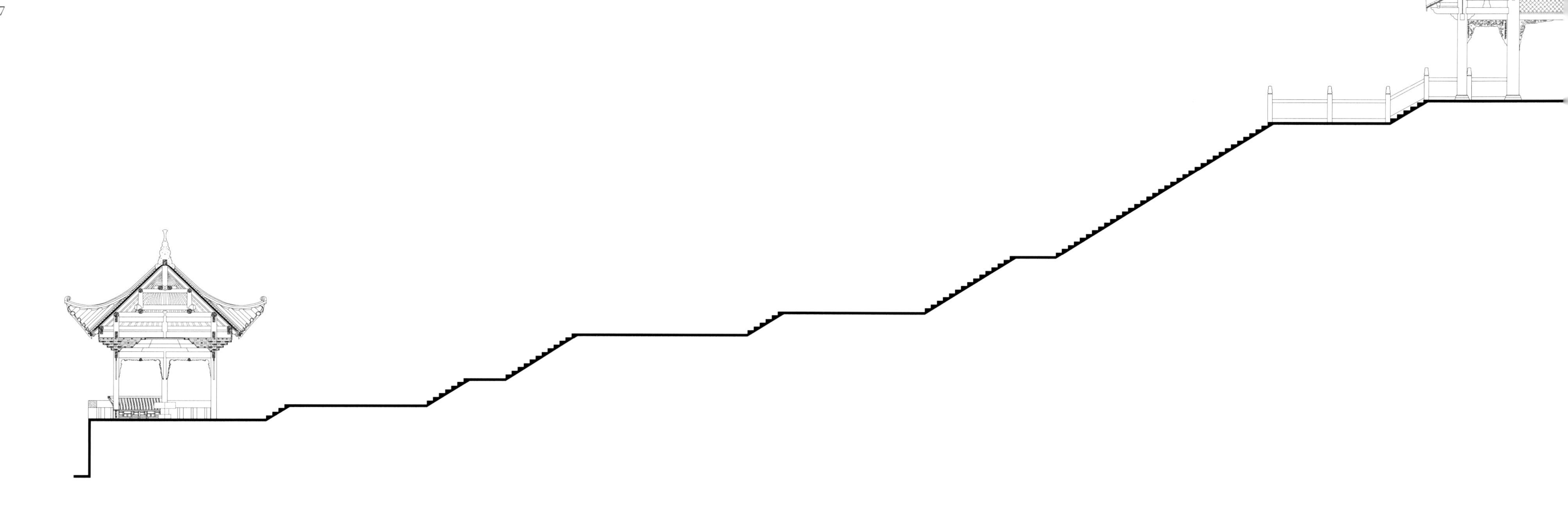

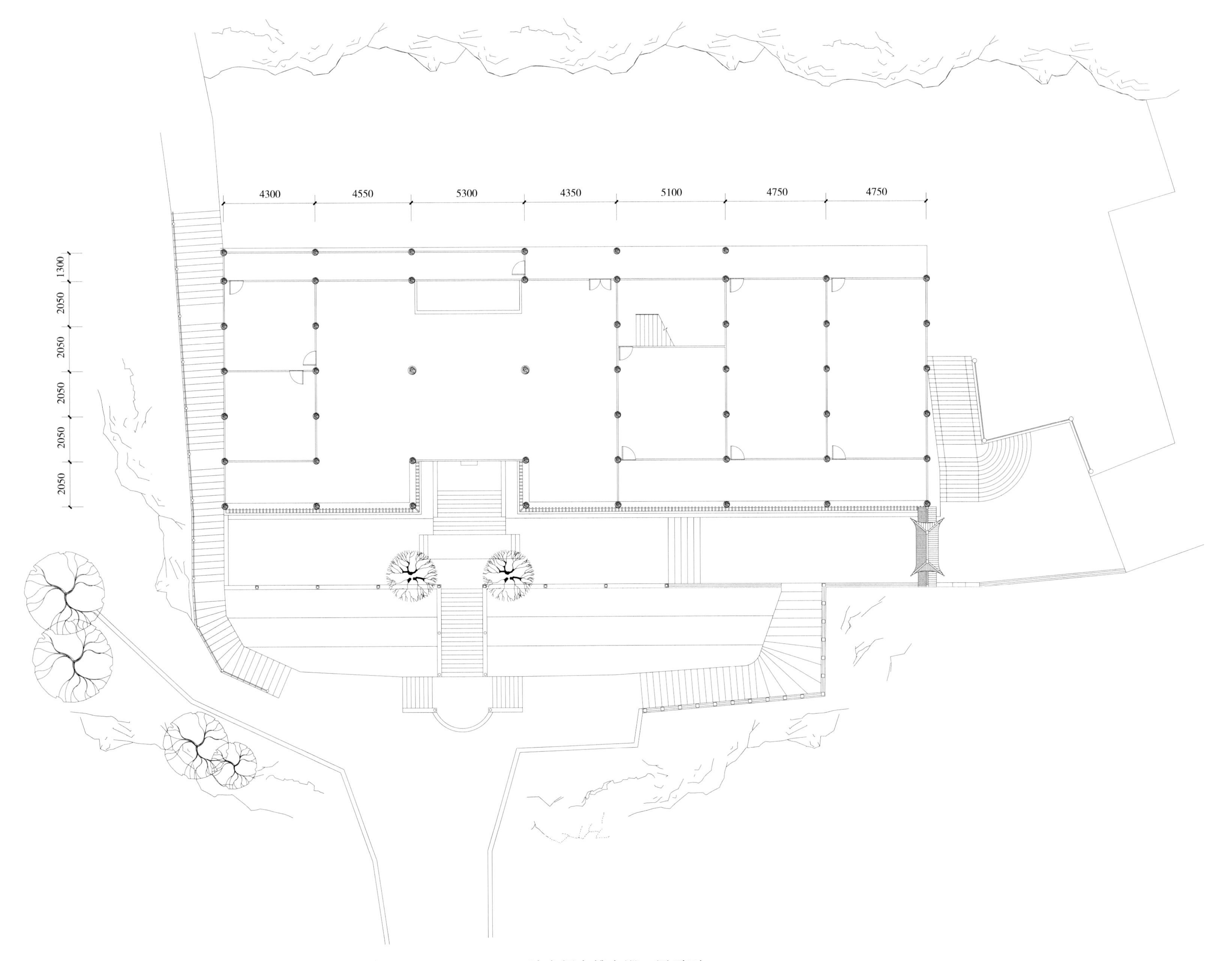
4300
4550
5300
4350
5100
4750
4750
1300
2050
2050
2050
2050
2050

清音阁大雄宝殿一层平面
Ground Floor Plan of Mahavira Hall, Qingyin Building

0 1 2 3 4 5m

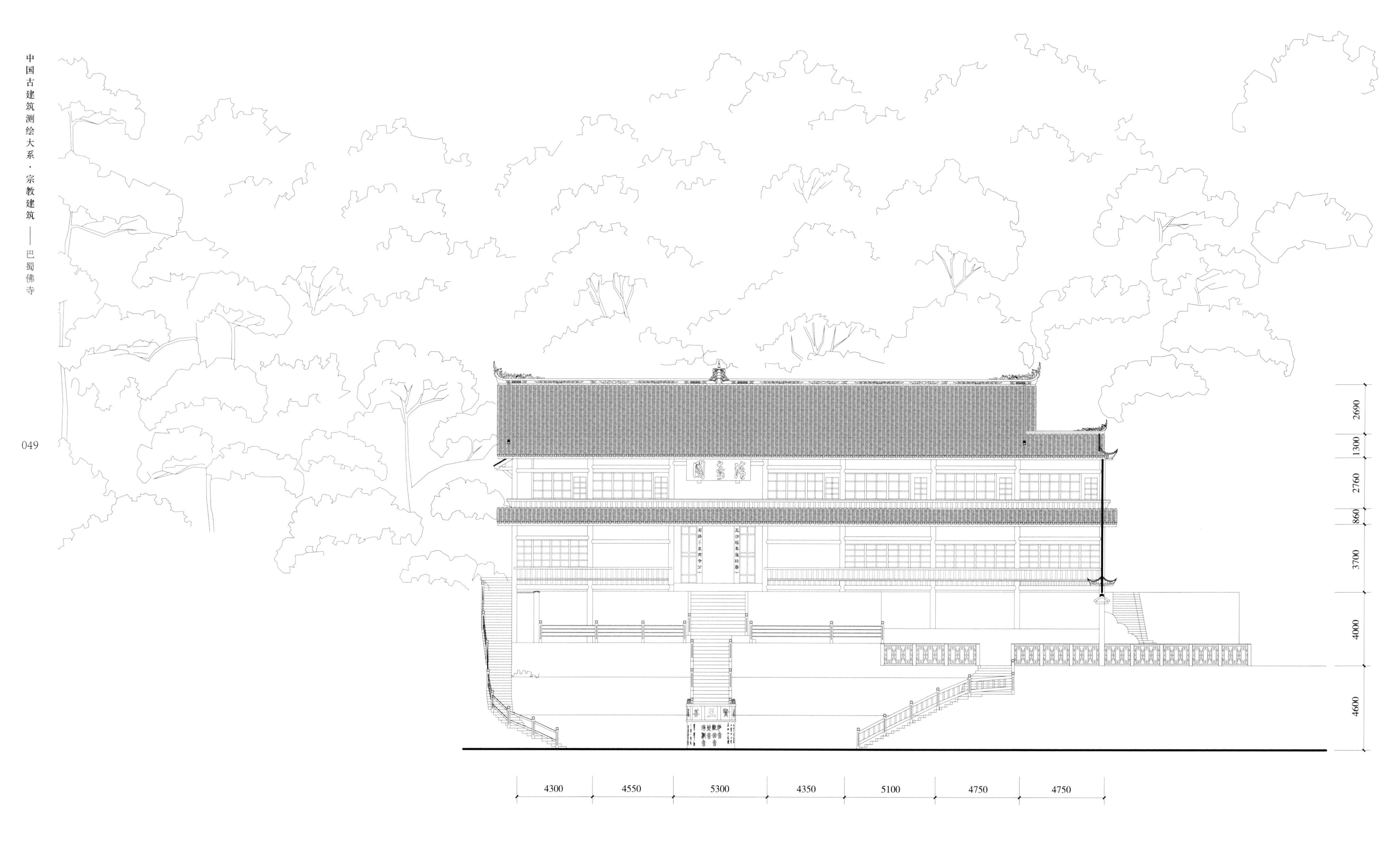
2690
1300
2760
860
3700
4000
4600
4300
4550
5300
4350
5100
4750
4750
0 1 2 3 4 5m

清音阁大雄宝殿正立面图
Front Elevation of Mahavira Hall, Qingyin Building

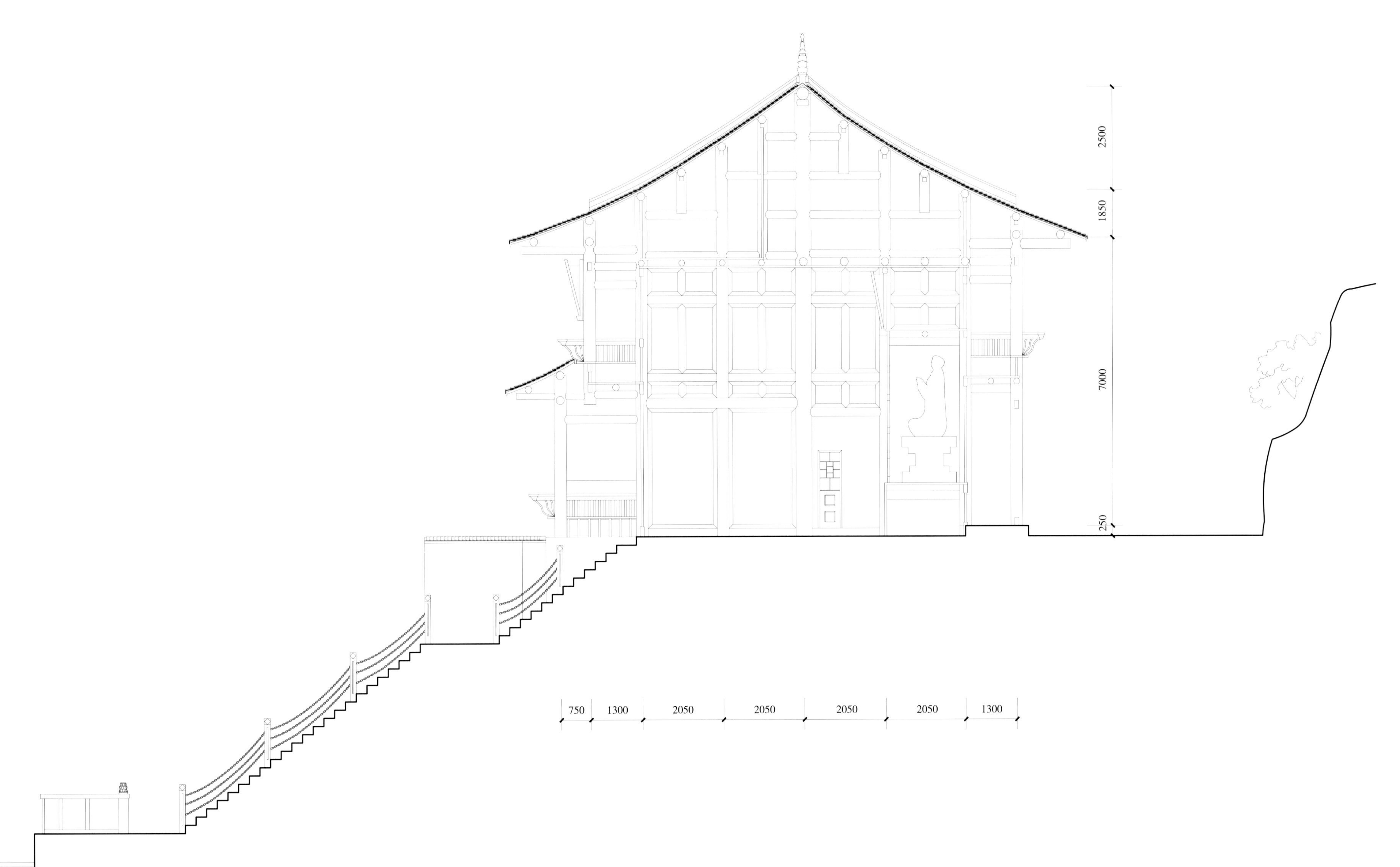
2500
1850
7000
250
750
1300
2050
2050
2050
2050
1300
0 1 2 3 4 5m

清音阁大雄宝殿横剖面图
Cross Section of Mahavira Hall, Qingyin Building

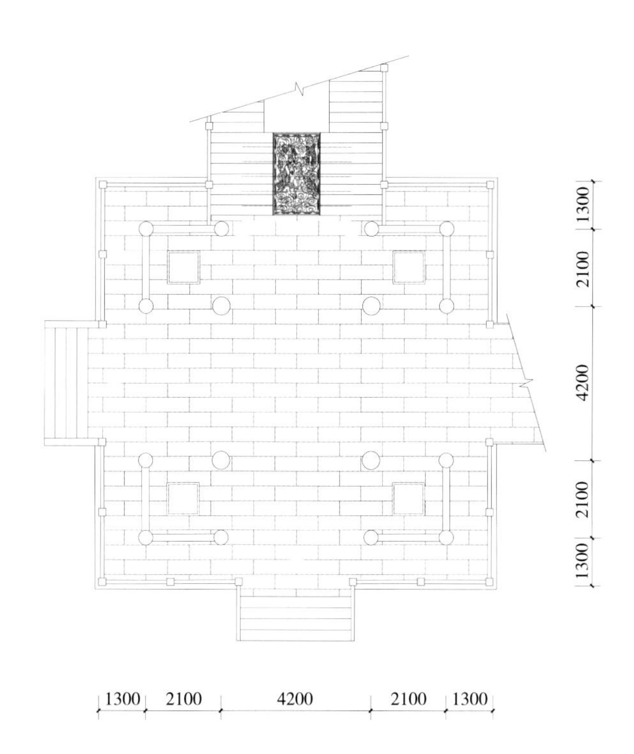
1300
2100
4200
2100
1300
1300 2100 4200 2100 1300
0 1 2 3m

清音阁双飞亭一层平面图
Ground Floor Plan of Shuangfei Pavilion, Qingyin Building

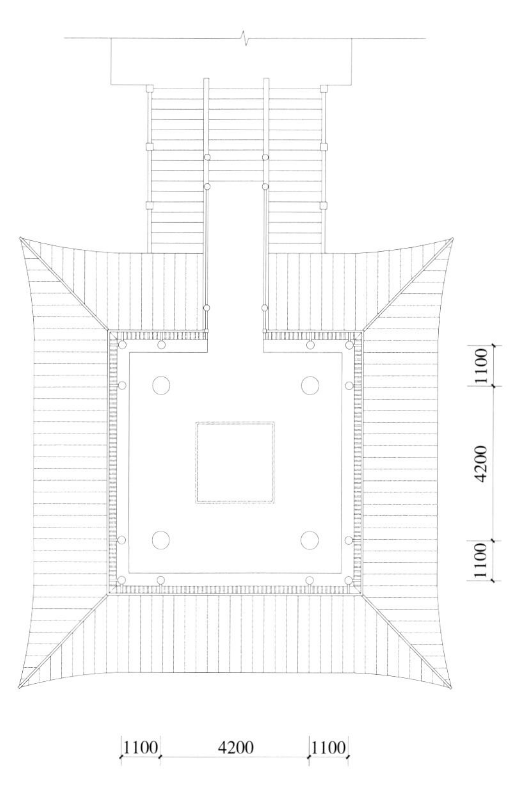
1100
4200
1100
1100 4200 1100
0 1 2 3m

清音阁双飞亭二层平面图
First Floor Plan of Shuangfei Pavilion, Qingyin Building

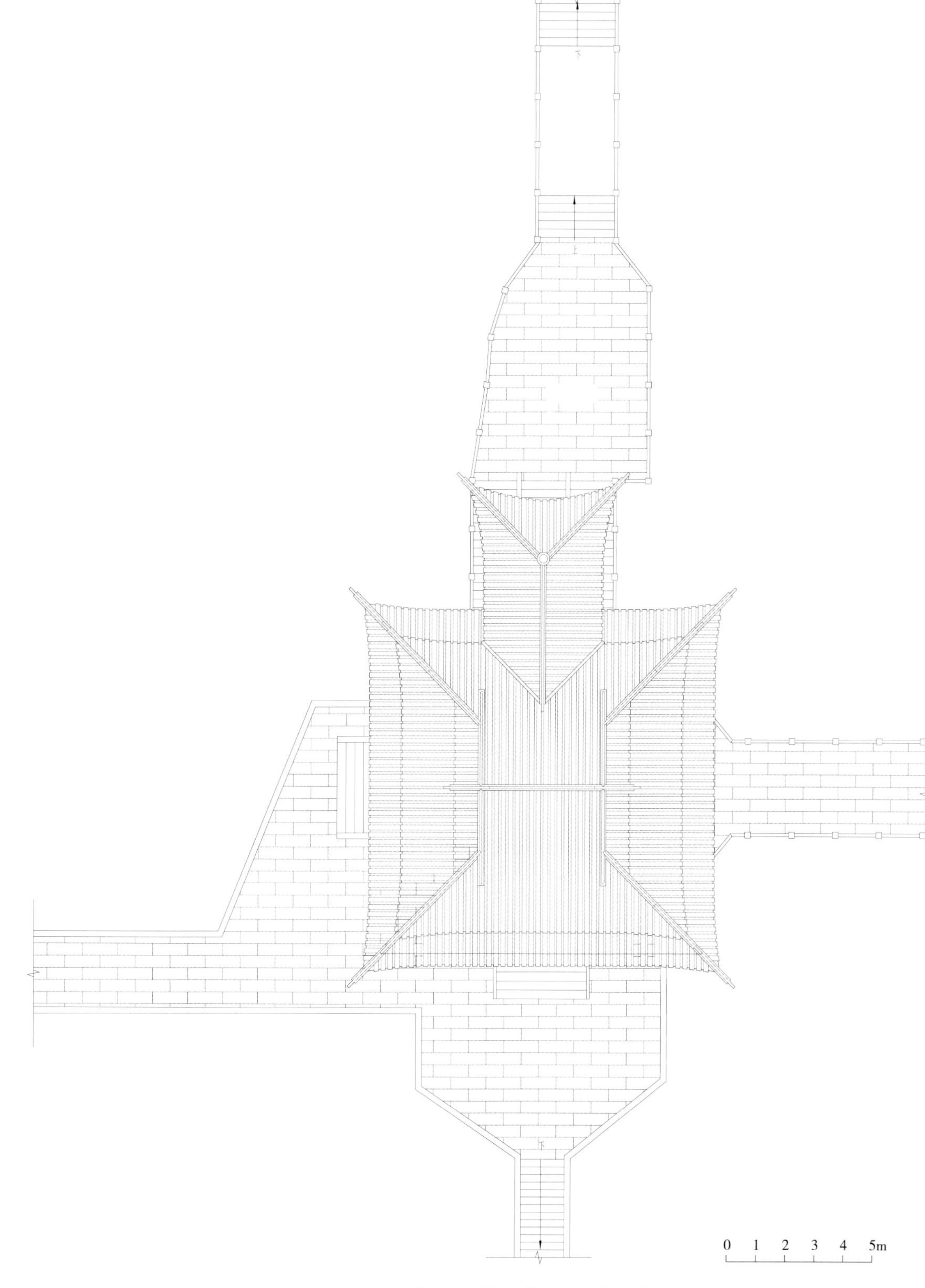
0 1 2 3 4 5m

清音阁双飞亭屋顶平面图
Roof Plan of Shuangfei Pavilion, Qingyin Building

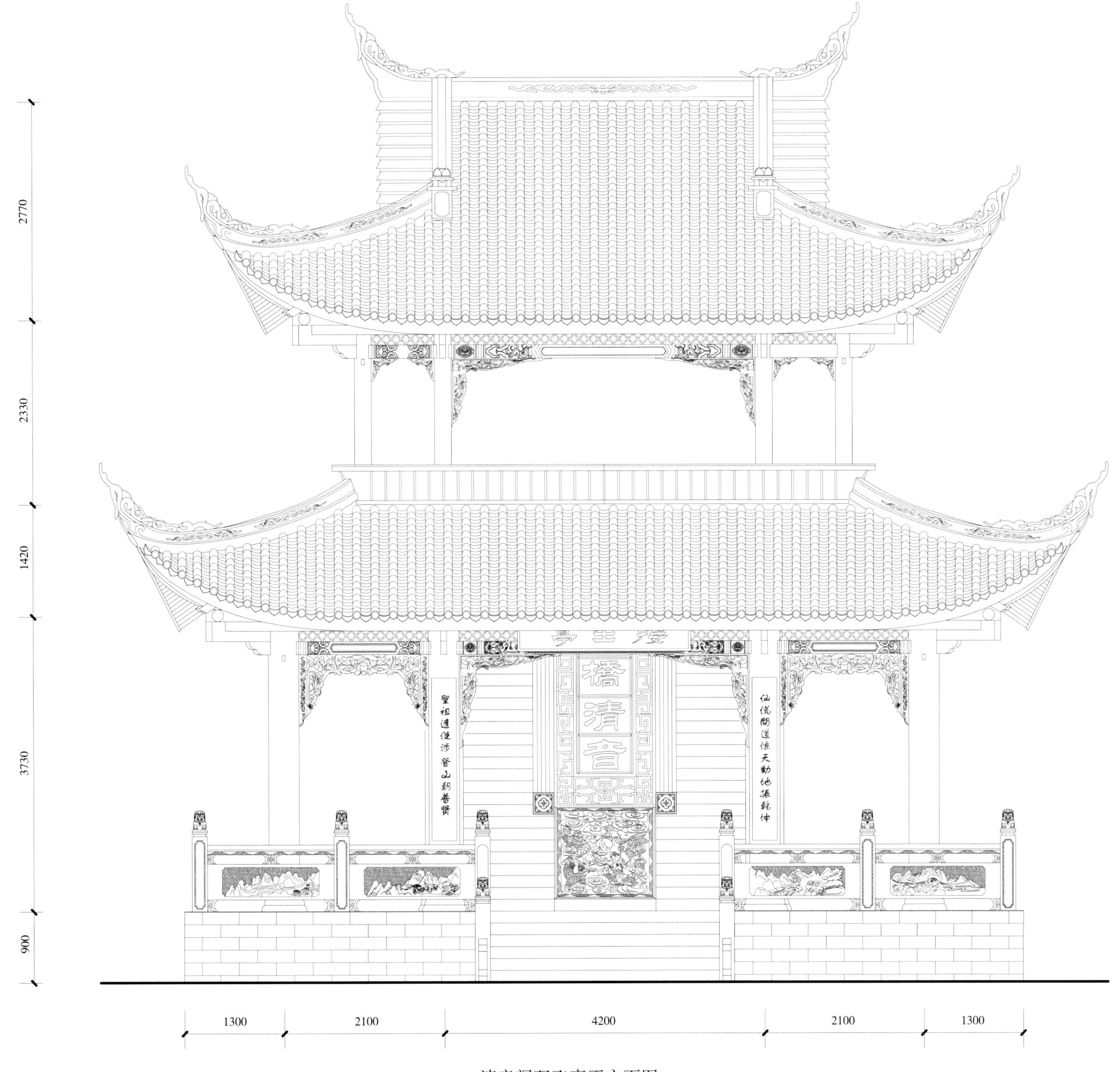
2770
2330
1420
3730
900
橋清音
1300
2100
4200
2100
1300
0
1
2
3m

清音阁双飞亭正立面图

Front Elevation of Shuangfei Pavilion, Qingyin Building

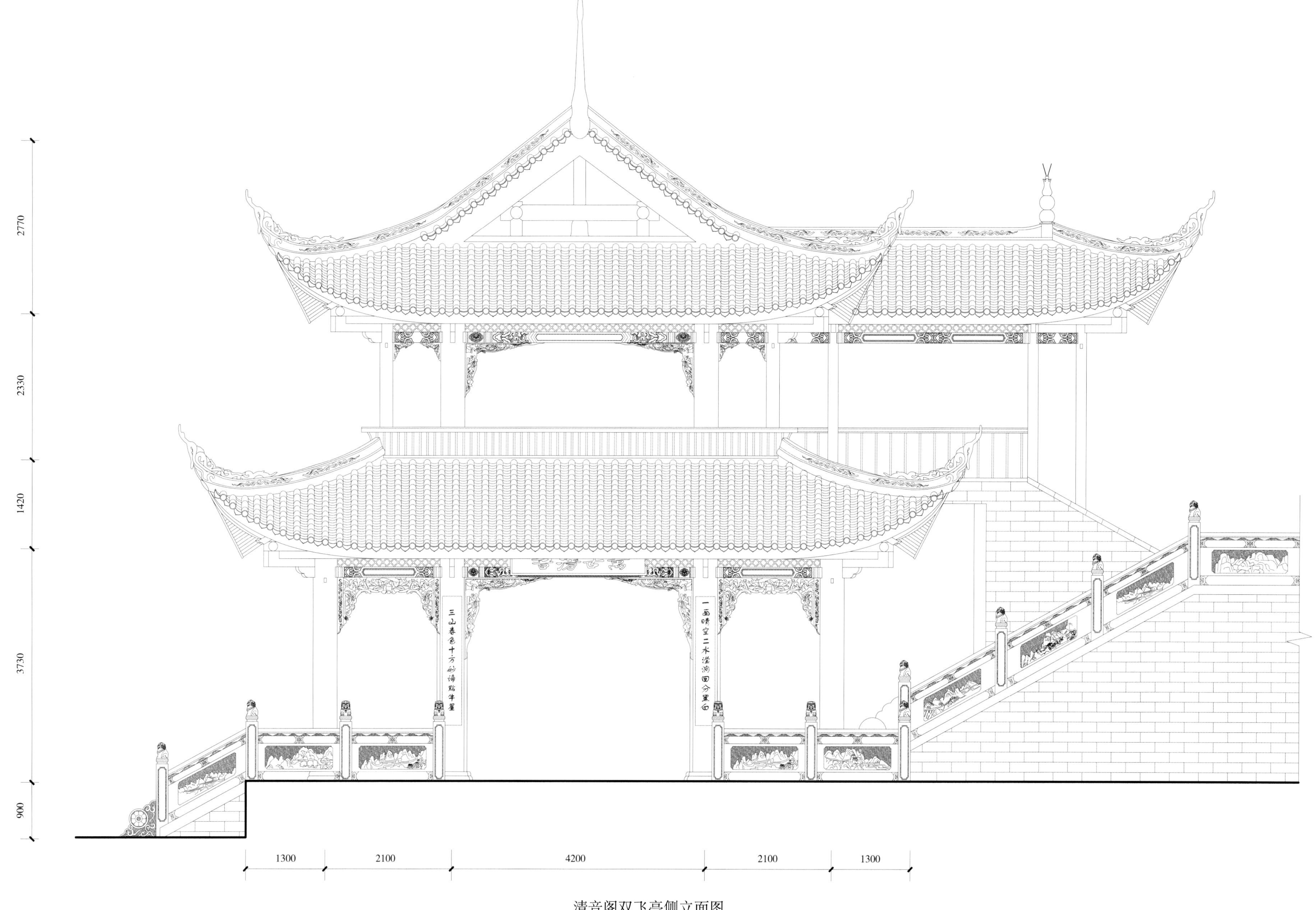
2770
2330
1420
3730
900
三山春色十方妙谛點牛星
一面晴空二水滢洄回分黑白
1300
2100
4200
2100
1300
0
1
2
3m

清音阁双飞亭侧立面图
Side Elevation of Shuangfei Pavilion, Qingyin Building

2770
2330
1420
3730
900
2100
4200
2100
0
1
2
3m

清音阁双飞亭剖面图

Section of Shuangfei Pavilion, Qingyin Building

洪椿坪

洪椿坪位于峨眉山天池峰山腰，海拔1120米。始建于宋代的千佛庵，又称『千佛禅院』，明朝以来不断扩建而初具规模。乾隆四十三年（1778年）毁于火，清乾隆四十七年至五十五年（1782—1790年）修复。因寺前有洪椿树而得名。主要佛教殿堂有山门、观音殿、大雄宝殿、普贤殿等。建筑群置于山腰台地上，三面临沟谷深壑，背面靠山是朝拜金顶的通道。建筑群空间组合巧妙利用地形，内部为四合院空间严谨对称，外部空间悬崖吊脚并适应地形自由错落。山门不在轴线上而通过道路一侧转折进入，具有浓郁的山地寺庙园林风格。在寺庙的右侧有宝掌峰，下有黑龙江，受小气候环境的影响，寺庙云雾环绕犹如仙境，『洪椿晓雨』是峨眉十景之一。

Hongchunping Temple

Hongchunping Temple is located on the mountainside of Tianchi Peak on Mount Emei, at an altitude of 1,120m. Originally built as the Thousand Buddha Convent in the Song Dynasty (AD960-1279), it was also known as the Thousand Buddha Zen Monastery. It was continually expanded from Ming Dynasty on and developed into an architectural complex of a significant scale. The monastery was destroyed by fire in the 43rd year of Qianlong's reign in Qing Dynasty (AD1778), and restored between the 47th and 55th year of Qianlong's reign in Qing Dynasty (AD1782-1790). It was named after the Ailanthus tree (Hongchun) in front of the monastery. Major Buddhist buildings include the entrance gate, the Avalokiteshvara Hall, the Great Buddha Hall, and the Samantabhadra Hall among others. The complex sits on a hillside terrace with deep valley on three sides, and the pilgrimage path towards the mountain top rises from its back. The spatial combination of the architectural complex is a clever adaptation to the terrain. Inside is a courtyard space of regular symmetry, while from the outside stilts over the cliff are adapted to the terrain in a free manner. The entrance gate is not on the axis, but is entered from one side of the path, which clearly indicates a mountain temple garden style. To the right of the monastery is Baozhang Peak, with the Heilong River running beneath. Thanks to the microclimate, the monastery is surrounded by cloud and fog like a wonderland, which makes a picture of "Hongchun Xiaoyu", one of the ten famous Emei Scenes.

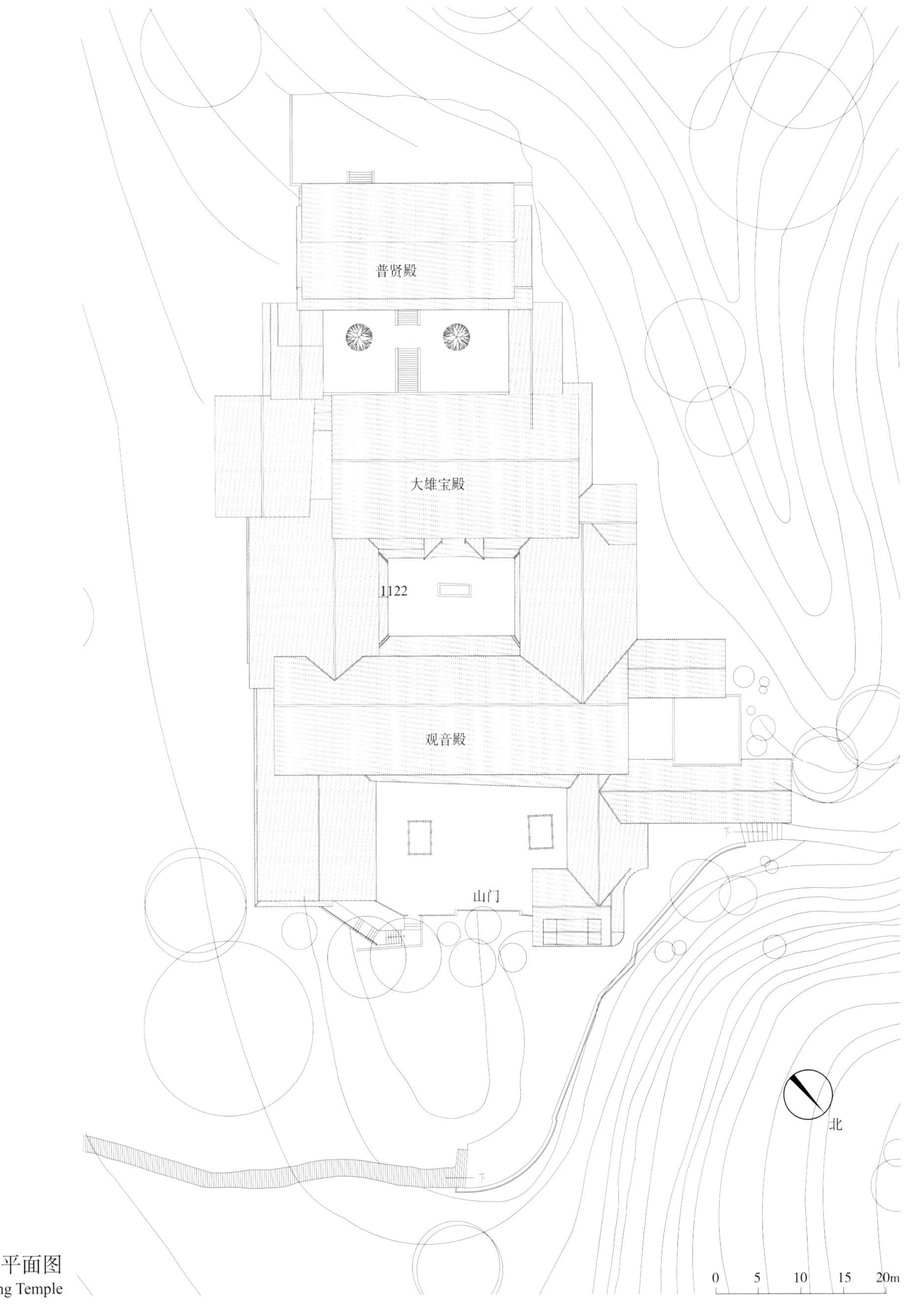
普贤殿
大雄宝殿
1122
观音殿
山门
北
0 5 10 15 20m

洪椿坪总平面图
Site Plan of Hongchunping Temple

图二　洪椿坪大雄宝殿

图一　洪椿坪鸟瞰图

图三　洪椿坪大雄宝殿内景

Fig.1 Bird's Eye View of Hongchunping Temple
Fig.2 Mahavira Hall, Hongchunping Temple
Fig.3 Interior View of Mahavira Hall, Hongchunping Temple

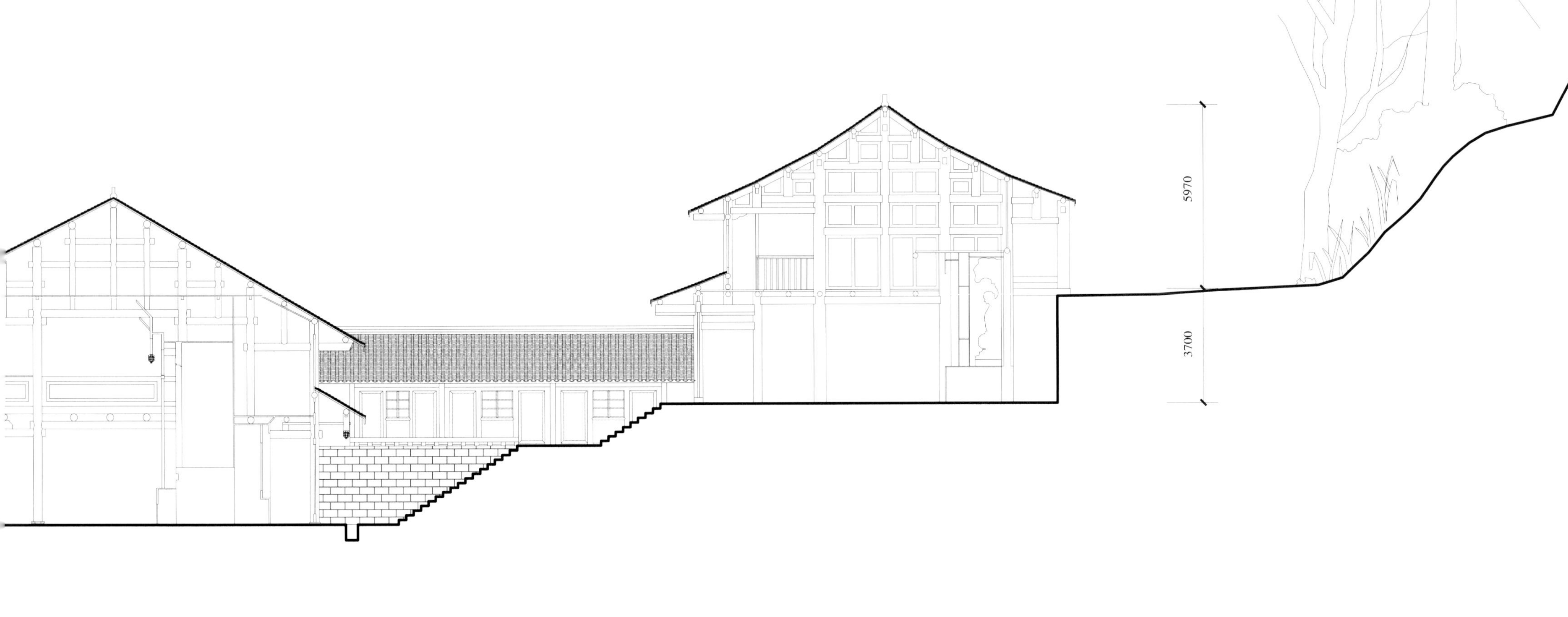
5970
3700
0
4850
4450
16050
2000
2100
4200
2100
1710

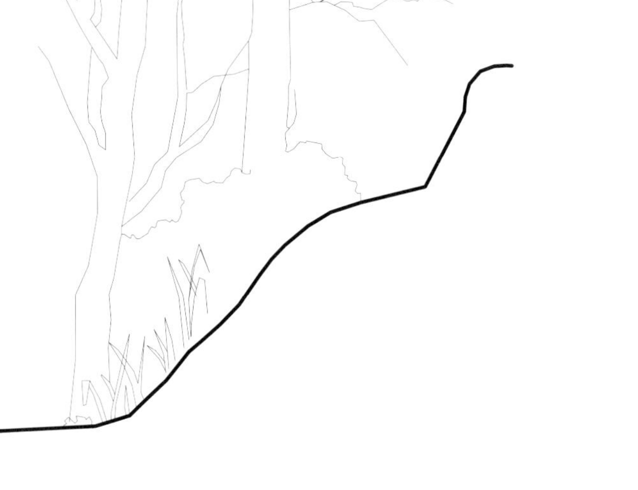

洪椿坪中轴剖面图
Cross Section on Axis of Hongchunping Temple

0 1 2 3 4 5m

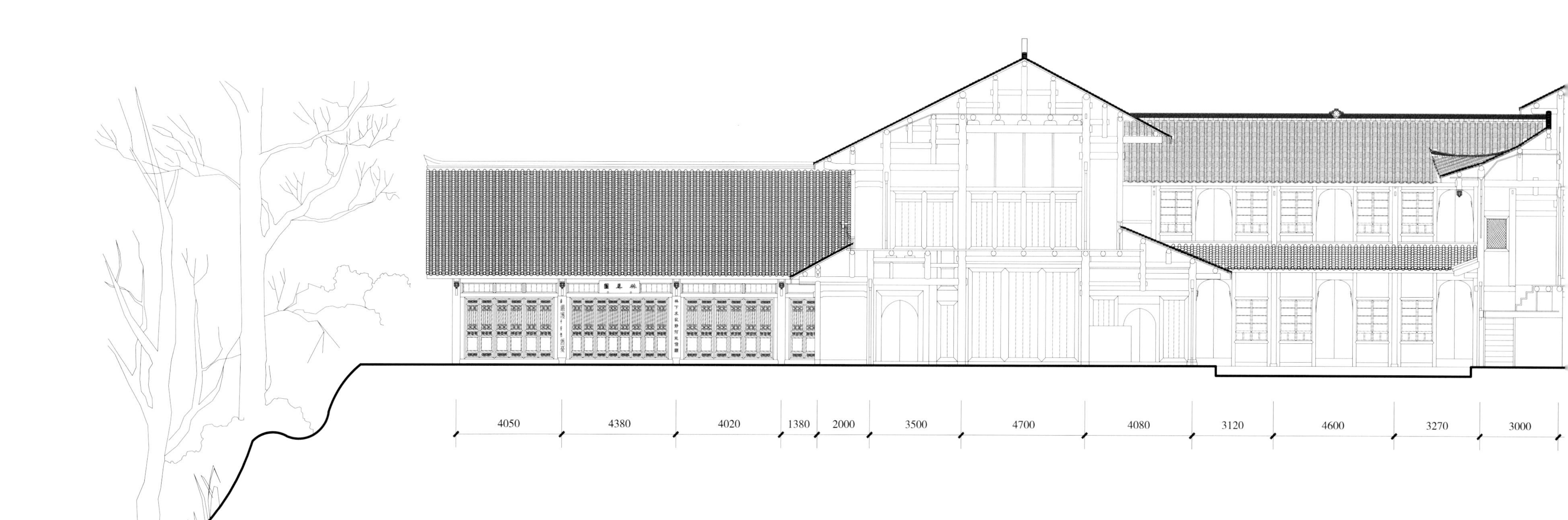
4050
4380
4020
1380
2000
3500
4700
4080
3120
4600
3270
3000

2000 3000 4700 5100 5350 5560 5350 5100 4090 3130
3600 3900 1880 3200 4150 4600 4100 2300 4700 1270 2290 3270 4020 4380 4050
2180 4400 4900 4400 1490 1030 3270 4560 1720 2240 2000 5550 5750 1750
1370
观音殿
1550 2280 2290 4720 4300 4080 6060 4060 3510 2120 2620 2220 1950 2760 950 3450 4250 4350 4130

洪椿坪一层平面图
Ground Floor Plan of Hongchunping Temple

0 1 2 3 4 5m

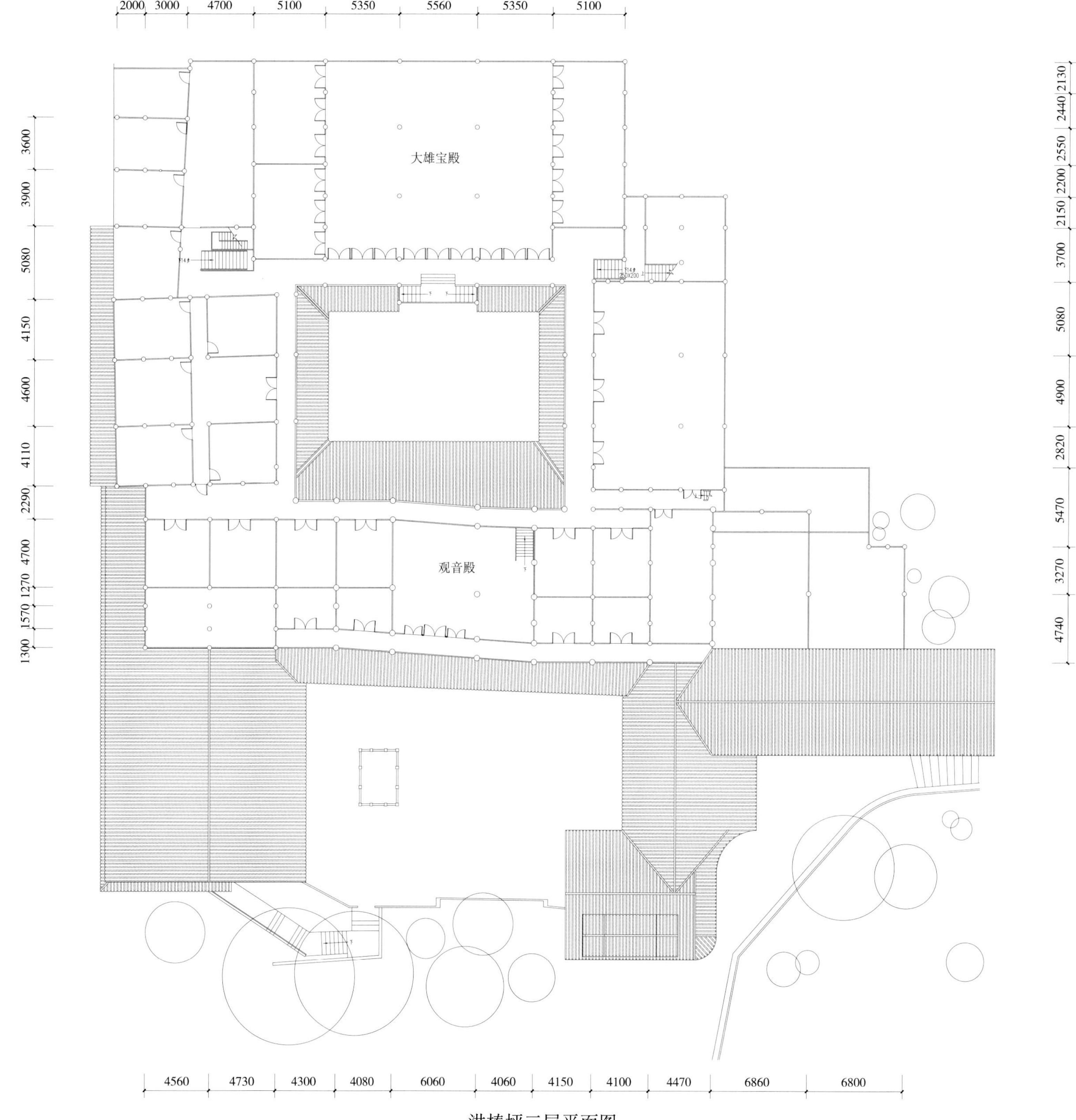
2000
3000
4700
5100
5350
5560
5350
5100
大雄宝殿
观音殿
3600
3900
5080
4150
4600
4110
2290
4700
1270
1570
1300
2130
2440
2550
2200
2150
3700
5080
4900
2820
5470
3270
4740
4560
4730
4300
4080
6060
4060
4150
4100
4470
6860
6800

洪椿坪二层平面图

First Floor Plan of Hongchunping Temple

0 1 2 3 4 5m

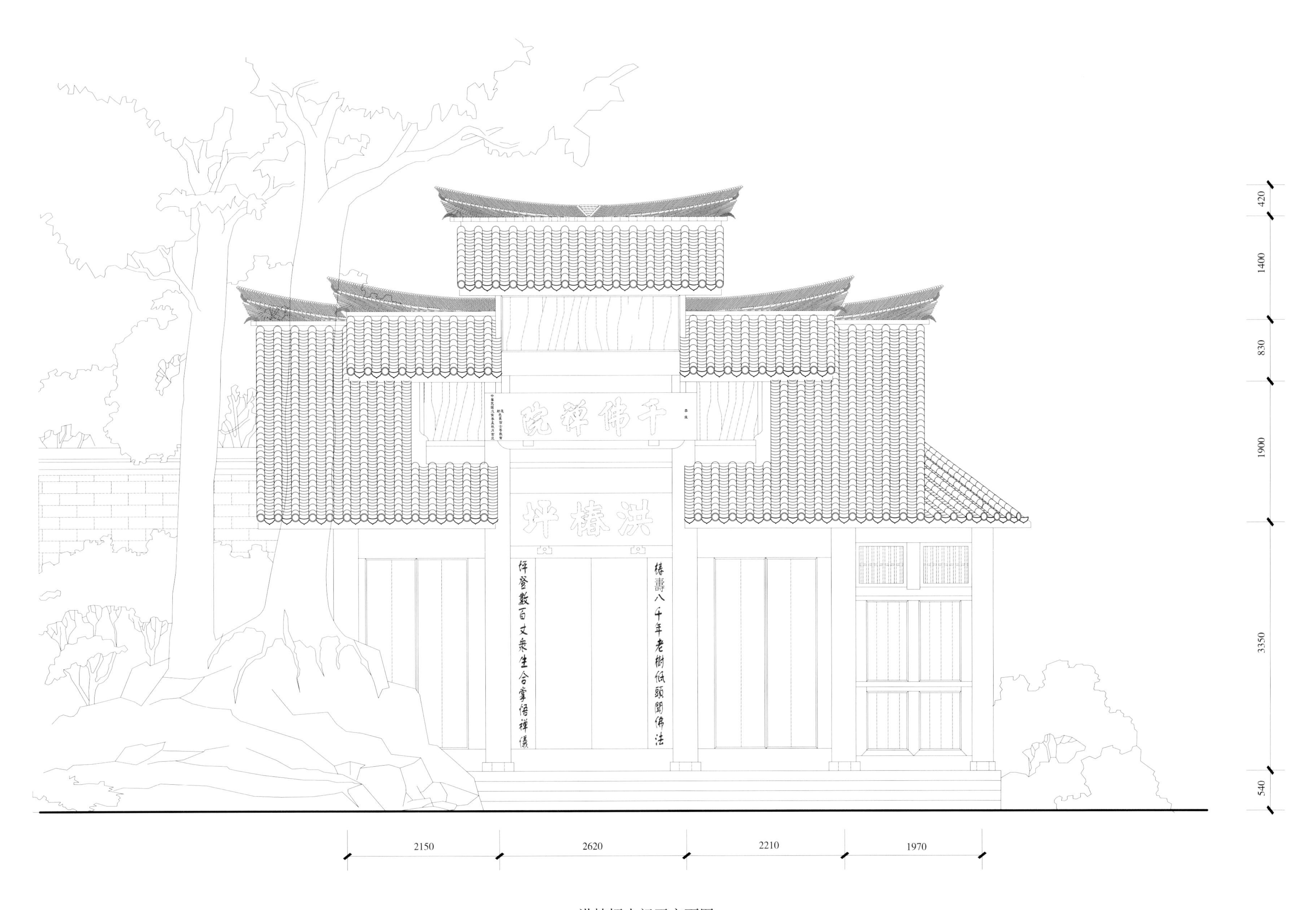
千佛禅院
洪椿坪
坪蓊數百丈衆生合掌悟禅機
椿壽八千年老樹低頭聞佛法
420
1400
830
1900
3350
540
2150
2620
2210
1970
0
1
2
3m

洪椿坪山门正立面图

Front Elevation of Main Gate (Shan Men), Hongchunping Temple

420
2230
1900
3350
540
2500
3580
1750
1090
2010
1200
0
1
2
3m

洪椿坪山门剖面图

Cross Section of Main Gate (Shan Men) , Hongchunping Temple

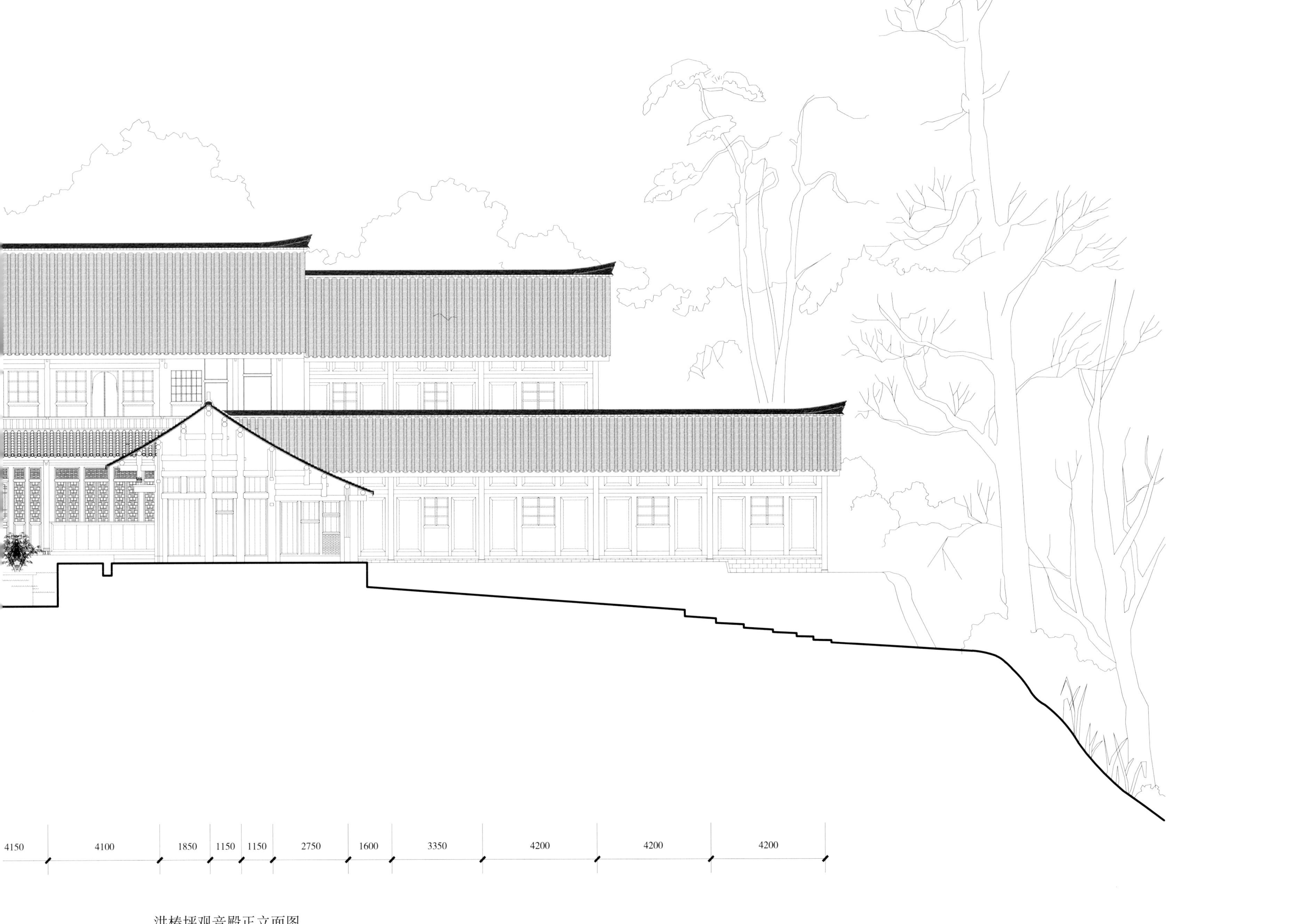
4150
4100
1850
1150
1150
2750
1600
3350
4200
4200
4200
0
2.5
5
7.5
10m

洪椿坪观音殿正立面图
Front Elevation of Guanyin Hall, Hongchunping Temple

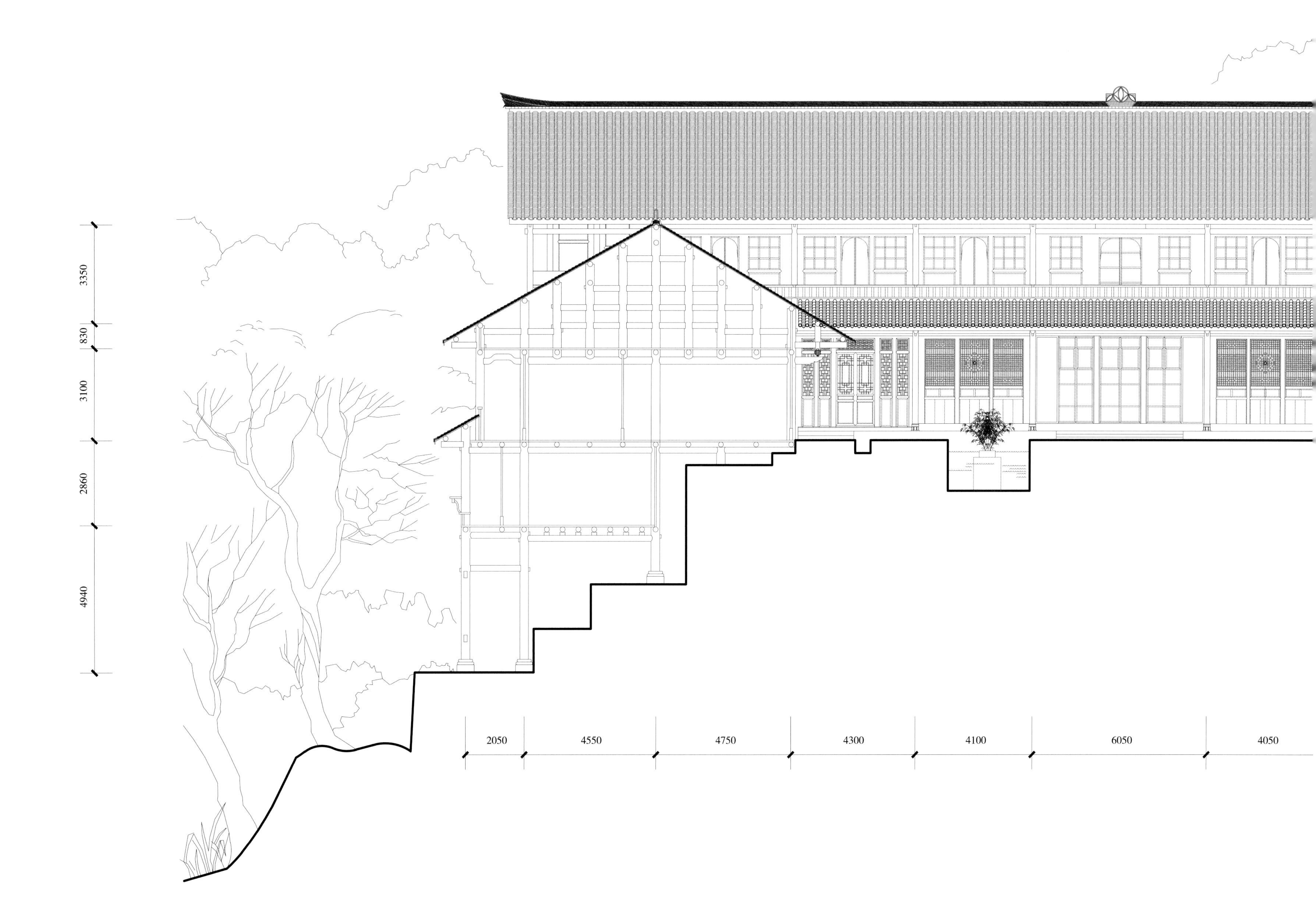
3350
830
3100
2860
4940
2050
4550
4750
4300
4100
6050
4050

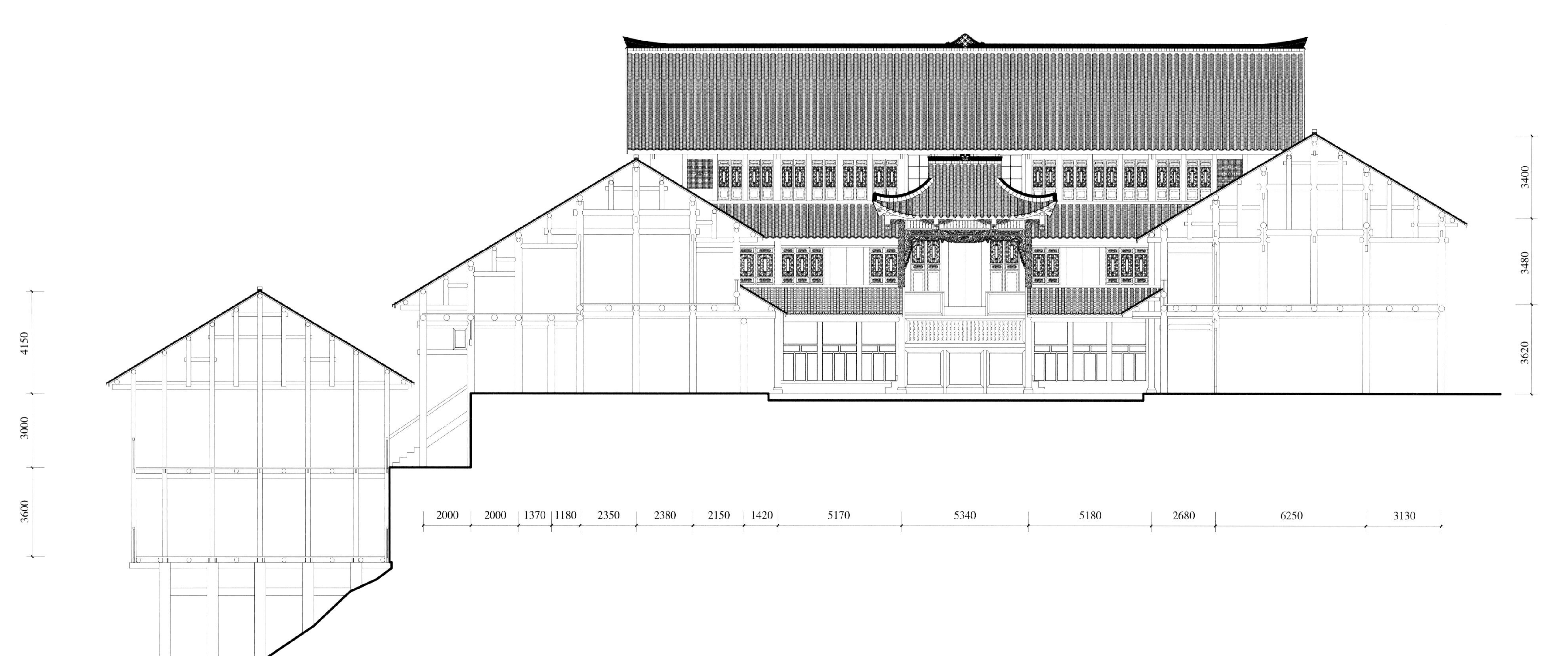
4150
3000
3600
3400
3480
3620
2000
2000
1370
1180
2350
2380
2150
1420
5170
5340
5180
2680
6250
3130
1300
2000
2000
2000
2000
1300
0 1 2 3 4 5m

洪椿坪大雄宝殿正立面图
Front Elevation of Mahavira Hall, Hongchunping Temple

3030
2940
3700
2000
2100
4220
2080
1900

0
1
2
3
4
5m

洪椿坪普贤殿横剖面图
Cross Section of Samantabhadra Hall, Hongchunping Temple

Xianfeng Temple

Xianfeng Temple sits next to the Xianfeng Rock on Mount Emei, at an altitude of 1,725m. In the 18th year of Zhiyuan period of Yuan Dynasty (AD1281) Ciyan Temple was built, which was also called Xianfeng Temple, commonly known as the Jiulao Cave. It was destroyed by fire during Tianshun period in Ming Dynasty, and expanded in the 40th year of Wanli period (AD1612). In the 17th year of Chongzhen period (AD1644) the temple was again destroyed by warfare. The existing building is mainly a reconstruction in the 44th year of Qianlong's reign in Qing Dynasty (AD1779). The temple complex, following the height differences in the terrain, gradually sets back. It is composed of a courtyard and the independent Sarira Hall, which makes up a triple-hall spatial pattern. These include, successively, the Hall to the God of Wealth, the Great Buddha Hall, and the Sarira Hall (the Bhaisajyaguru Hall). The Maitreya Hall and the Great Buddha Hall create a large courtyard along with two side buildings. The Sarira Hall was built on a terrace behind the Great Buddha Hall. Due to frequent rain and fog over the year, roofing materials are made of tin and lead, which creates a unique architectural style on Mount Emei. "Jiulao Xianfu" is one of the ten famous Emei Scenes.

仙峰寺

仙峰寺位于峨眉山仙峰岩旁，海拔1725米。元代至元十八年（1281年）创建慈延寺，亦名『仙峰寺』，俗称『九老洞』。明天顺年间毁于火，明万历四十年（1612年）扩建；明崇祯十七年（1644年）又毁于兵火，现存建筑主体为清乾隆四十四年（1779年）重建。寺庙利用地形高差往后层层退台，由一四合院与独立的舍利殿组合而成，构成三重殿堂的空间格局，依次分别是财神殿、大雄宝殿和舍利殿（药师殿）。弥勒殿和大雄宝殿与两侧厢房围合为大型的院落。舍利殿建于大雄宝殿后的台地上。因常年多数时间笼罩在雨雾之中，屋面材料采用锡和铅皮制作，构成独特的峨眉山建筑风格。『九老仙府』是峨眉十景之一。

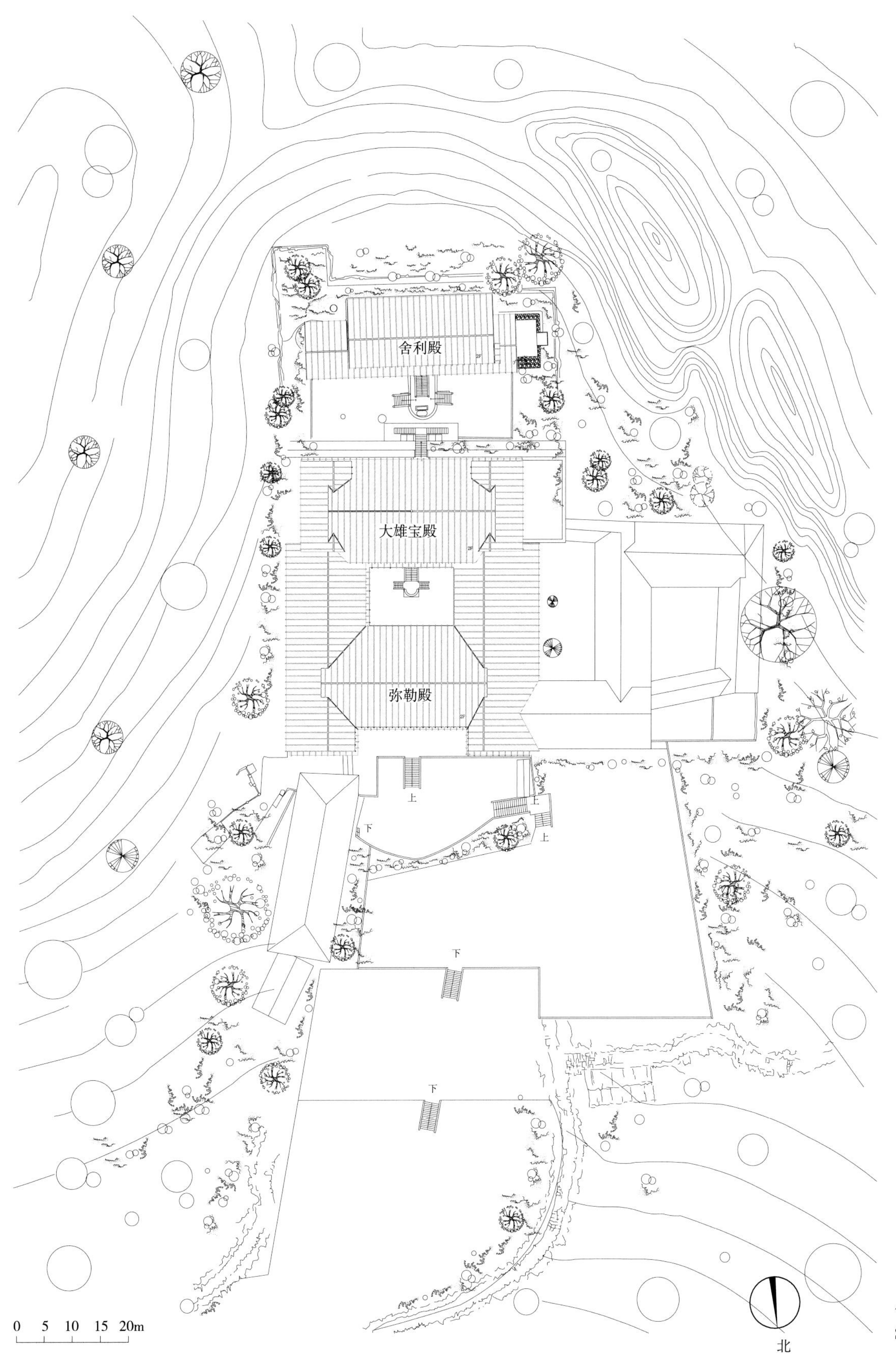
舍利殿
大雄宝殿
弥勒殿
上
下
0 5 10 15 20m
北

仙峰寺总平面图
Site Plan of Xianfeng Temple

图一 仙峰寺鸟瞰图

图二 仙峰寺总平面图

Fig.1 Bird's Eye View of Xianfeng Temple
Fig.2 Site Plan of Xianfeng Temple

图三 仙峰寺鸟瞰图

图四 仙峰寺立面图

Fig.3 Bird's Eye View of Xianfeng Temple
Fig.4 Front Elevation of Xianfeng Temple

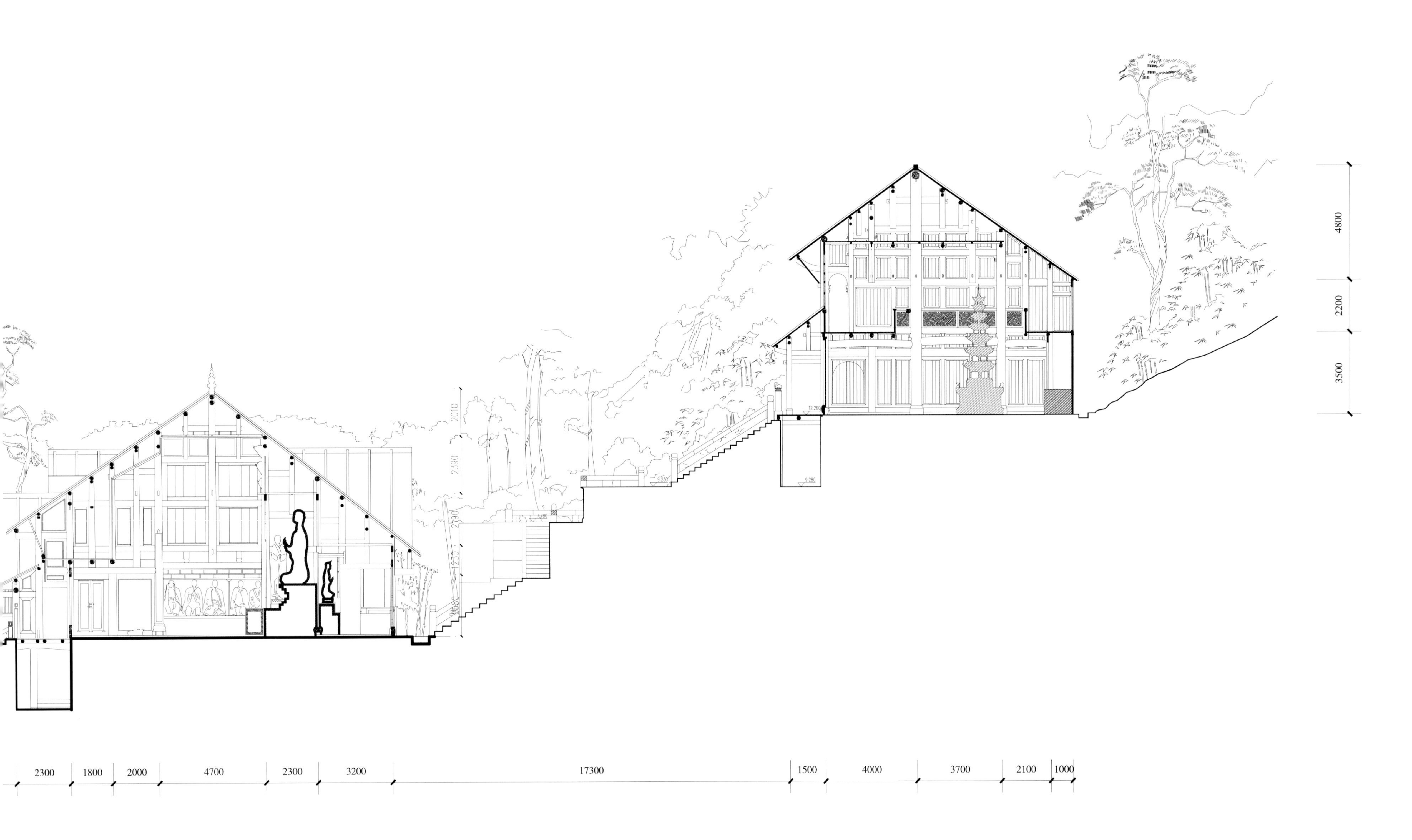
4800
2200
3500
2010
2390
2190
1230
2300
1800
2000
4700
2300
3200
17300
1500
4000
3700
2100
1000

仙峰寺中轴剖面图
Cross Section on Axis of Xianfeng Temple

0 5 10m

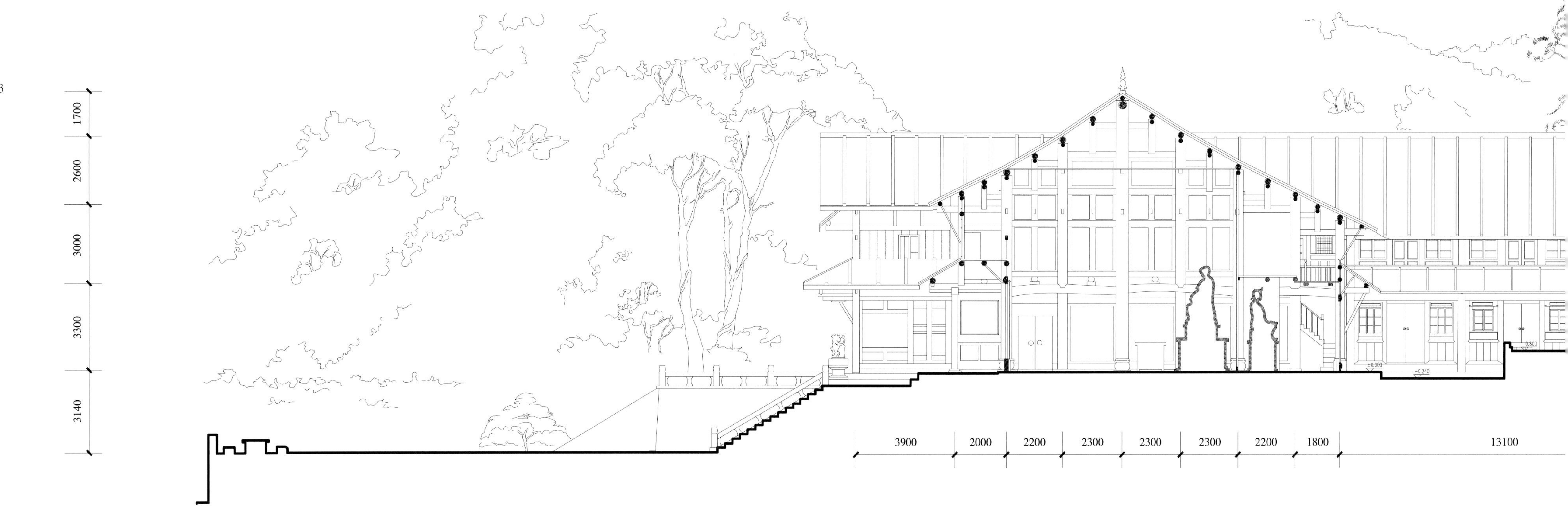
1700
2600
3000
3300
3140
3900
2000
2200
2300
2300
2300
2200
1800
13100

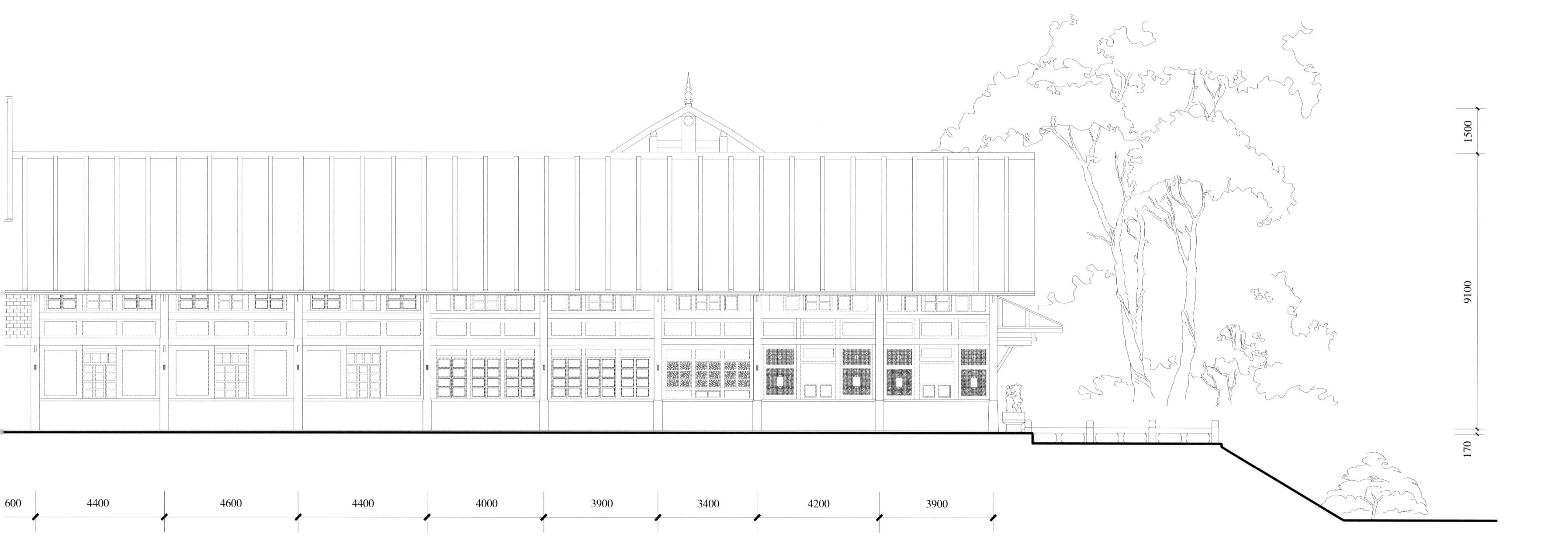
1500
9100
170
600
4400
4600
4400
4000
3900
3400
4200
3900
0
5
10m

仙峰寺侧立面图
Side Elevation of Xianfeng Temple

1100
3300
2400
1950
2100
2020
16600
5500
4650
38

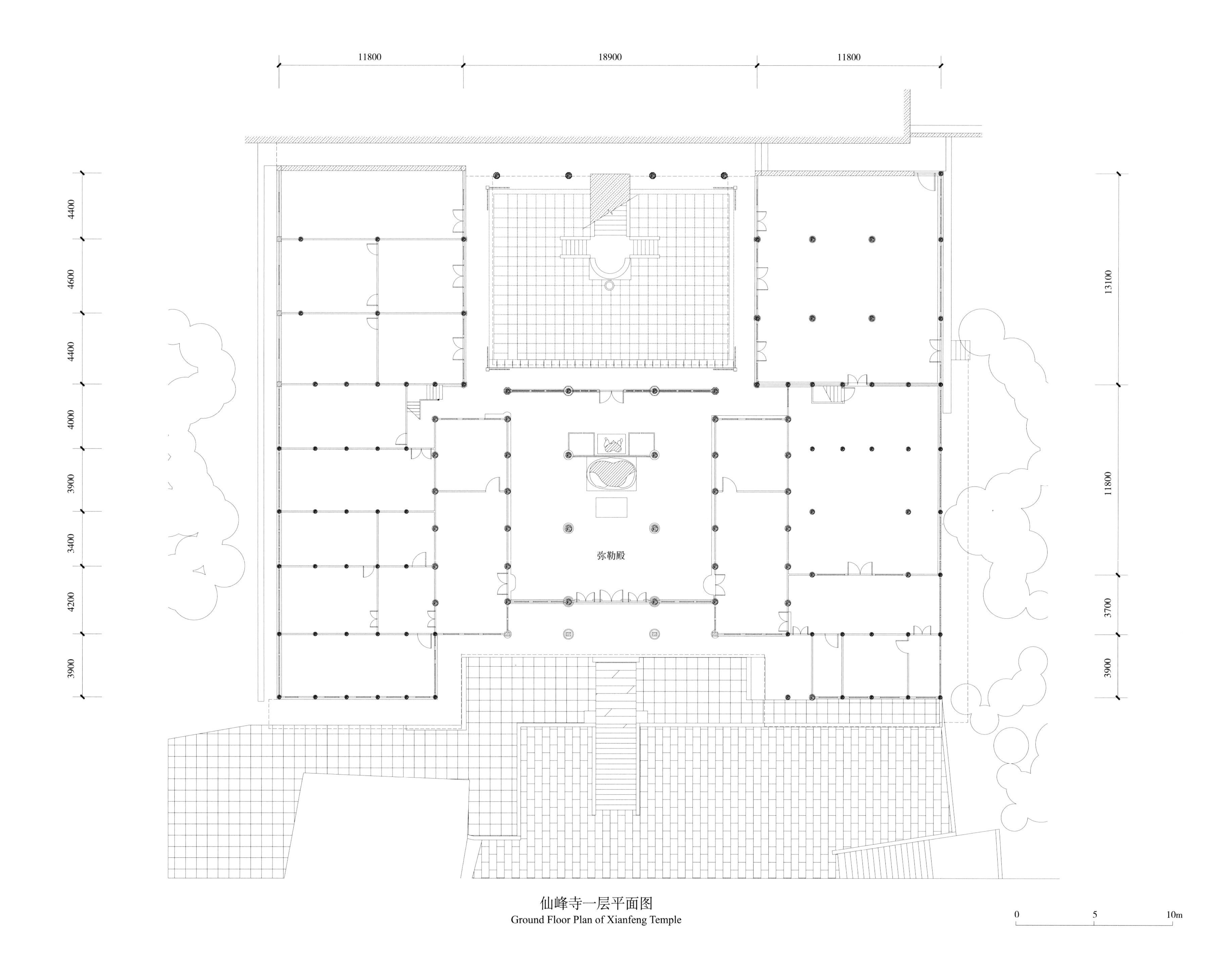
11800
18900
11800
4400
4600
4400
4000
3900
3400
4200
3900
13100
11800
3700
3900
弥勒殿
0
5
10m

仙峰寺一层平面图
Ground Floor Plan of Xianfeng Temple

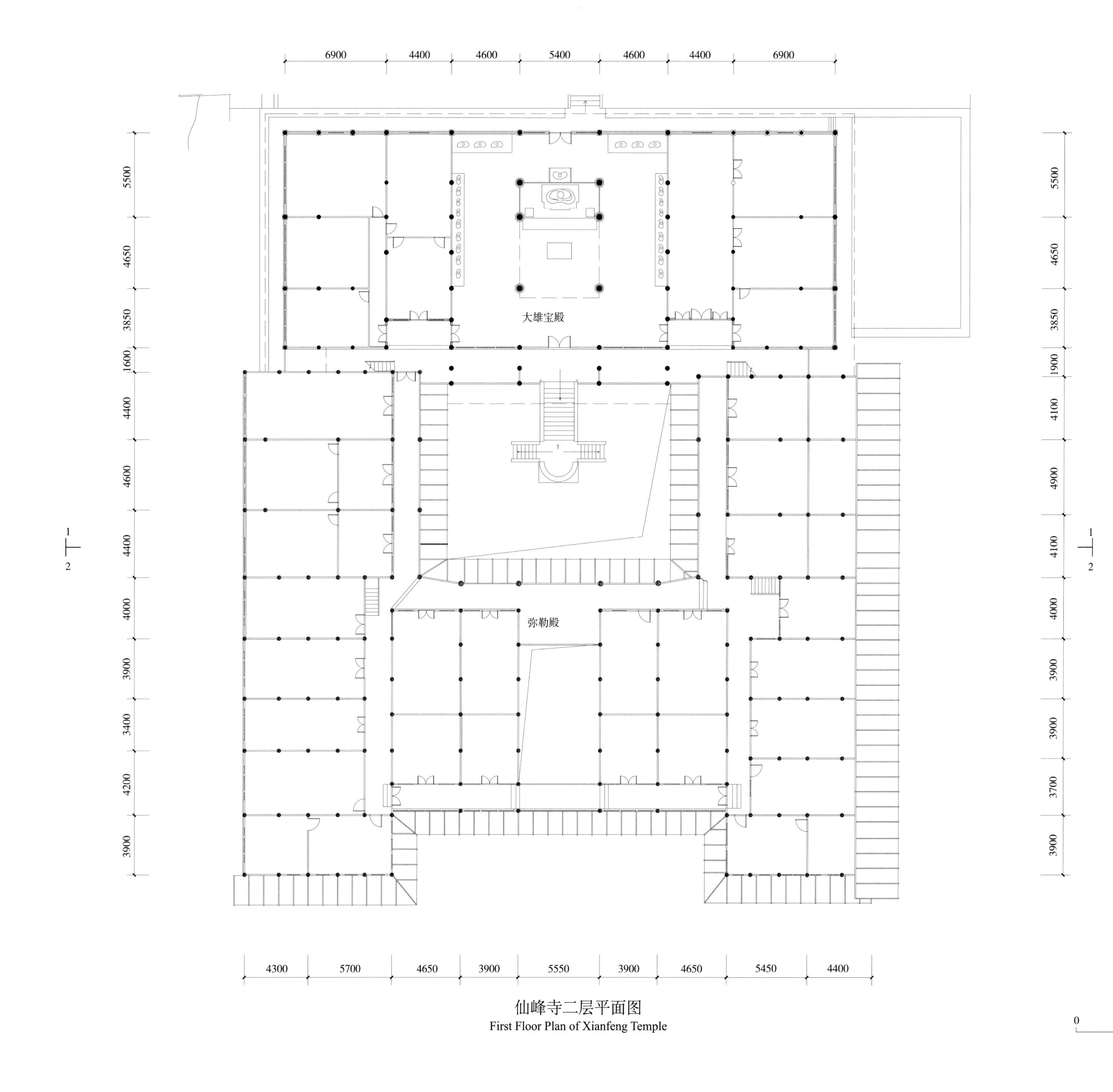
6900
4400
4600
5400
4600
4400
6900
5500
4650
3850
1600
4400
4600
4400
4000
3900
3400
4200
3900
1900
4100
4900
4100
3700
大雄宝殿
弥勒殿
1
2
4300
5700
4650
3900
5550
5450
4400
0
5
10m

仙峰寺二层平面图
First Floor Plan of Xianfeng Temple

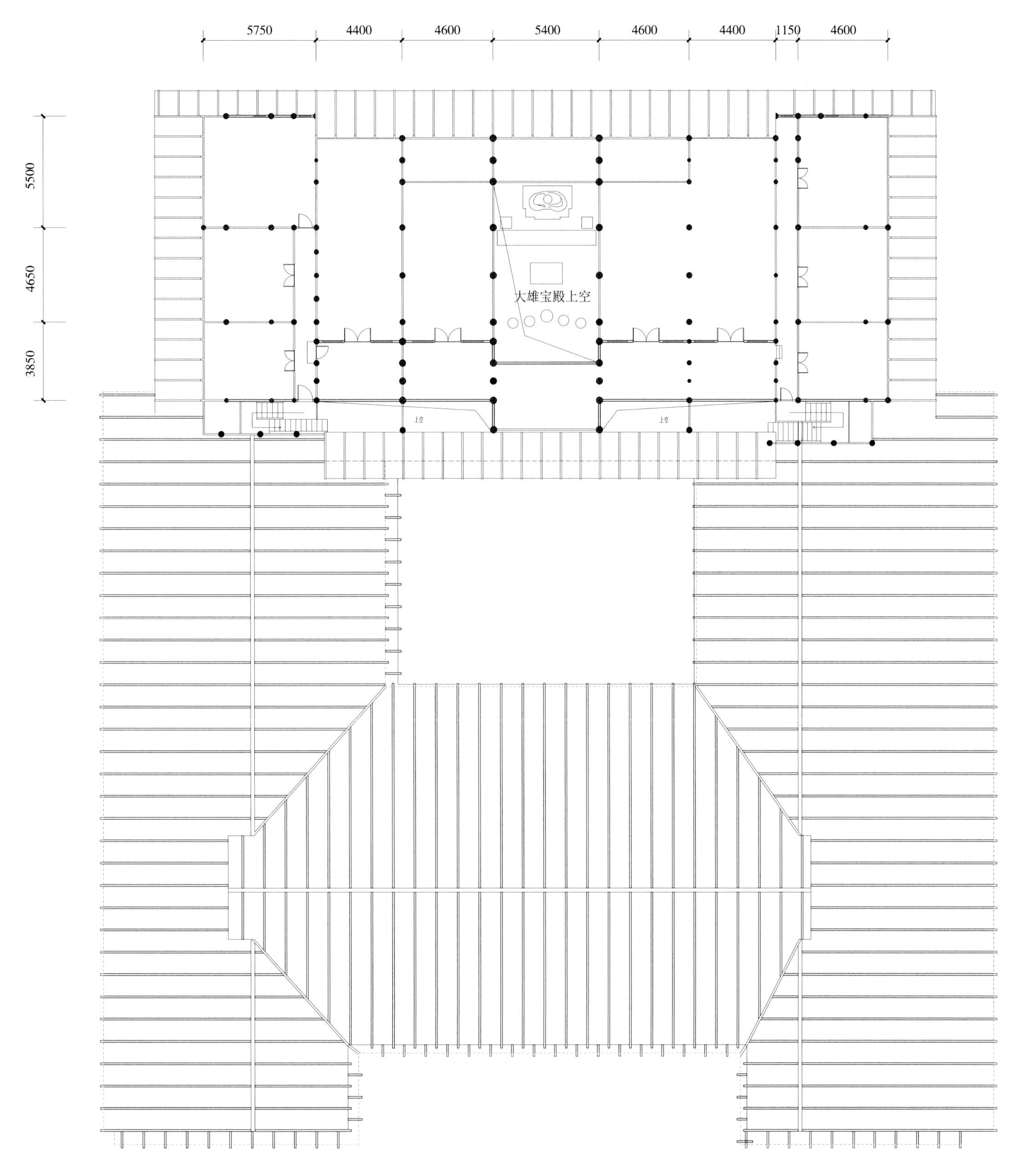
5750
4400
4600
5400
4600
4400
1150
4600
5500
4650
3850
大雄宝殿上空
上
上

仙峰寺三层平面图

Second Floor Plan of Xianfeng Temple

0 5 10m

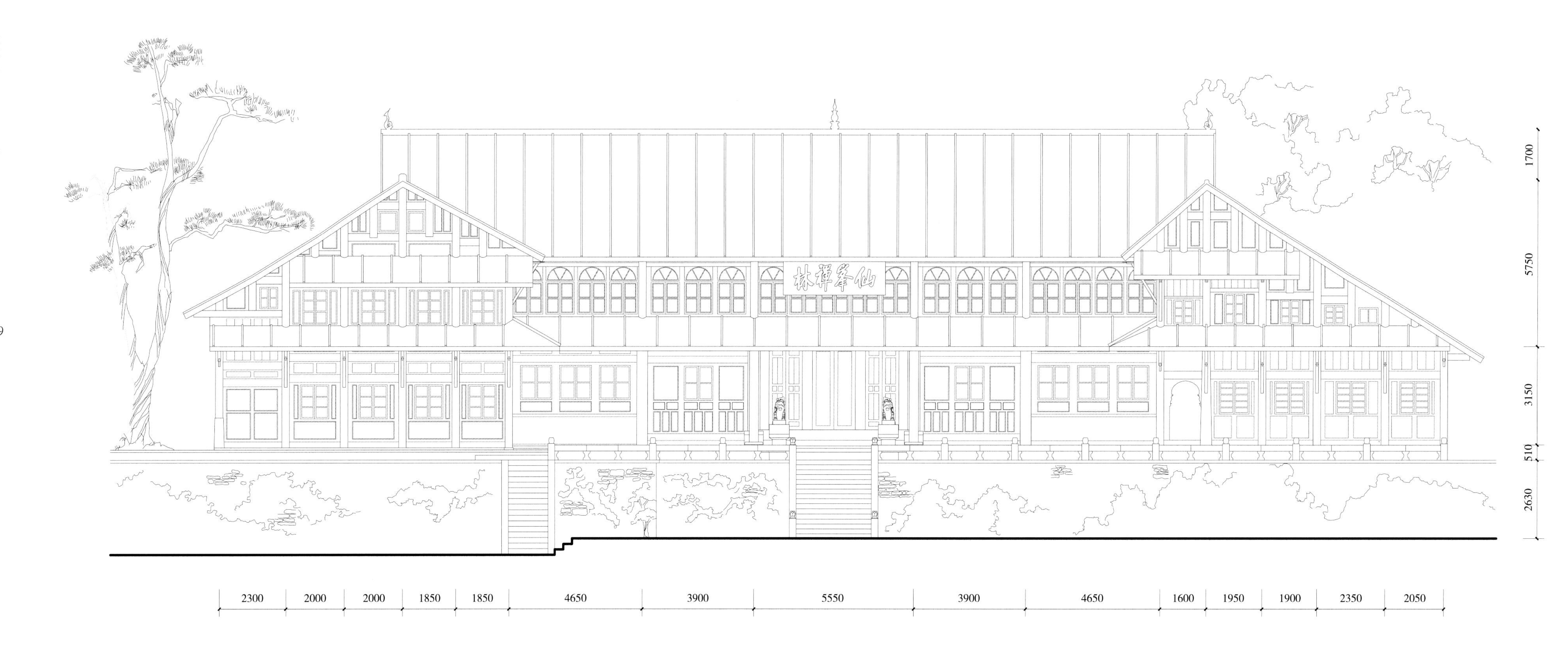
仙峰禅林
1700
5750
3150
510
2630
2300
2000
2000
1850
1850
4650
3900
5550
3900
4650
1600
1950
1900
2350
2050

仙峰寺弥勒殿正立面图
Front Elevation of Maitreya Hall, Xianfeng Temple

0 1 2 3 4 5m

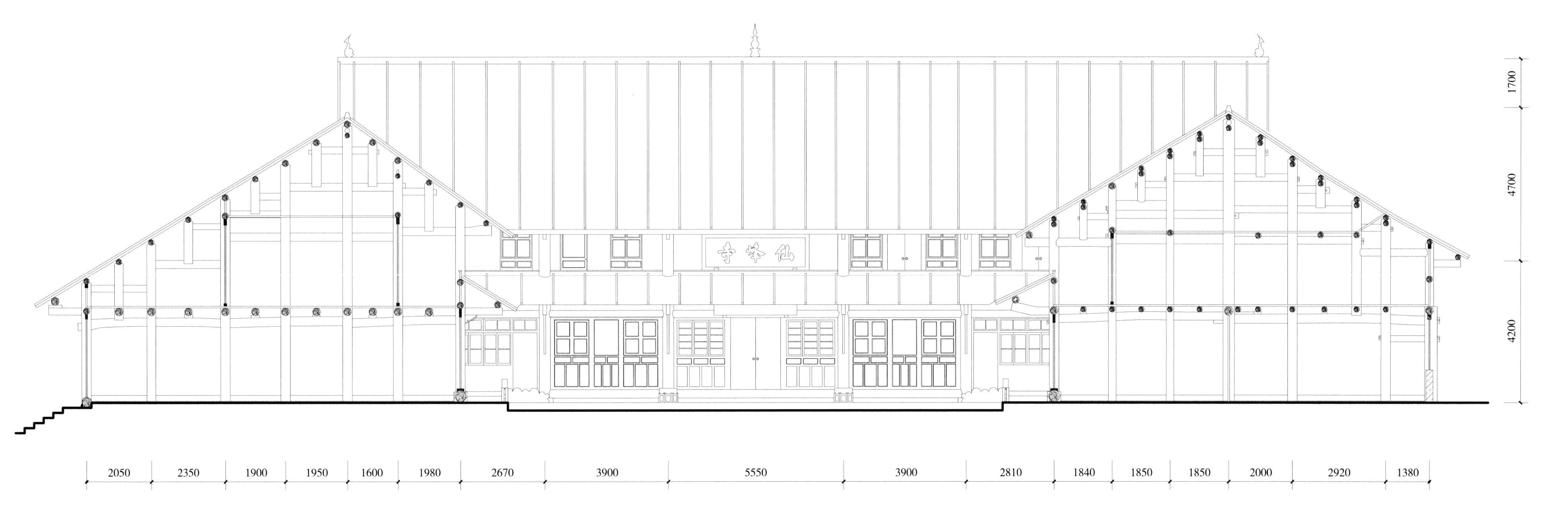
仙峰寺
1700
4700
4200
2050
2350
1900
1950
1600
1980
2670
3900
5550
3900
2810
1840
1850
1850
2000
2920
1380
0 1 2 3 4 5m

仙峰寺弥勒殿背立面图
Back Elevation of Maitreya Hall, Xianfeng Temple

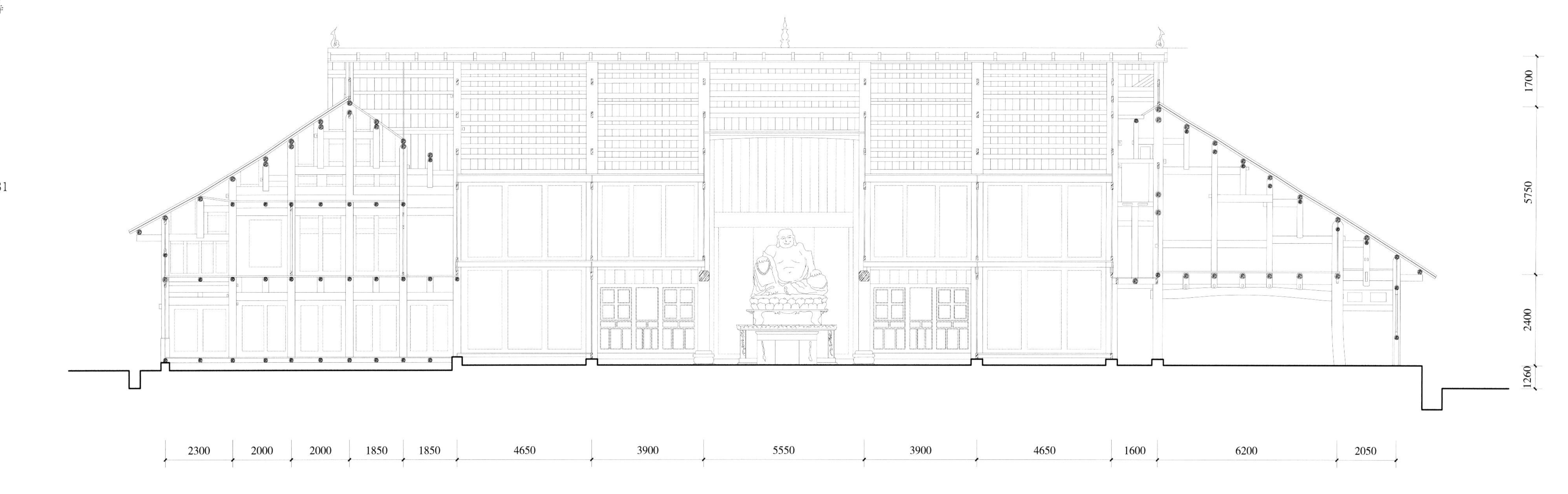
1700
5750
2400
1260
2300
2000
2000
1850
1850
4650
3900
5550
3900
4650
1600
6200
2050
0 1 2 3 4 5m

仙峰寺弥勒殿纵剖面图
Longitudinal Section of Maitreya Hall, Xianfeng Temple

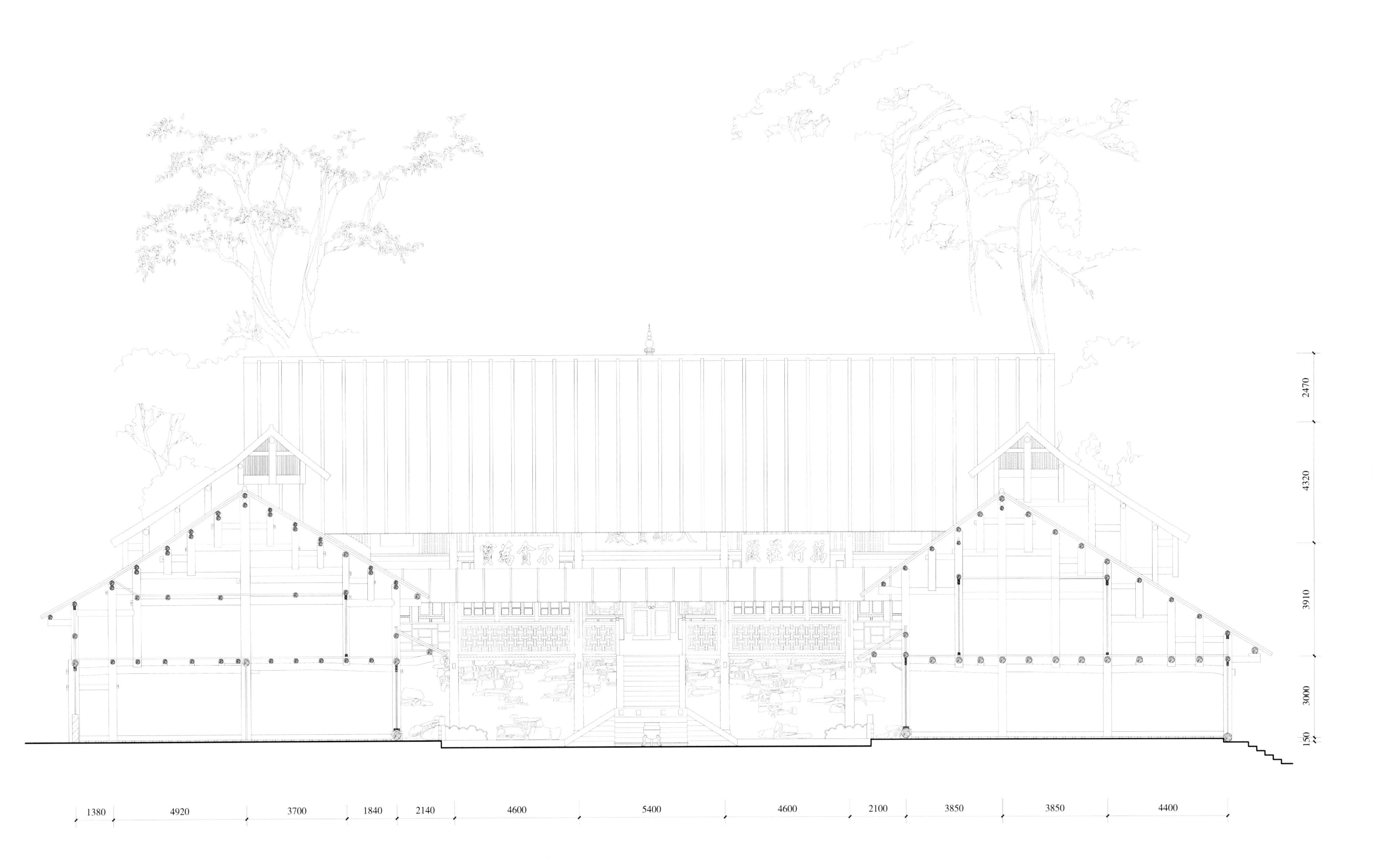

2470
4320
3910
3000
150
1380
4920
3700
1840
2140
4600
5400
4600
2100
3850
3850
4400

仙峰寺大雄宝殿正立面图
Front Elevation of Mahavira Hall, Xianfeng Temple

0 1 2 3 4 5m

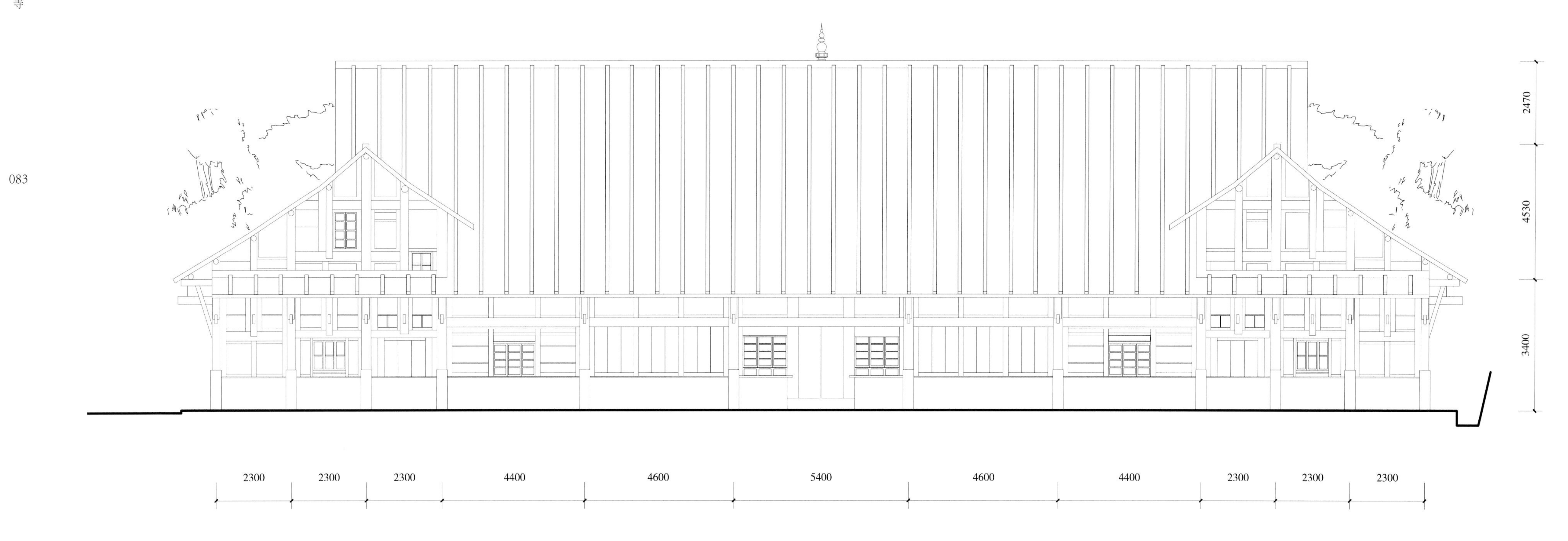
2470
4530
3400
2300
2300
2300
4400
4600
5400
4600
4400
2300
2300
2300

仙峰寺大雄宝殿背立面图
Back Elevation of Mahavira Hall, Xianfeng Temple

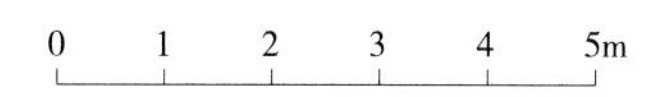
0 1 2 3 4 5m

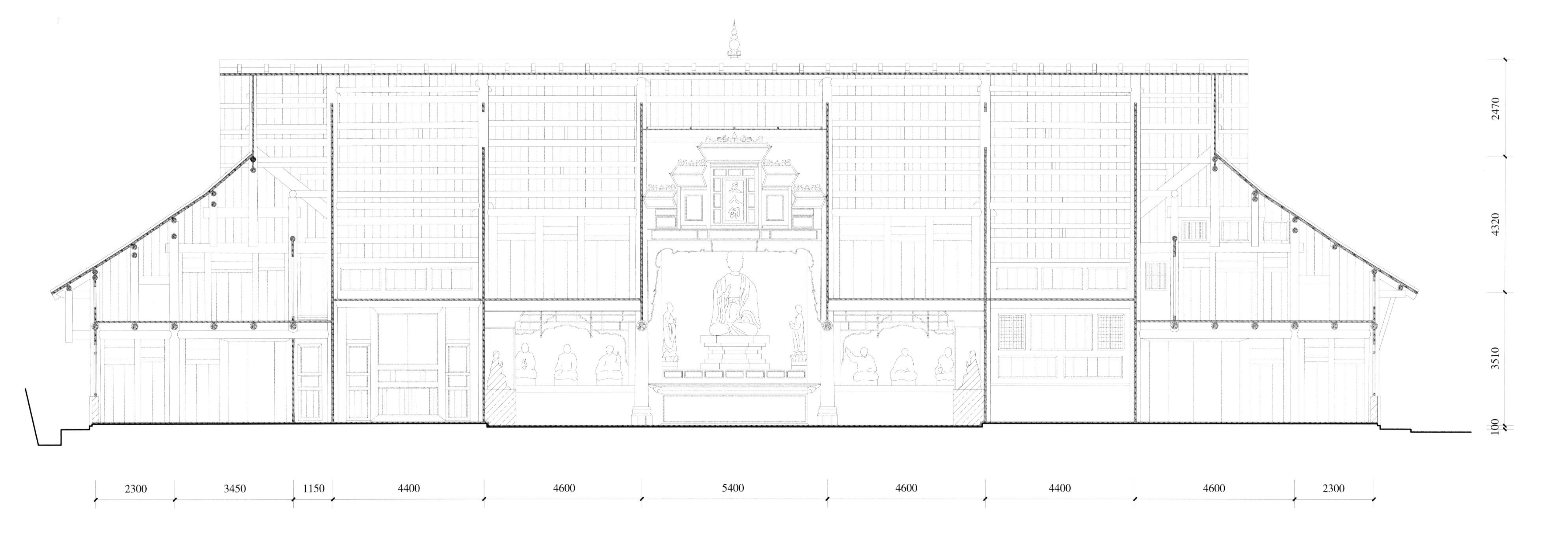
2470
4320
3510
100
2300
3450
1150
4400
4600
5400
4600
4400
4600
2300
0 1 2 3 4 5m

仙峰寺大雄宝殿纵剖面图
Longitudinal Section of Mahavira Hall, Xianfeng Temple

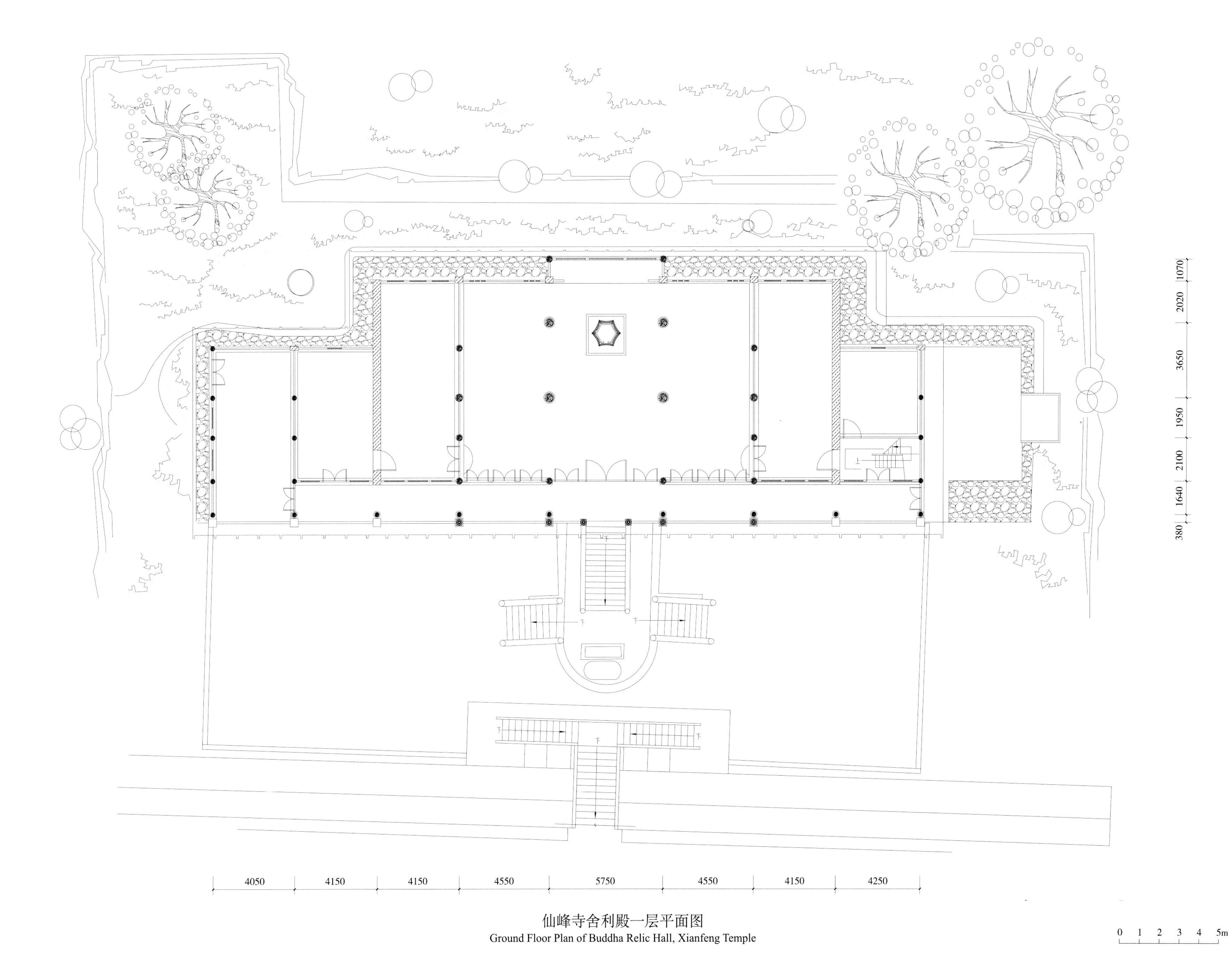
1070
2020
3650
1950
2100
1640
380
4050
4150
4150
4550
5750
4550
4150
4250
0 1 2 3 4 5m

仙峰寺舍利殿一层平面图
Ground Floor Plan of Buddha Relic Hall, Xianfeng Temple

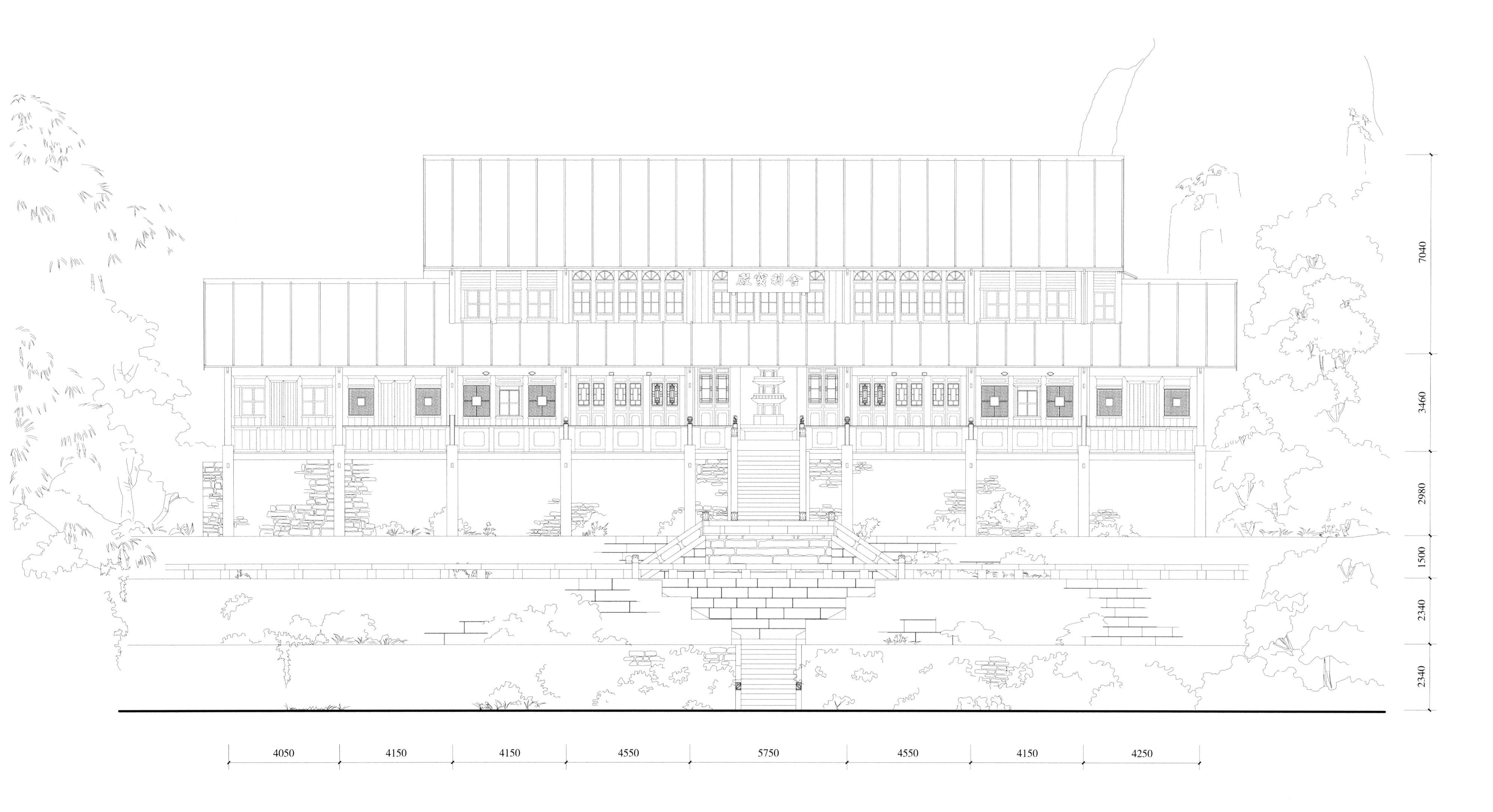
舍利宝殿
7040
3460
2980
1500
2340
2340
4050
4150
4150
4550
5750
4550
4150
4250

仙峰寺舍利殿正立面图
Front Elevation of Buddha Relic Hall, Xianfeng Temple

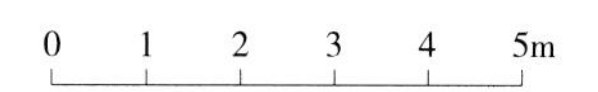
0 1 2 3 4 5m

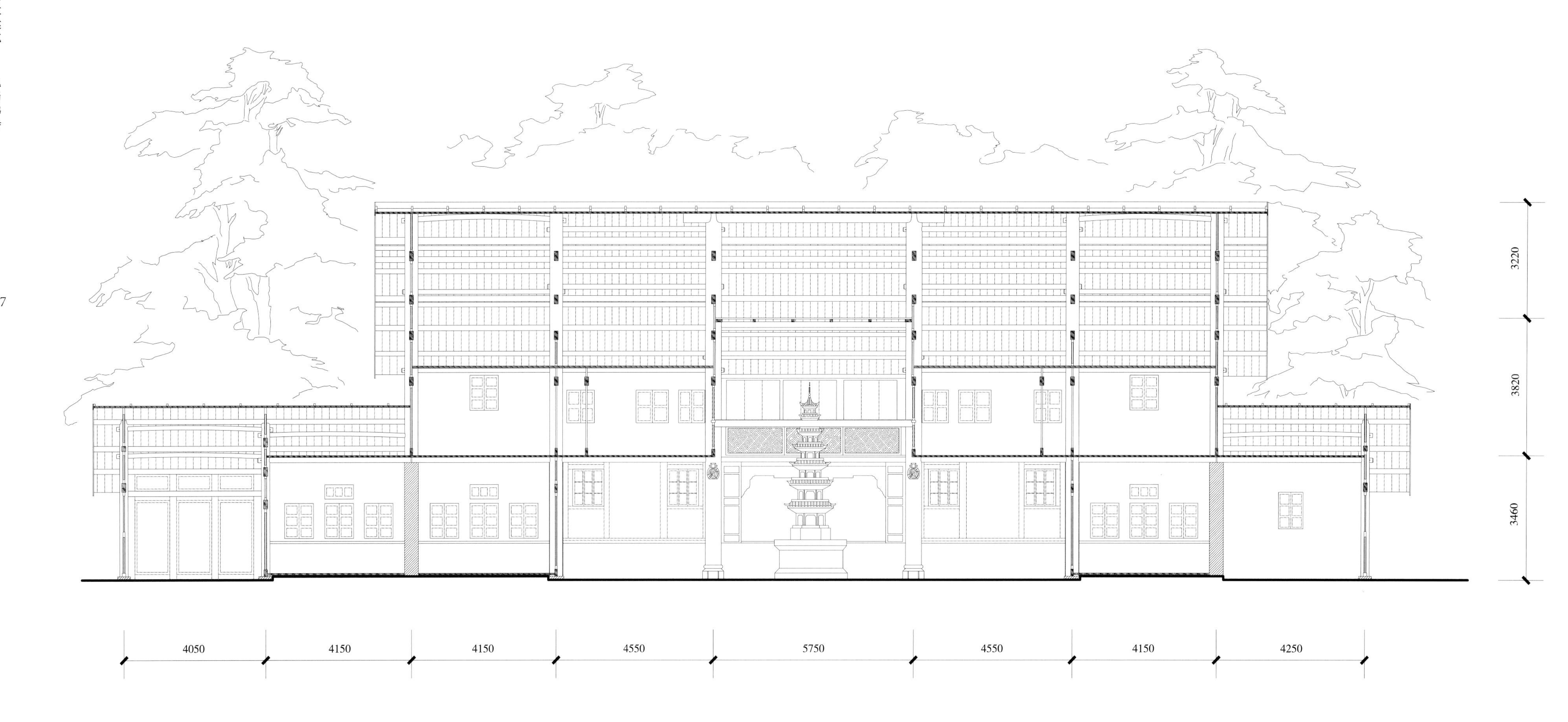
3220
3820
3460
4050
4150
4150
4550
5750
4550
4150
4250
0
1
2
3
4
5m

仙峰寺舍利殿纵剖面图
Longitudinal Section of Buddha Relic Hall, Xianfeng Temple

Xixiang Pool

The Xixiang (literally, washing the elephant) Pool is located on the Zuantian Slope of Mount Emei, at an altitude of 2,070m. Originally built as Chuxi Pavilion during Zhengde period in Ming Dynasty (AD1506-1510), it was later named Chuxi Convent. In the 38th year Kangxi's reign of Qing Dynasty (AD1699), it was expanded into a compound named Tianhua Zen Courtyard, commonly known as the Xixiang Pool. This name was officially recognized for the hexagon pool built in front of the complex as an interpretation of the legend "when Samantabhadra rode on an elephant for travelling up the mountain, the bodhisattva stopped by the pool to draw water for washing the elephant". The complex, facing north, occupies an elevated terrace of Zuantian Slope in a steep topography. Most of the existing buildings were remodeled in the late years of Guangxu period in Qing Dynasty. The trio of the Maitreya Hall, the Great Buddha Hall and the Avalokiteshvara Hall occupies two courtyards in the front and back. The internal space is compact and the axial spatial sequence clear. In the rugged external terrain, a steep climbing slope is in the north, while in the east and west are steep cliffs and deep valleys. The buildings are adapted to the terrain for spatial organization. In the east, platforms are built for the uphill pilgrimage, whereas in the west overhangs are used for greater internal space. Due to frequent rain and fog over the year, buildings are covered by tin tiles and lead rooftops. There are also steep slopes for rainwater drainage, which make up the unique architectural appearance of Xixiang Pool. Xixiang Pool is an important passage for the pilgrimage up the mountain, which also embodies the interpretation of the elephant washing legend. This is known as "Yueye Xixiang", one of the ten famous Emei Scenes.

洗象池

洗象池位于峨眉山钻天坡上，海拔2070米。始建于明正德年间（1506—1510年），名初喜亭，后名『初喜庵』，清康熙三十八年（1699年）扩建为大院，名天花禅院，俗称『洗象池』。因寺前一侧建六方水池，以附会『普贤骑象登山，曾在池中汲水洗象』的传说，后正式命名为洗象池。寺踞钻天坡高岗台地，形势险奇，坐南向北。现存建筑，大部分为清光绪末年改建。殿宇三重，按弥勒、大雄、观音殿，由前后两重院落空间构成。内部空间紧凑而轴线空间序列明确。外部地形环境十分险要，北面是陡峭的爬山坡道，东西两侧均是险峻悬崖深壑。建筑外观利用地形组织空间，东面通过筑台构成上山朝拜的通道，西面利用吊脚出挑手法，建筑凌空吊脚以争取更大内部空间。因常年多数时间笼罩在雨雾之中，建筑的屋面采用锡瓦和铅皮覆盖，屋顶坡度陡峭以利排水，构成洗象池独特的建筑风貌。洗象池是上山朝拜金顶的重要通道，并有洗象的附会，是峨眉十景之一的『月夜洗象』。

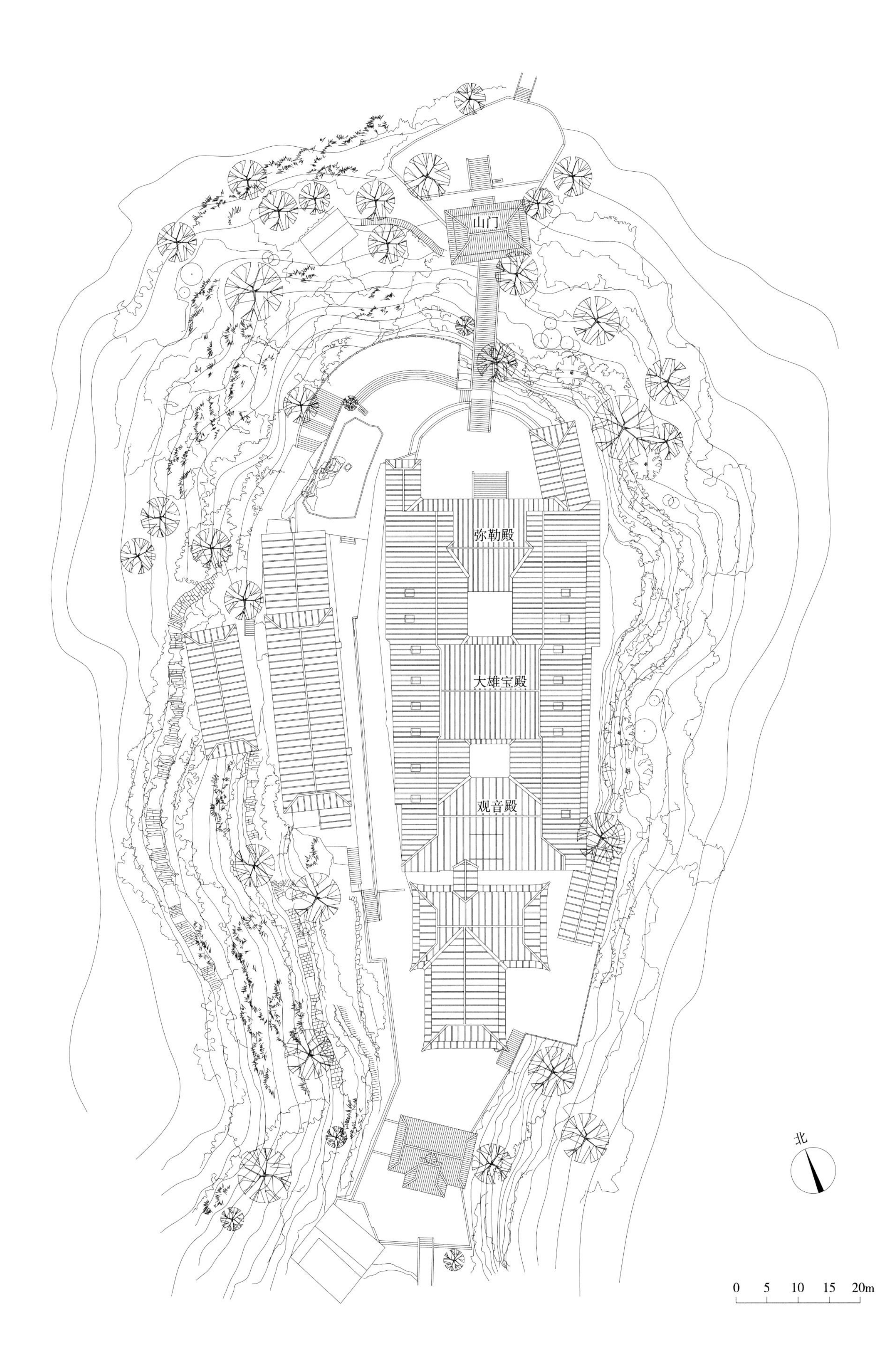
山门
弥勒殿
大雄宝殿
观音殿
北
0 5 10 15 20m

洗象池总平面图
Site Plan of Xixiang Pool

图一 洗象池鸟瞰图

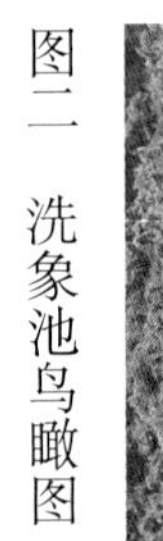

图二 洗象池鸟瞰图

Fig.1 Bird's Eye View of Xixiang Pool
Fig.2 Bird's Eye View of Xixiang Pool

图三 洗象池鸟瞰图

图四 洗象池正立面

Fig.3 Bird's Eye View of Xixiang Pool
Fig.4 Front Elevation of Xixiang Pool

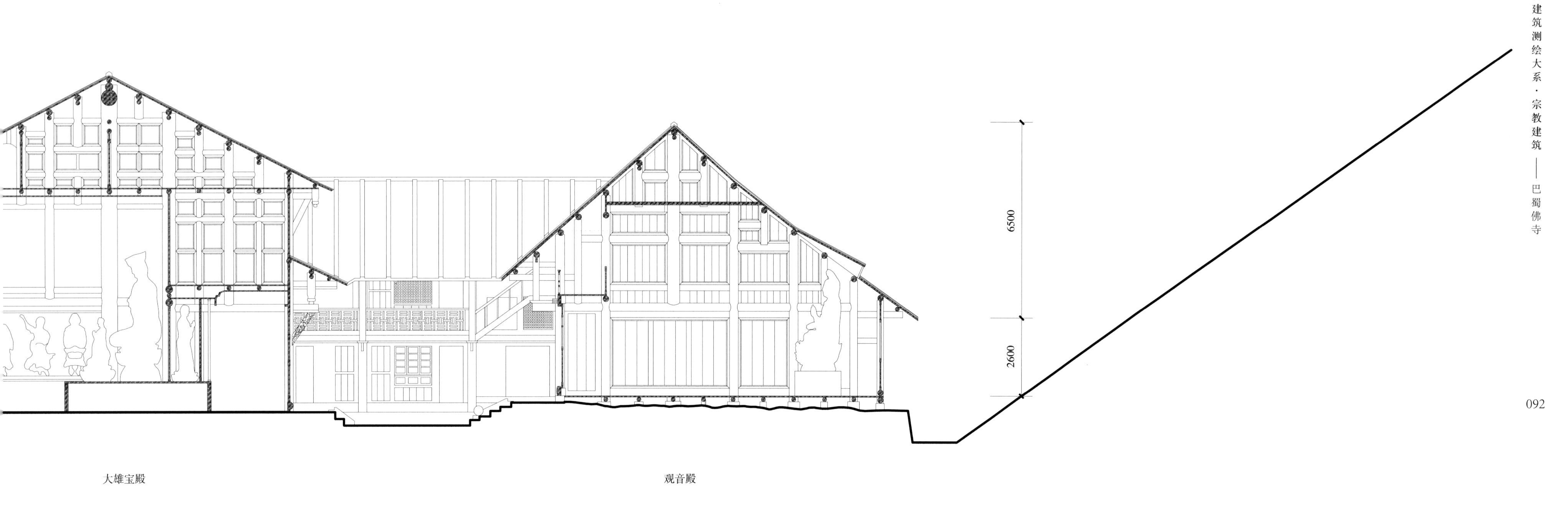

6500
2600
大雄宝殿
观音殿

洗象池中轴剖面图
Cross Section on Axis of Xixiang Pool

0 1 2 3 4 5m

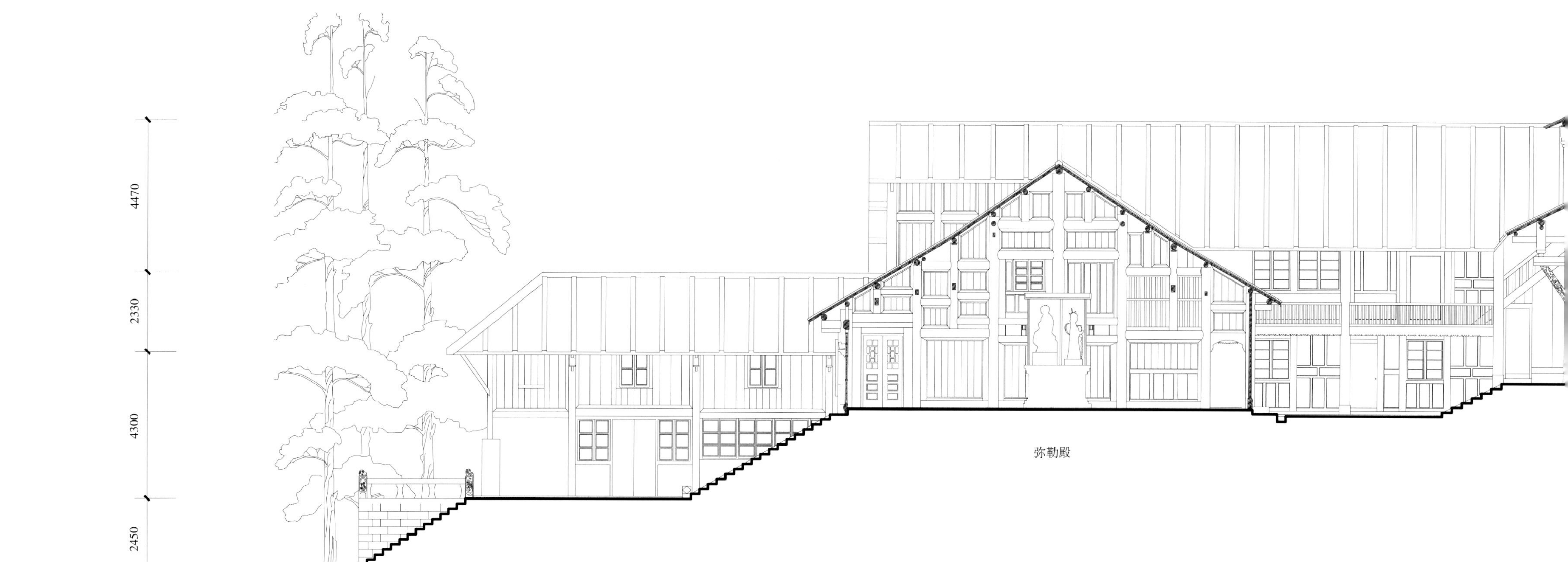
4470
2330
4300
2450
弥勒殿

6250
3900
1300
2050
2300
1950
1950
2600
4700
4700
1000
3500
4100
4100
4600
3700
3000
1500
4500
4200
5790
3300
4200
2100
1200
3000
4200
4200
1200
1100
4350
3900
4150
5300
4150
3600
1800
3300
弥勒殿
大雄宝殿
观音殿
0
5
10
15
20m

洗象池一层平面图

Ground Floor Plan of Xixiang Pool

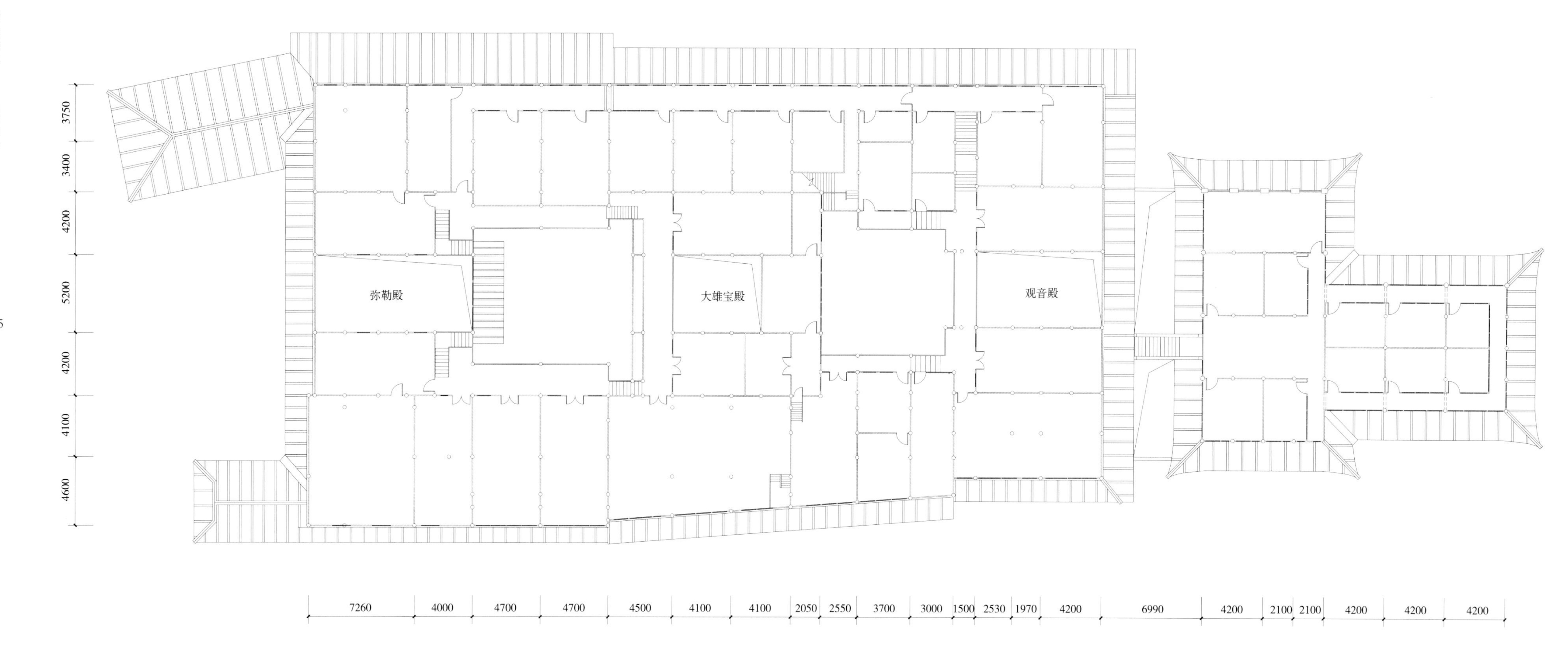
弥勒殿
大雄宝殿
观音殿
3750
3400
4200
5200
4200
4100
4600
7260
4000
4700
4700
4500
4100
4100
2050
2550
3700
3000
1500
2530
1970
4200
6990
4200
2100
2100
4200
4200
4200
0
5
10m

洗象池二层平面图
First Floor Plan of Xixiang Pool

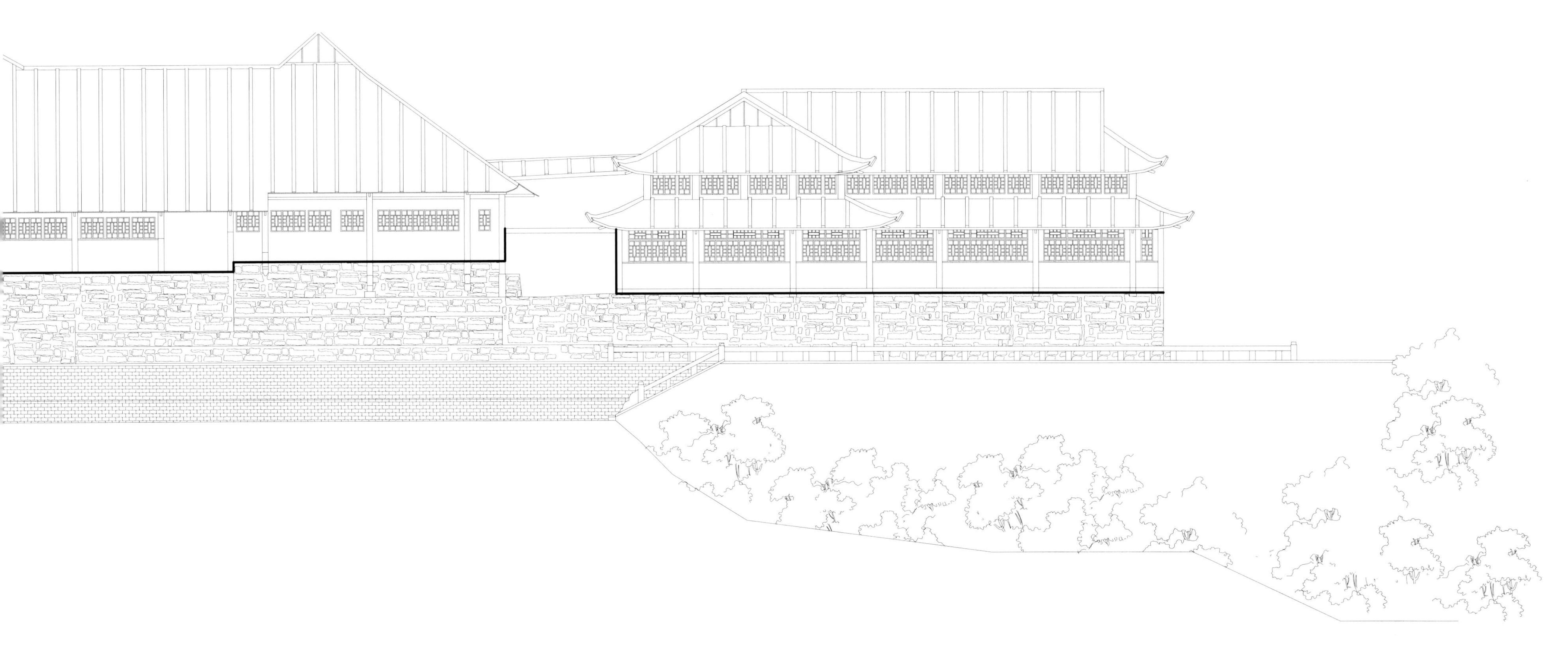

洗象池西立面图
West Elevation of Xixiang Pool

0 5 10m

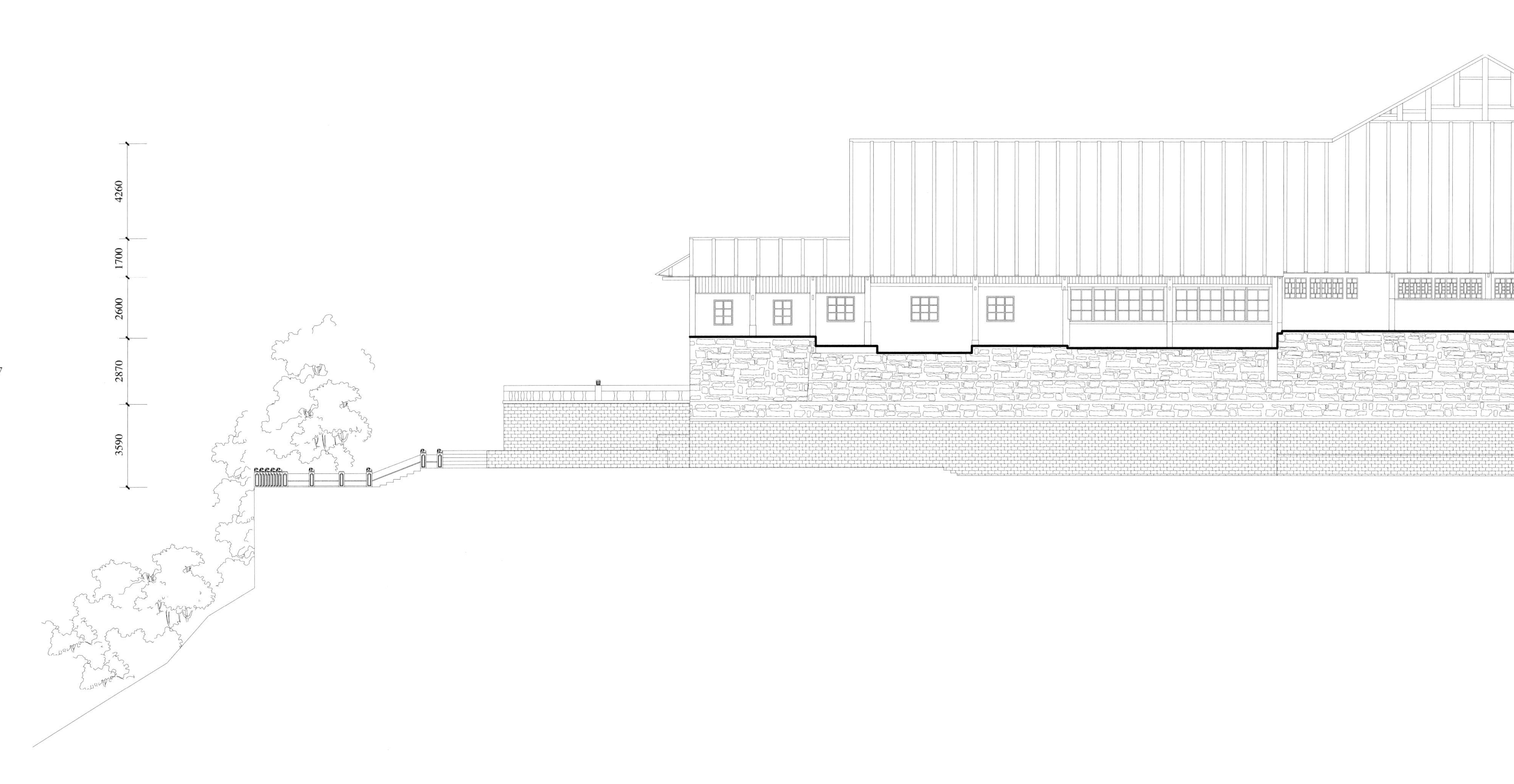
4260
1700
2600
2870
3590

洗象池东立面图
East Elevation of Xixiang Pool

0
5
10m

3400
2300
2500
3500

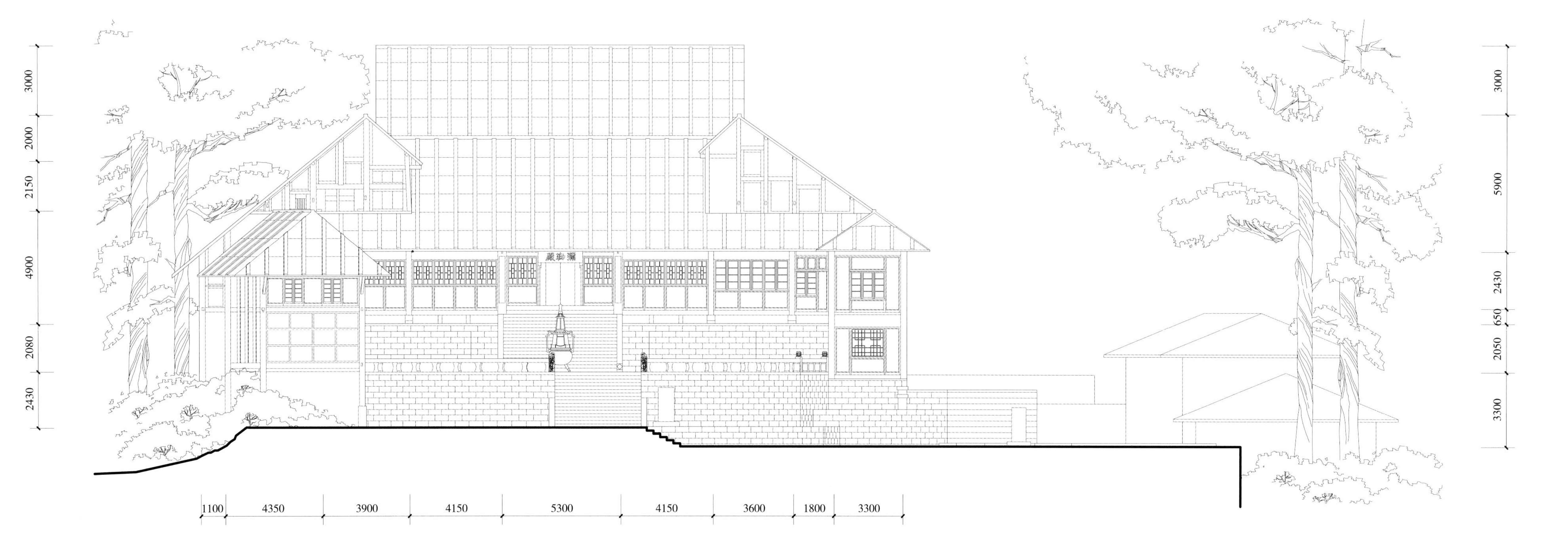

3000
2000
2150
4900
2080
2430
3000
5900
2430
650
2050
3300
1100
4350
3900
4150
5300
4150
3600
1800
3300

洗象池弥勒殿正立面图

Front Elevation of Maitreya Hall, Xixiang Pool

0 1 2 3 4 5m

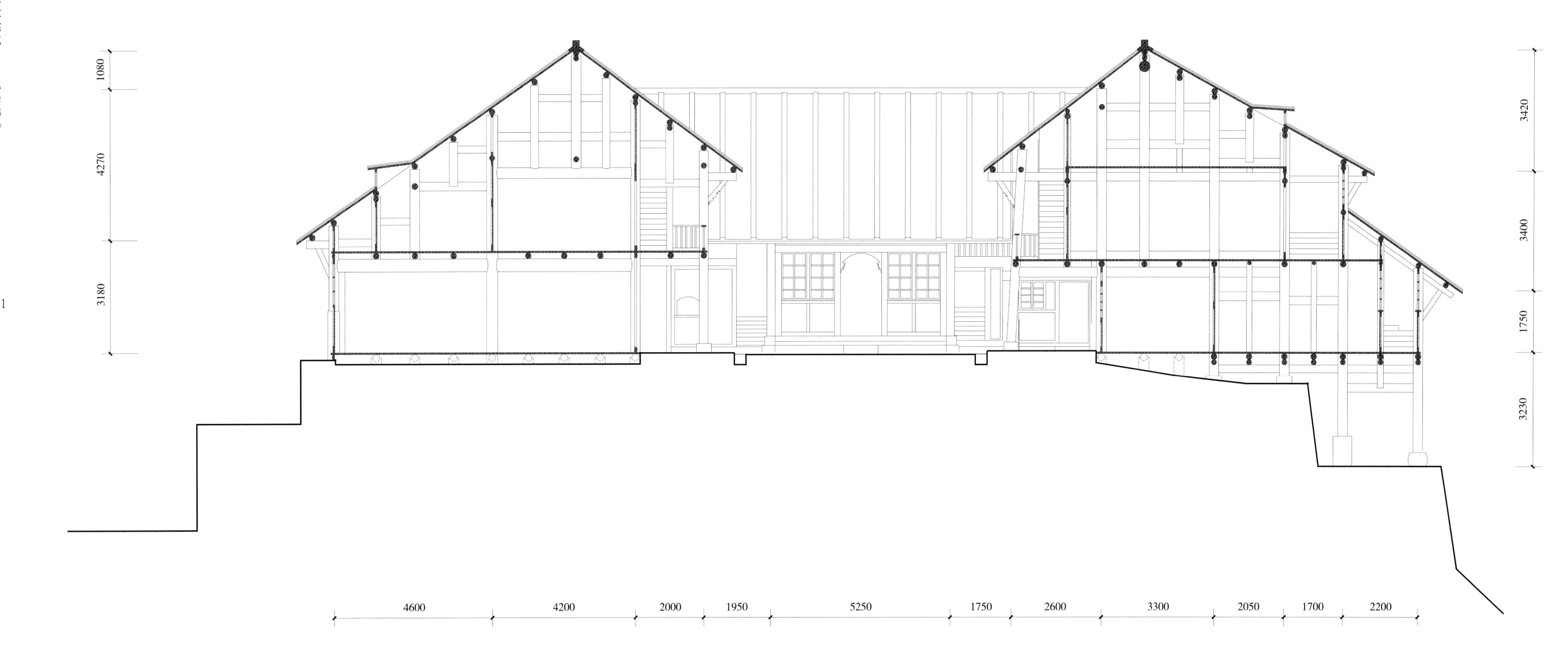
1080
4270
3180
3420
3400
1750
3230
4600
4200
2000
1950
5250
1750
2600
3300
2050
1700
2200
0 1 2 3 4 5m

洗象池弥勒殿背立面图

Back Elevation of Maitreya Hall, Xixiang Pool

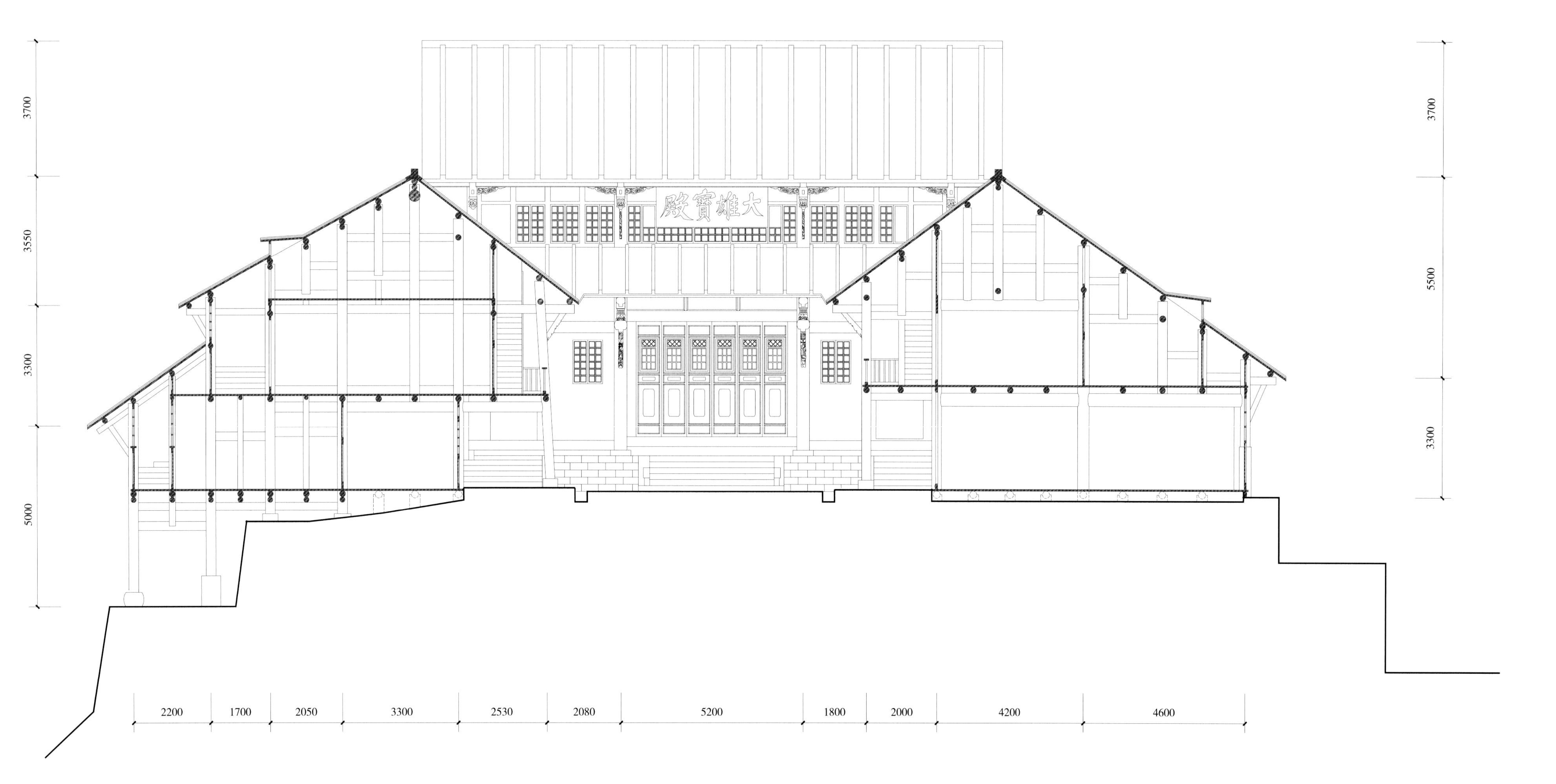
大雄寶殿
3700
3550
3300
5000
3700
5500
3300
2200
1700
2050
3300
2530
2080
5200
1800
2000
4200
4600

洗象池大雄宝殿正立面图
Front Elevation of Mahavira Hall, Xixiang Pool

0 1 2 3 4 5m

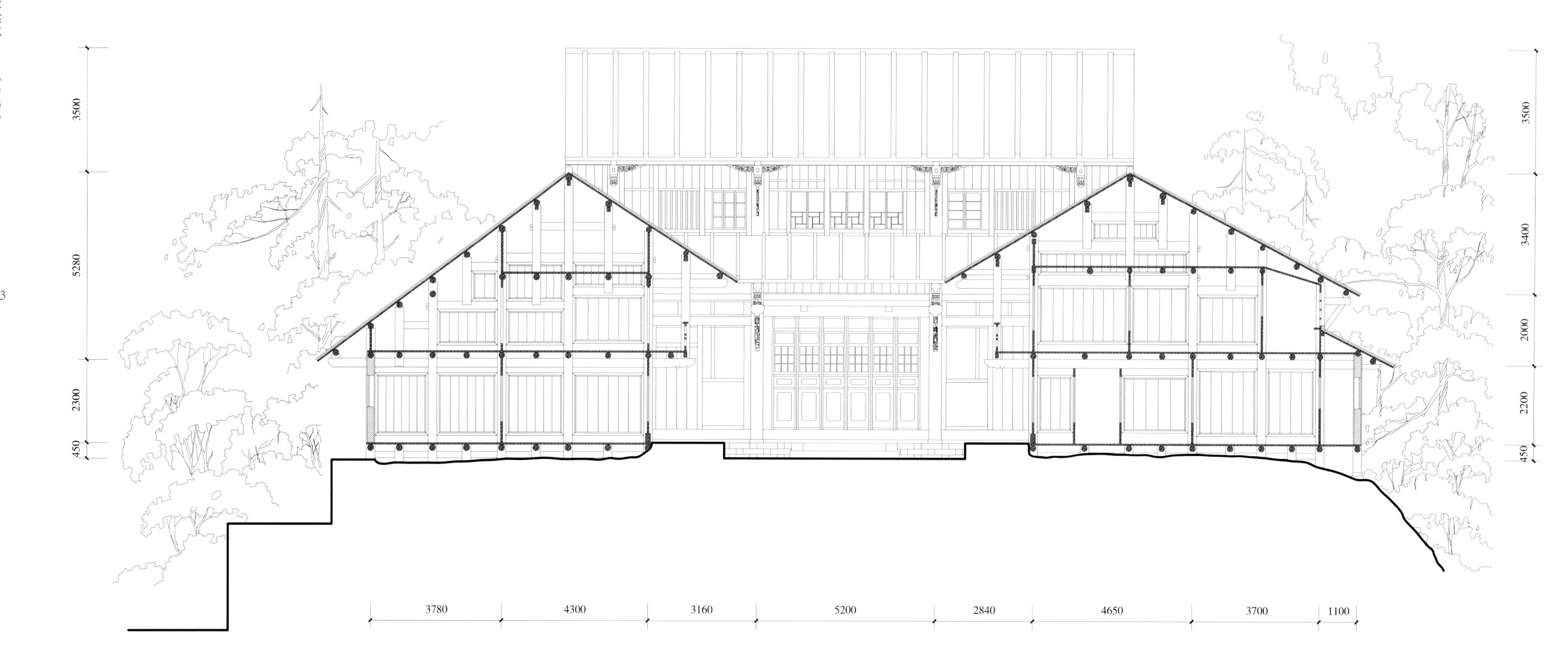
3500
5280
2300
450
3500
3400
2000
2200
450
3780
4300
3160
5200
2840
4650
3700
1100
0 1 2 3 4 5m

洗象池大雄宝殿背立面图
Back Elevation of Mahavira Hall, Xixiang Pool

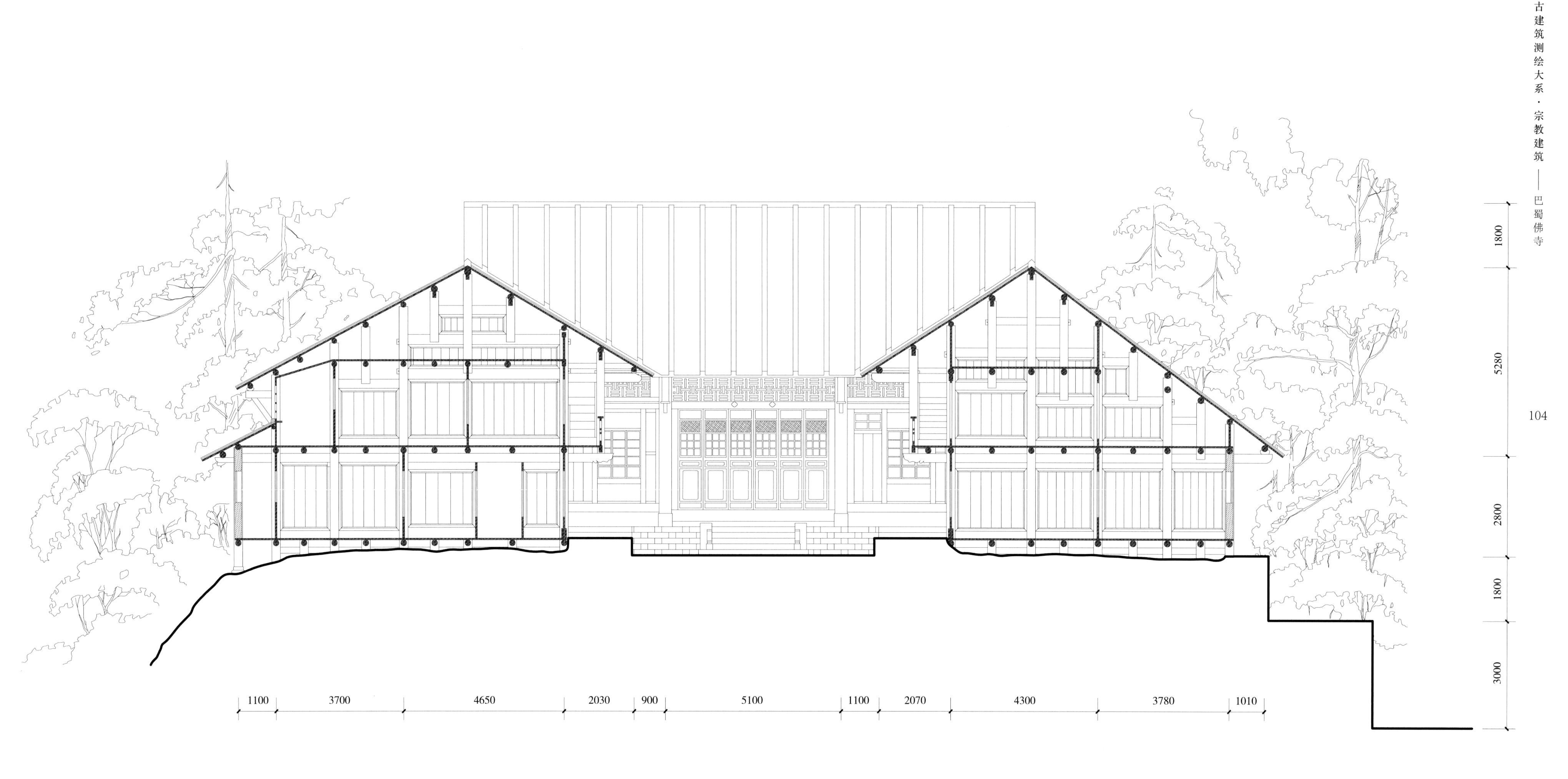
1800
5280
2800
1800
3000
1100
3700
4650
2030
900
5100
1100
2070
4300
3780
1010

洗象池观音殿正立面图
Front Elevation of Guanyin Hall, Xixiang Pool

0 1 2 3 4 5m

Bao'en Temple in Pingwu

Bao'en Temple is located in Pingwu County, Sichuan Province. Pingwu County belonged historically to Long'an Prefecture. The temple was originally built in the 4th year of Zhengtong period in Ming Dynasty (AD1439), and completed by the 4th year of Tianshun period (AD1460). The structure, though damaged in the 2008 Wenchuan earthquake, was restored, and still retained the originals of the Ming Dynasty.

The main building of the Bao'en Temple faces east, and the building complex, following the mountain landscape, is composed of three courtyards. In front of the entrance gate, a pair of stone Dharani dhvaja pillars are symmetrically arranged. On the central axis were successively the entrance gate, the Hall of Heavenly Kings, the Great Buddha's Hall and the Ten-Thousand-Buddha Building. In the first courtyard a two-story bell tower stands in the north, while the Huayan Hall and Dabei Hall occupies the south and north of the second courtyard. In the third courtyard, on either side of the axis were symmetrically a pair of Imperial Stele Pavilions. The entrance gate and the Hall of Heavenly Kings were connected with three bridges. Major components of the *Dashi* style buildings in the whole complex were constructed with top quality Nanmu timber, with unadorned bracket sets and regular building layout. Inside the halls, there are exquisite timber fixtures, polychromatic paintings, and murals. The whole building complex, from the layout to architectural form, demonstrate strong characteristics of the northern official architectural style.

平武报恩寺

平武报恩寺位于四川省平武县城内，平武古属龙安府。明正统四年（1439年）年始建，到明天顺四年（1460年）全部建成。2008年汶川大地震遭到损毁后曾进行修缮，但仍保持明代原物。

报恩寺主体建筑坐西朝东，建筑群依山就势由三进院落构成。山门前对称布置陀罗尼经幢，中轴线依次为山门、天王殿、大雄宝殿和万佛阁，第一进院落的北侧是二层楼阁的钟楼，第二进院落的南北侧分别是华严殿和大悲殿，第三进院落轴线两侧对称布局御碑亭，山门与天王殿之间还有金水桥三道相互联系。整个建筑群的建筑是大木大式建筑，建筑主要用材均为上等楠木，斗栱形制古朴，建筑布局严谨，殿堂内的小木作、彩画、壁画华丽而精美。整个建筑群从布局到建筑形制都具有浓厚的北方官式建筑风格特色。

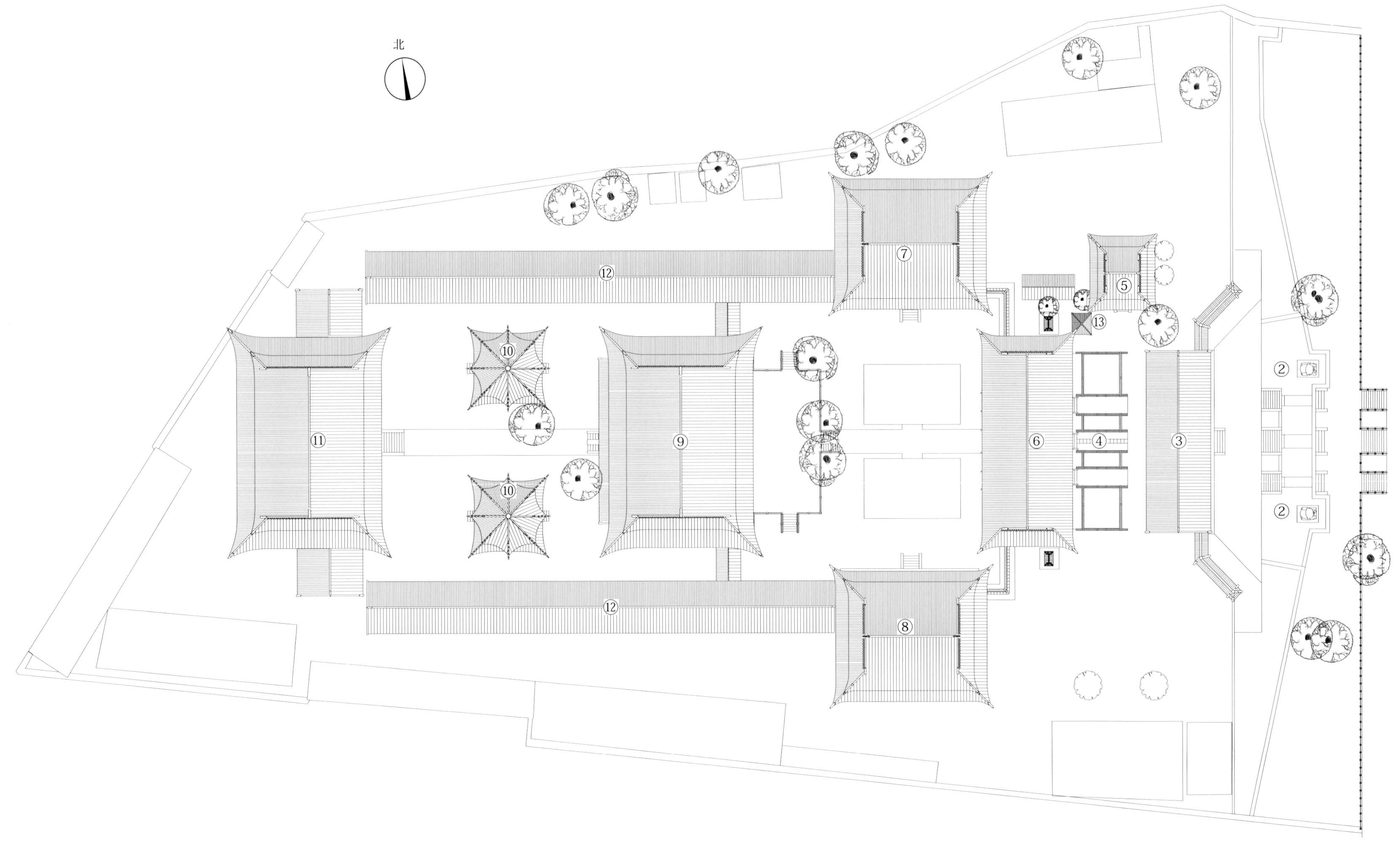

北
①
②
③
④
⑤
⑥
⑦
⑧
⑨
⑩
⑪
⑫
⑬
① 经幢
② 石狮
③ 山门
④ 金水桥
⑤ 钟楼
⑥ 天王殿
⑦ 大悲殿
⑧ 华严殿
⑨ 大雄宝殿
⑩ 御碑亭
⑪ 万佛阁
⑫ 回廊
⑬ 范公井
0 5 10 15 20m

平武报恩寺总平面图

Site Plan of Bao'en Temple, Pingwu

图三　平武报恩寺山门背立面图

图一　平武报恩寺山门

知恩報
AAAA
全国重点文物保护单位
图二　平武报恩寺山门正立面图

Fig.1 Main Gate (Shan Men), Bao’en Temple, Pingwu
Fig.2 Front Elevation of Main Gate (Shan Men), Bao’en Temple, Pingwu
Fig.3 Back Elevation of Main Gate (Shan Men), Bao’en Temple, Pingwu

天音醒世

图四 平武报恩寺钟楼

大雄寶殿

图五 平武报恩寺大雄宝殿

大雄寶殿

图六 平武报恩寺大雄宝殿

Fig.4 Bell Tower, Bao’en Temple, Pingwu
Fig.5 Mahavira Hall, Bao’en Temple, Pingwu
Fig.6 Mahavira Hall, Bao’en Temple, Pingwu

藏
嚴
華

图九 平武报恩寺华严殿

图七 平武报恩寺天王殿

天王殿

图八 平武报恩寺天王殿门额

Fig.7 Tianwang Hall, Bao'en Temple, Pingwu
Fig.8 Detail, Door Lintel, Tianwang Hall, Bao'en Temple, Pingwu
Fig.9 Huayan Hall, Bao'en Temple, Pingwu

大悲殿

图十 平武报恩寺大悲殿

图十一 平武报恩寺碑亭

图十二 平武报恩寺碑廊

Fig.10 Dabei Hall, Bao’en Temple, Pingwu
Fig.11 Stele Pavilion, Bao’en Temple, Pingwu
Fig.12 Stele Corridor, Bao’en Temple, Pingwu

北

① 经幢
② 石狮
③ 山门
④ 金水桥
⑤ 钟楼
⑥ 天王殿
⑦ 大悲殿
⑧ 华严殿
⑨ 大雄宝殿
⑩ 御碑亭
⑪ 万佛阁
⑫ 回廊
⑬ 范公井

0 5 10 15 20m

平武报恩寺建筑群平面图

Ground Floor Plan of Entire Compound, Bao'en Temple, Pingwu

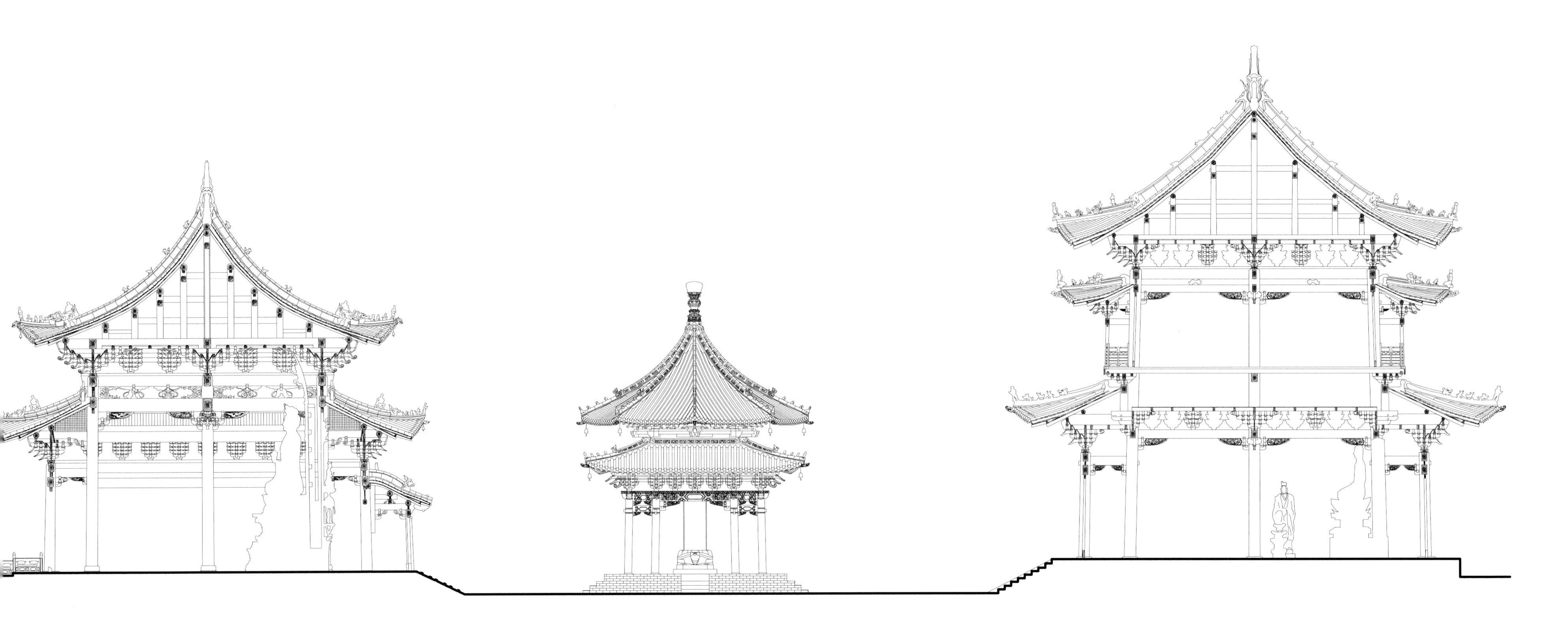
大雄宝殿
御碑亭
万佛阁
0
5
10m

平武报恩寺建筑群剖面图 1-1
1-1 Section of Entire Compound, Bao'en Temple, Pingwu

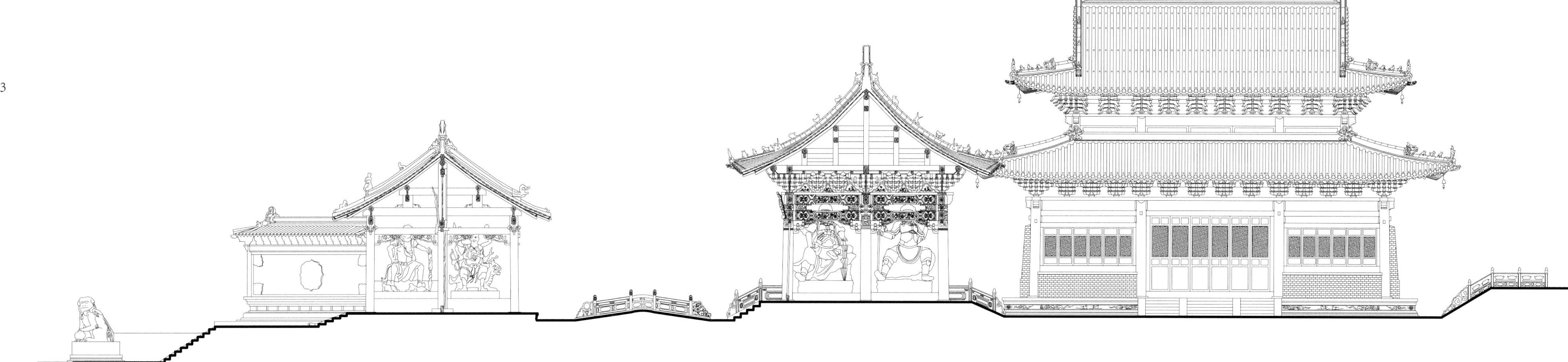

山门　　天王殿　　华严殿

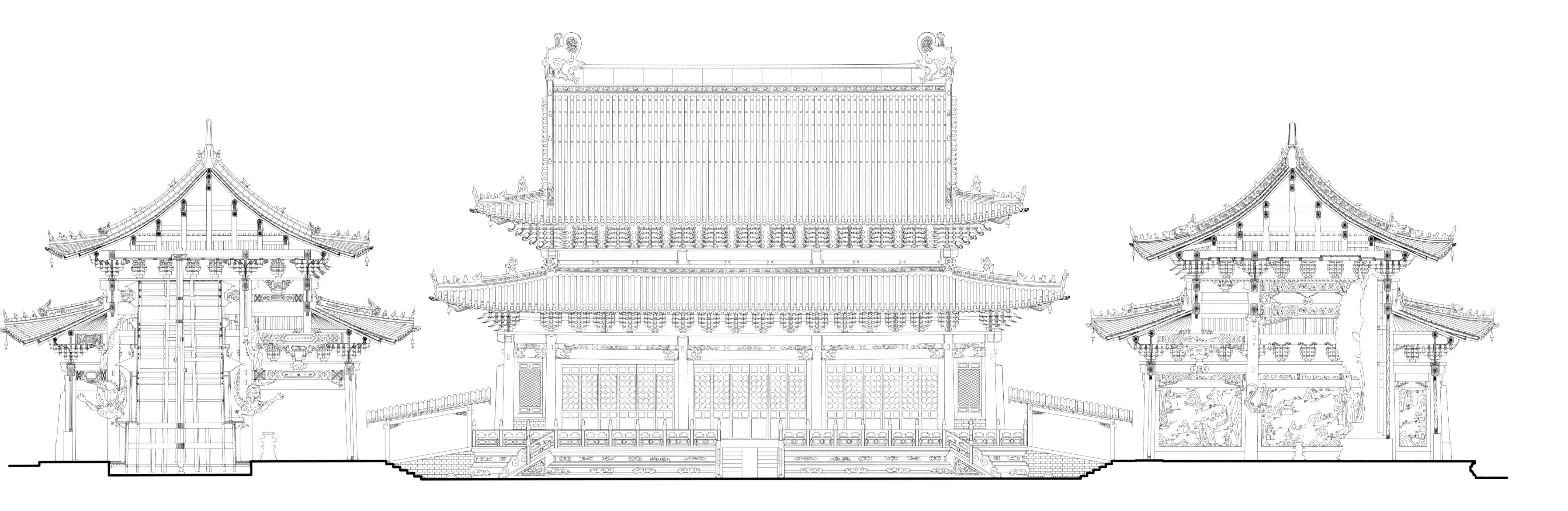
华严殿
大雄宝殿
大悲殿

平武报恩寺建筑群剖面图 2−2
2-2 Section of Entire Compound, Bao'en Temple, Pingwu

0 1 2 3 4 5m

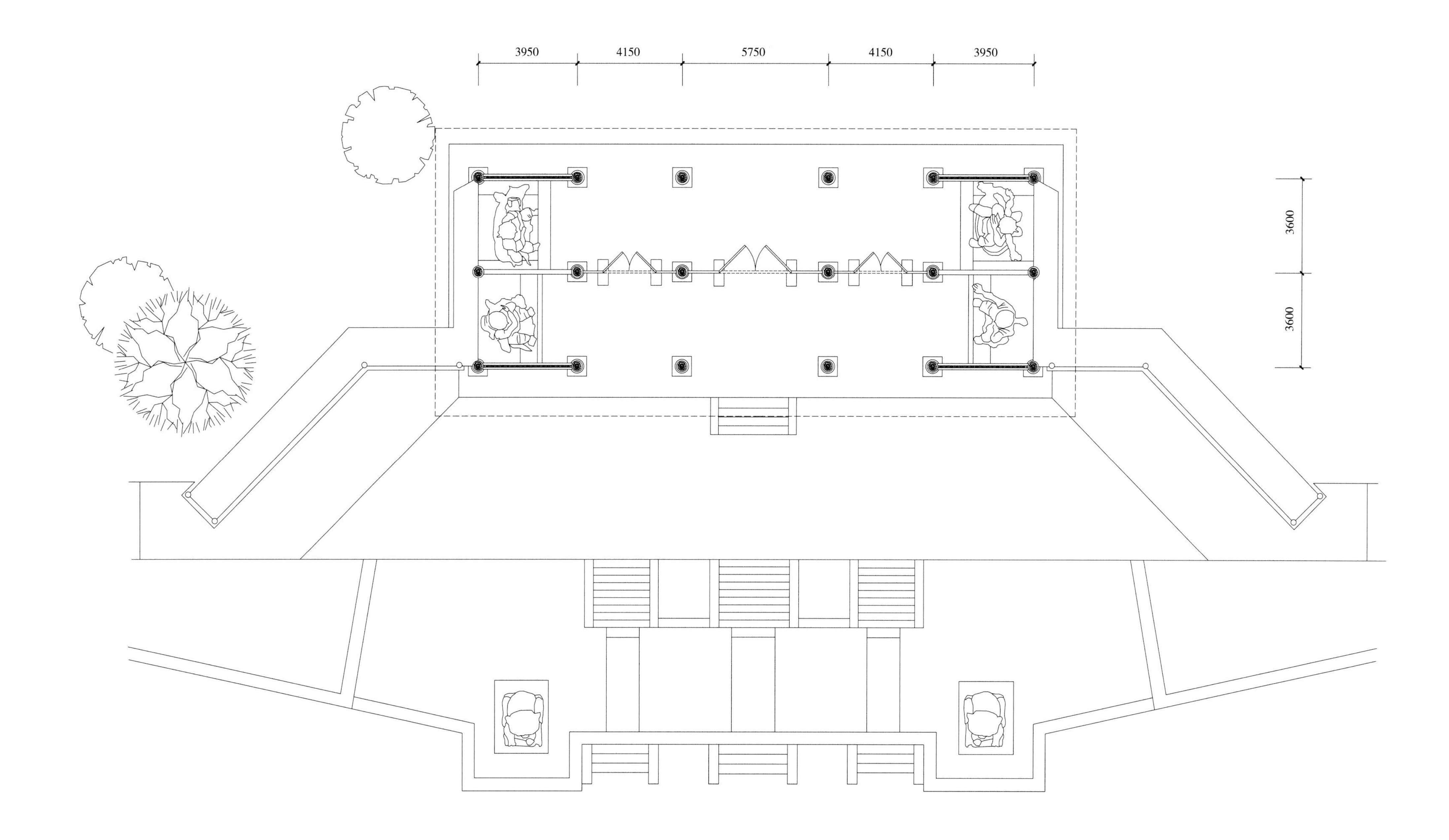
3950
4150
5750
4150
3950
3600
3600

平武报恩寺山门平面图
Ground Floor Plan of Main Gate (Shan Men), Bao’en Temple, Pingwu County

0 1 2 3 4 5m

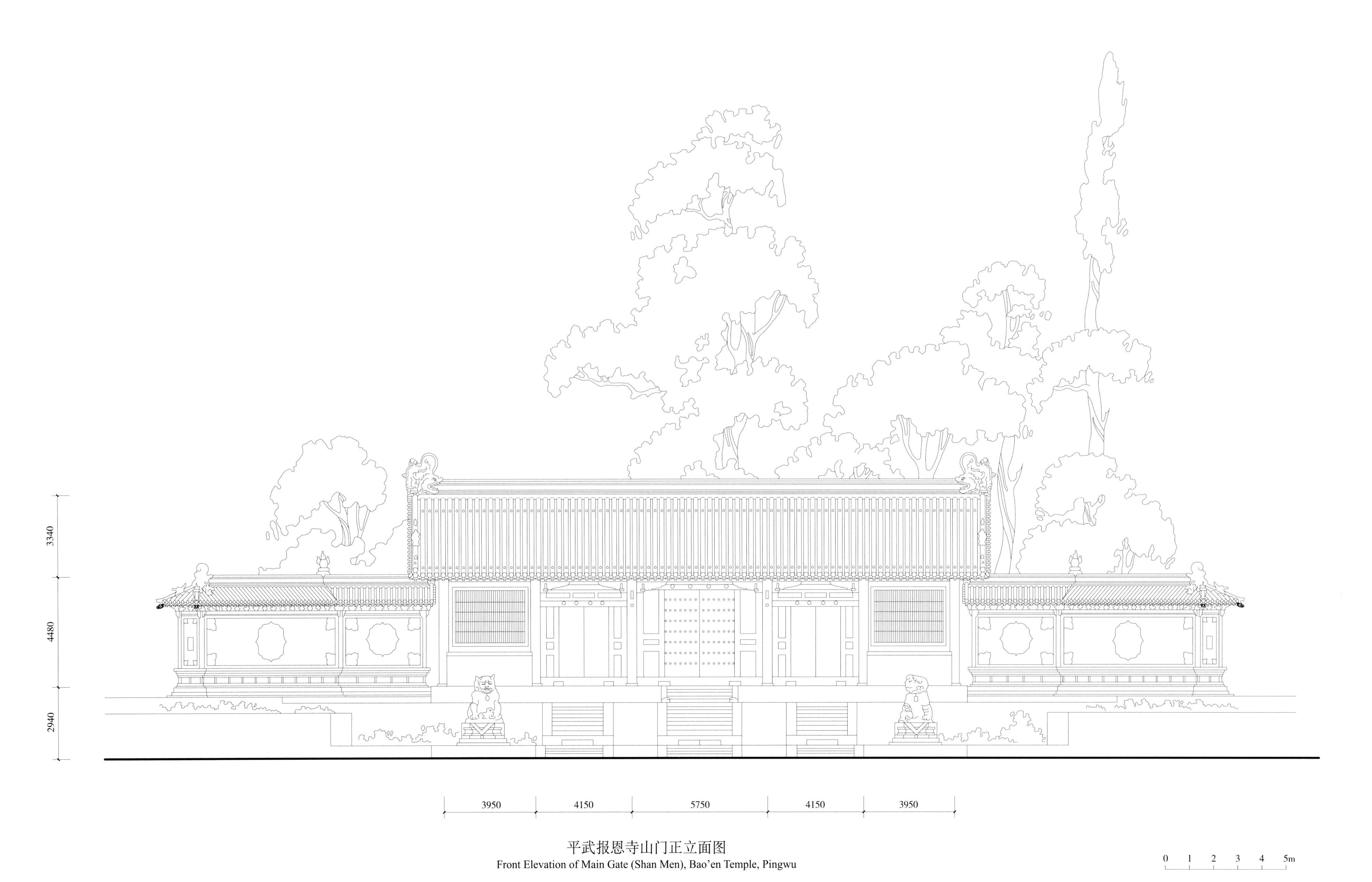
3340
4480
2940
3950
4150
5750
4150
3950
0 1 2 3 4 5m

平武报恩寺山门正立面图
Front Elevation of Main Gate (Shan Men), Bao'en Temple, Pingwu

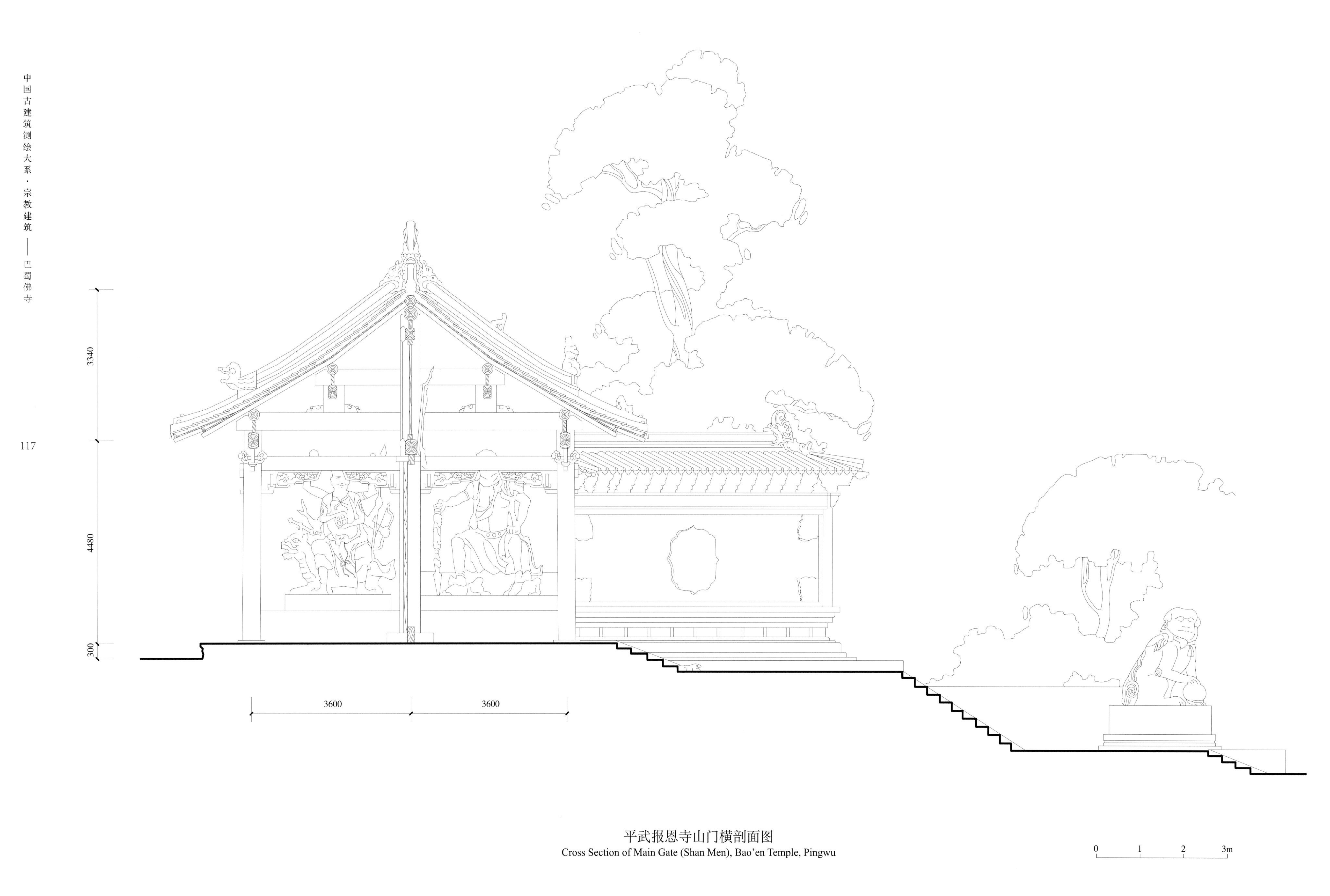
3340
4480
300
3600
3600
0
1
2
3m

平武报恩寺山门横剖面图

Cross Section of Main Gate (Shan Men), Bao'en Temple, Pingwu

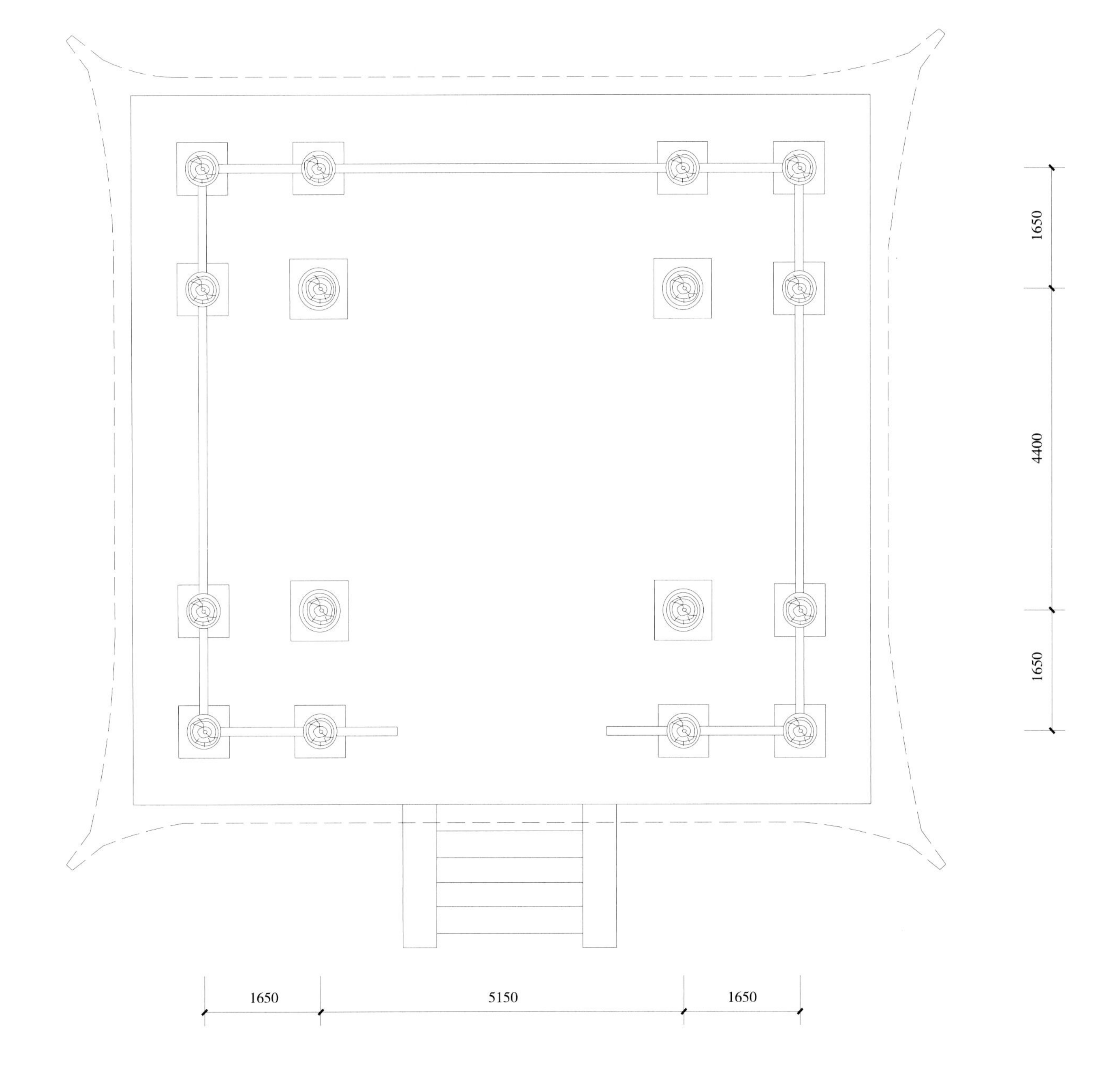
1650
4400
1650
1650
5150
1650

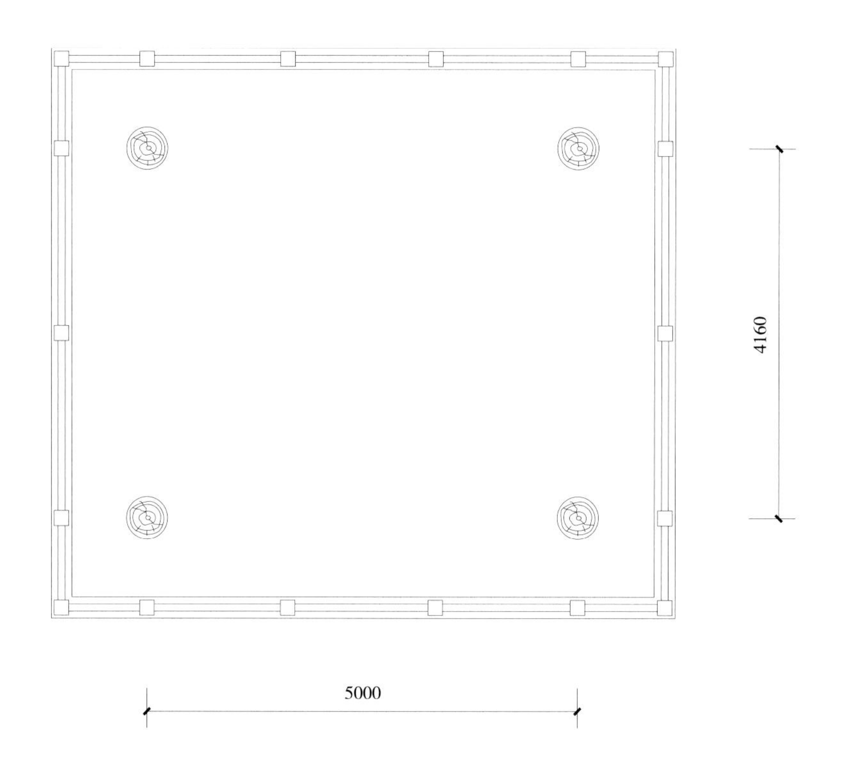
4160
5000

平武报恩寺钟楼一层平面图和楼层平面图

Ground Floor Plan and First Floor Plan of Bell Tower, Bao'en Temple, Pingwu

0 1 2 3m

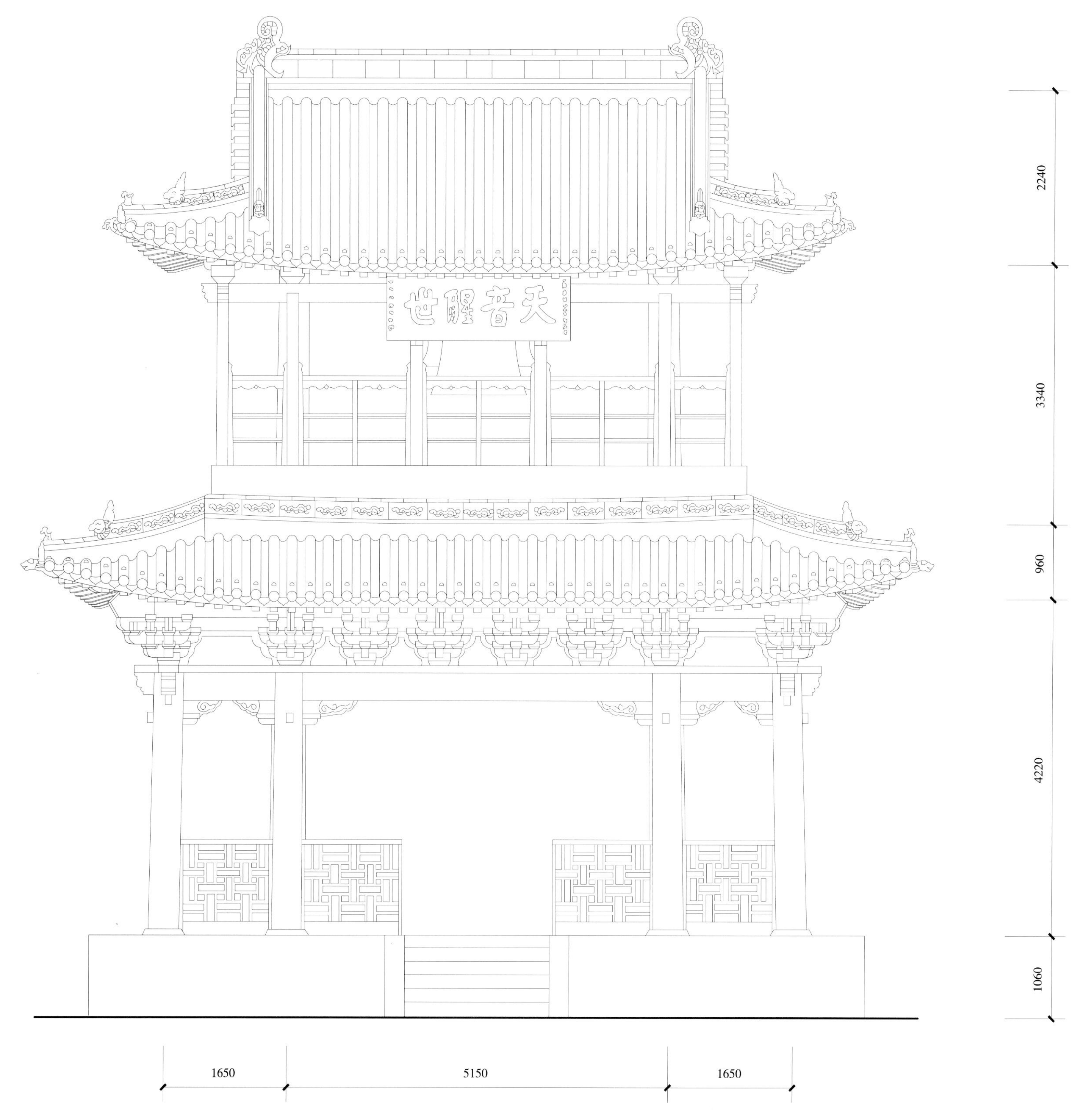
天香醒世
2240
3340
960
4220
1060
1650
5150
1650

平武报恩寺钟楼正立面图
Front Elevation of Bell Tower, Bao'en Temple, Pingwu

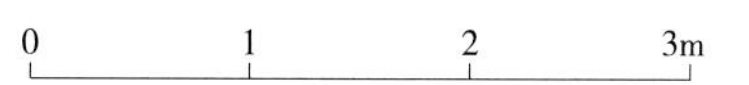
0
1
2
3m

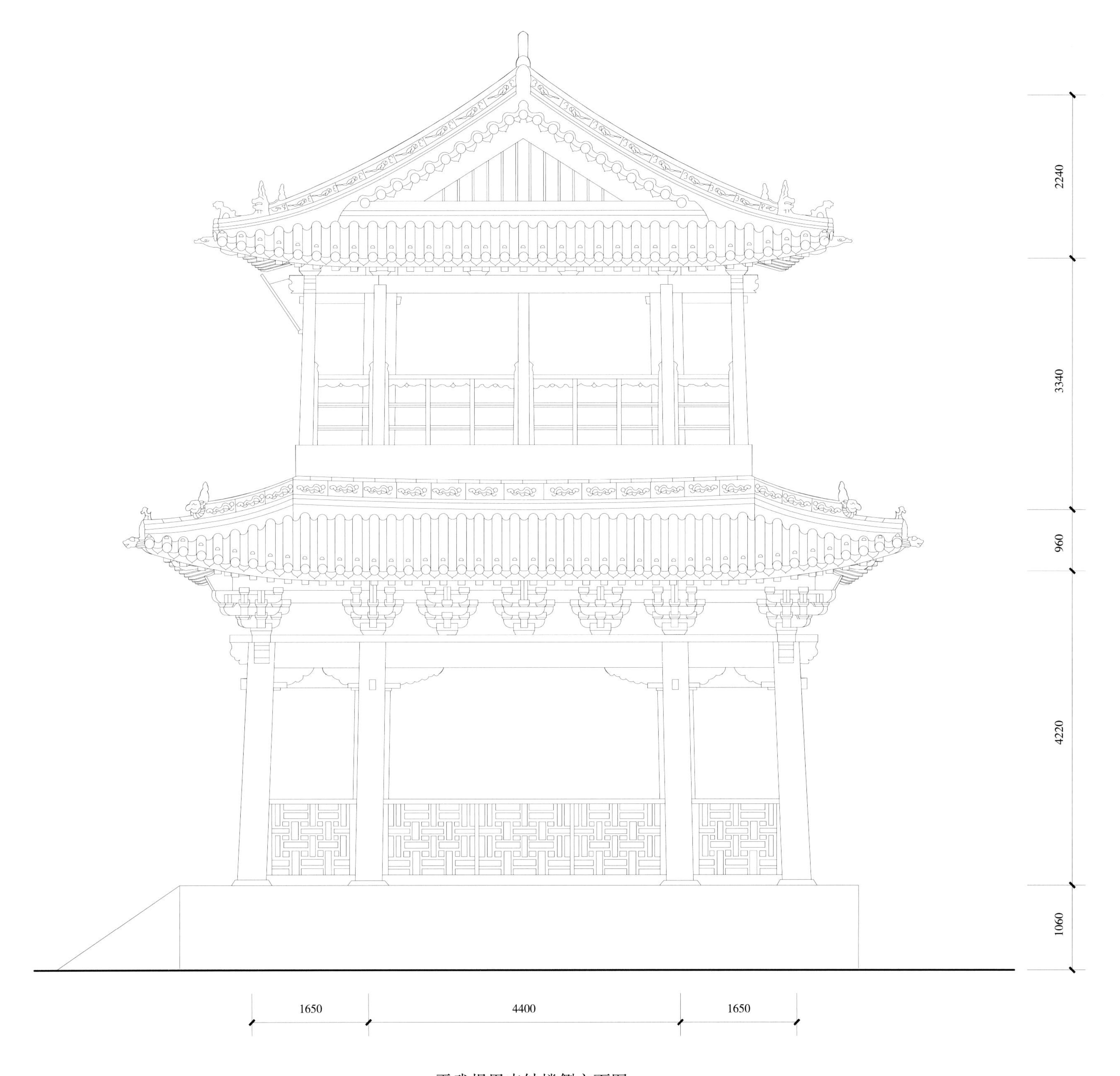
2240
3340
960
4220
1060
1650
4400
1650
0
1
2
3m

平武报恩寺钟楼侧立面图
Side Elevation of Bell Tower, Bao'en Temple, Pingwu

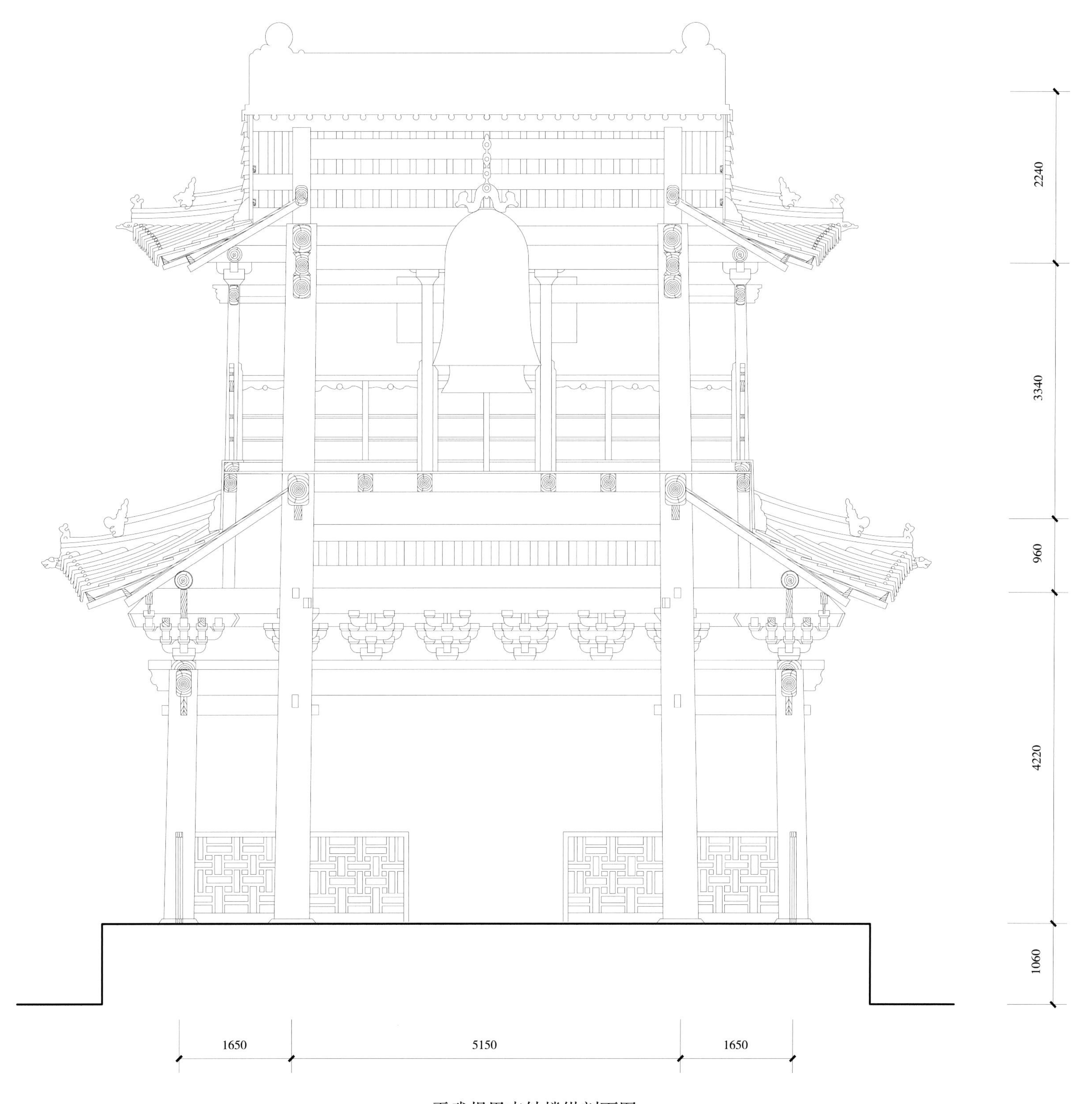
2240
3340
960
4220
1060
1650
5150
1650
0
1
2
3m

平武报恩寺钟楼纵剖面图
Longitudinal Section of Bell Tower, Bao'en Temple, Pingwu

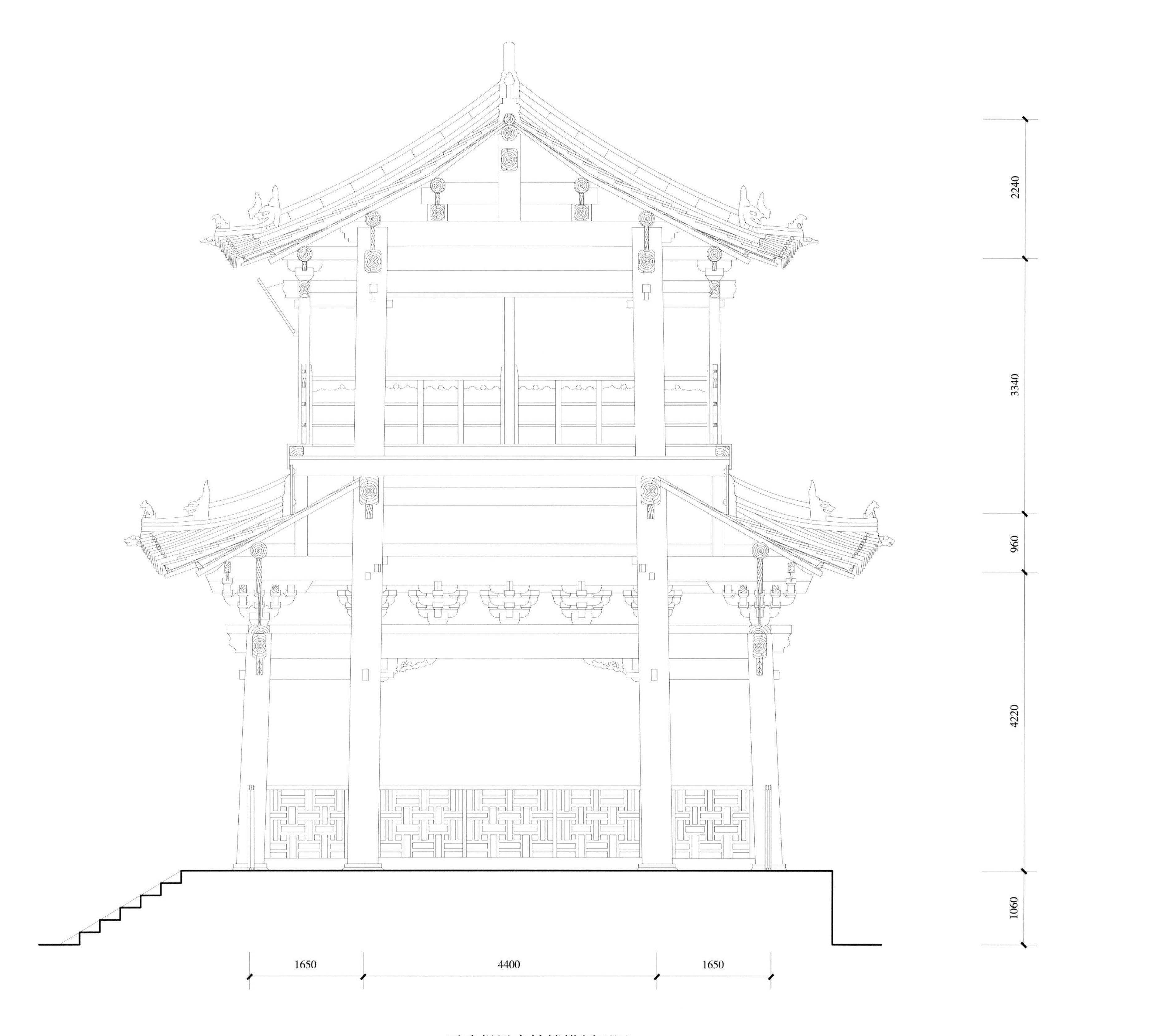
2240
3340
960
4220
1060
1650
4400
1650
0
1
2
3m

平武报恩寺钟楼横剖面图

Cross Section of Bell Tower, Bao'en Temple, Pingwu

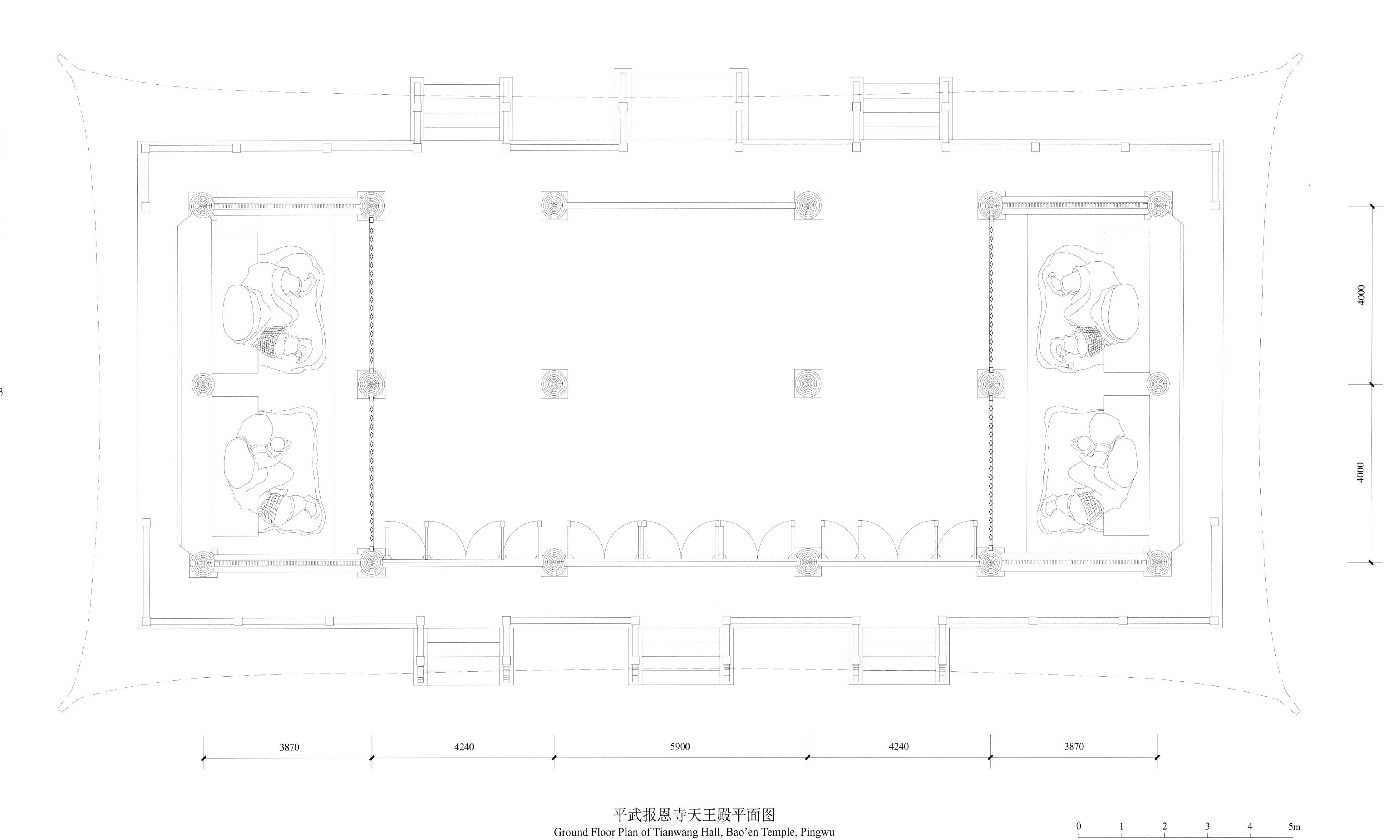
4000
4000
3870
4240
5900
4240
3870
0
1
2
3
4
5m

平武报恩寺天王殿平面图
Ground Floor Plan of Tianwang Hall, Bao’en Temple, Pingwu

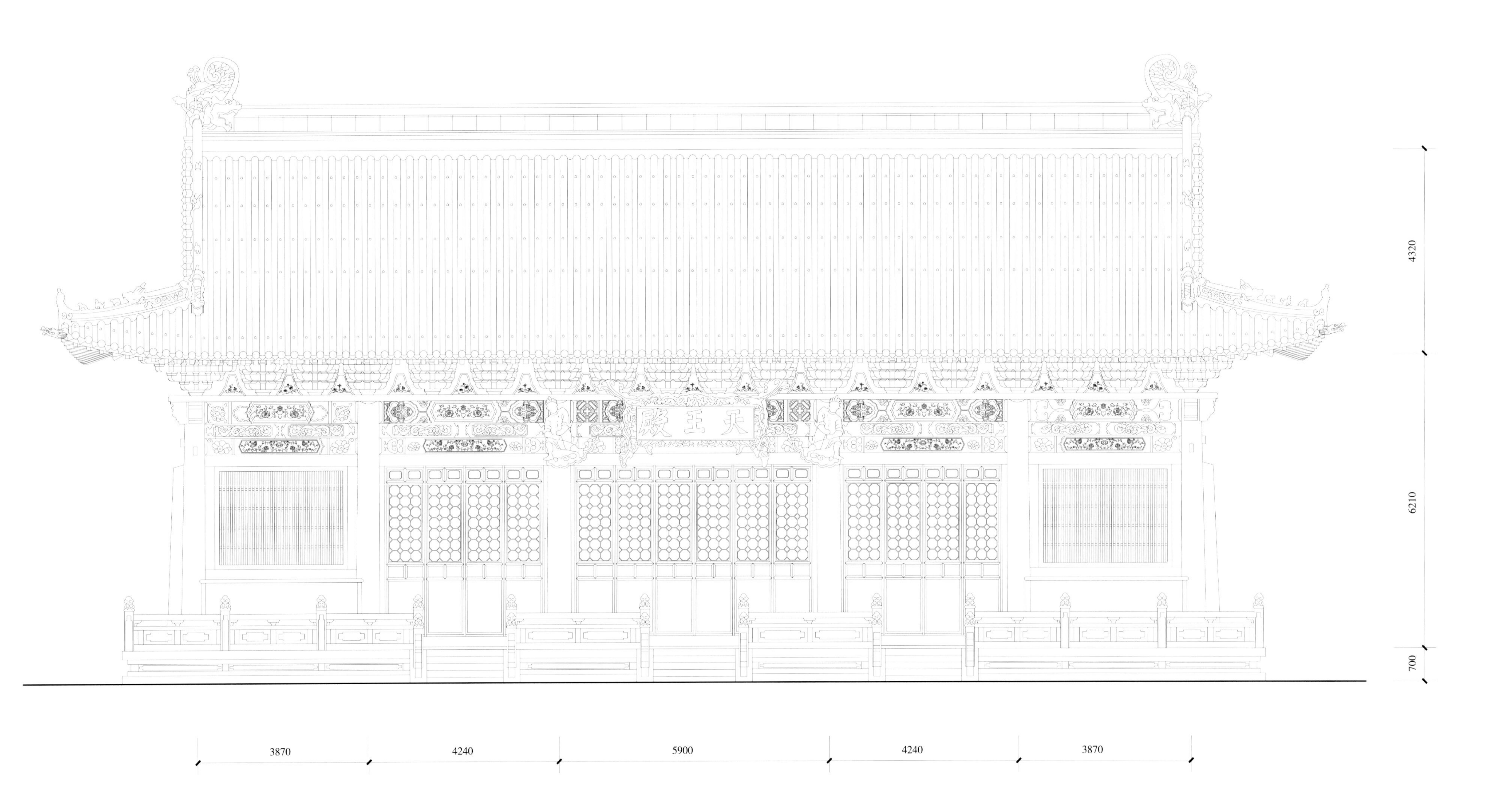
天王殿
4320
6210
700
3870
4240
5900
4240
3870

平武报恩寺天王殿正立面图
Front Elevation of Tianwang Hall, Bao'en Temple, Pingwu

0 1 2 3 4 5m

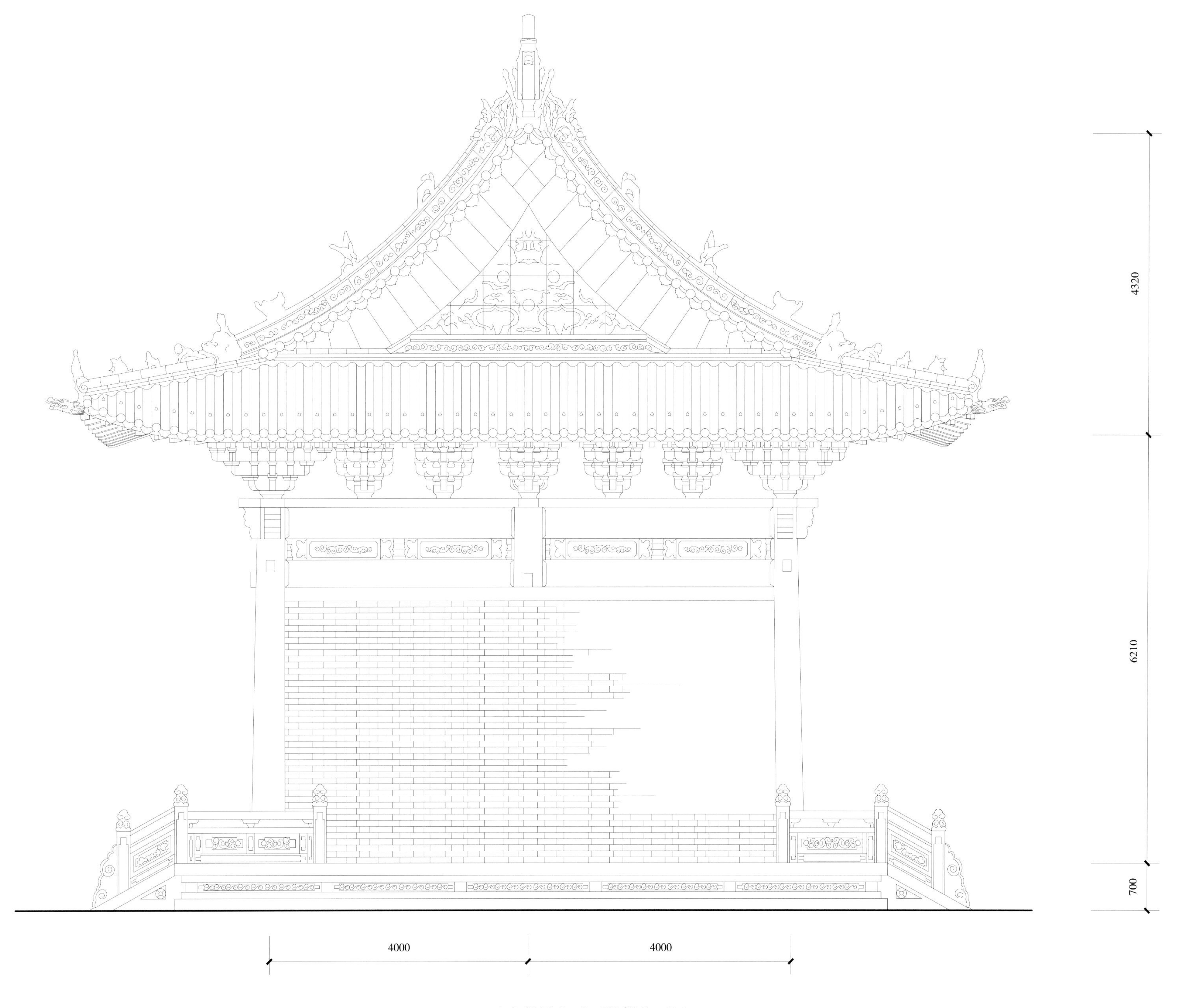
4320
6210
700
4000
4000
0
1
2
3m

平武报恩寺天王殿侧立面图
Side Elevation of Tianwang Hall, Bao'en Temple, Pingwu

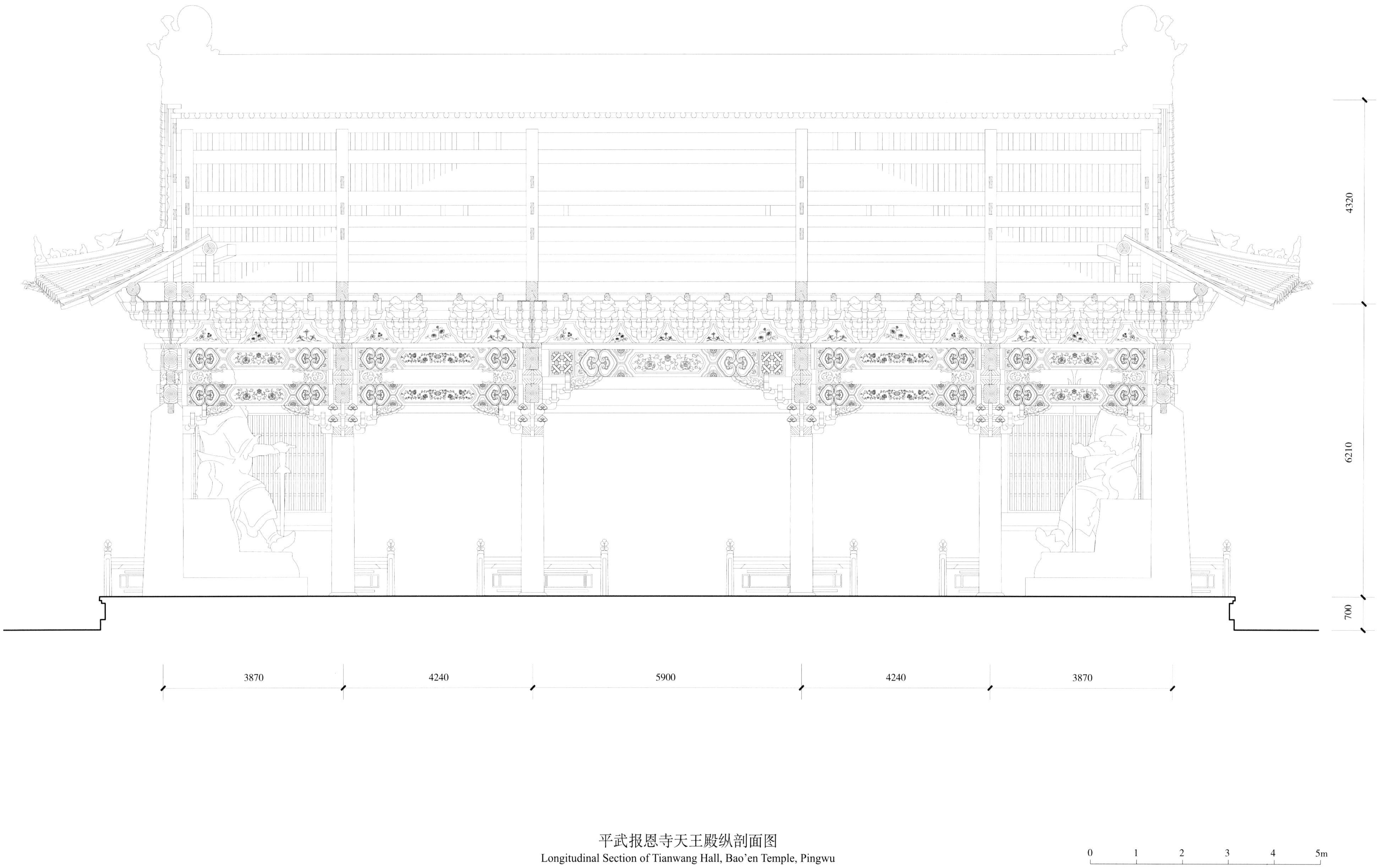

4320

6210

700

3870 4240 5900 4240 3870

平武报恩寺天王殿纵剖面图
Longitudinal Section of Tianwang Hall, Bao'en Temple, Pingwu

0 1 2 3 4 5m

4320
6210
700
4000
4000
0
1
2
3m

平武报恩寺天王殿横剖面图
Cross Section of Tianwang Hall, Bao'en Temple, Pingwu

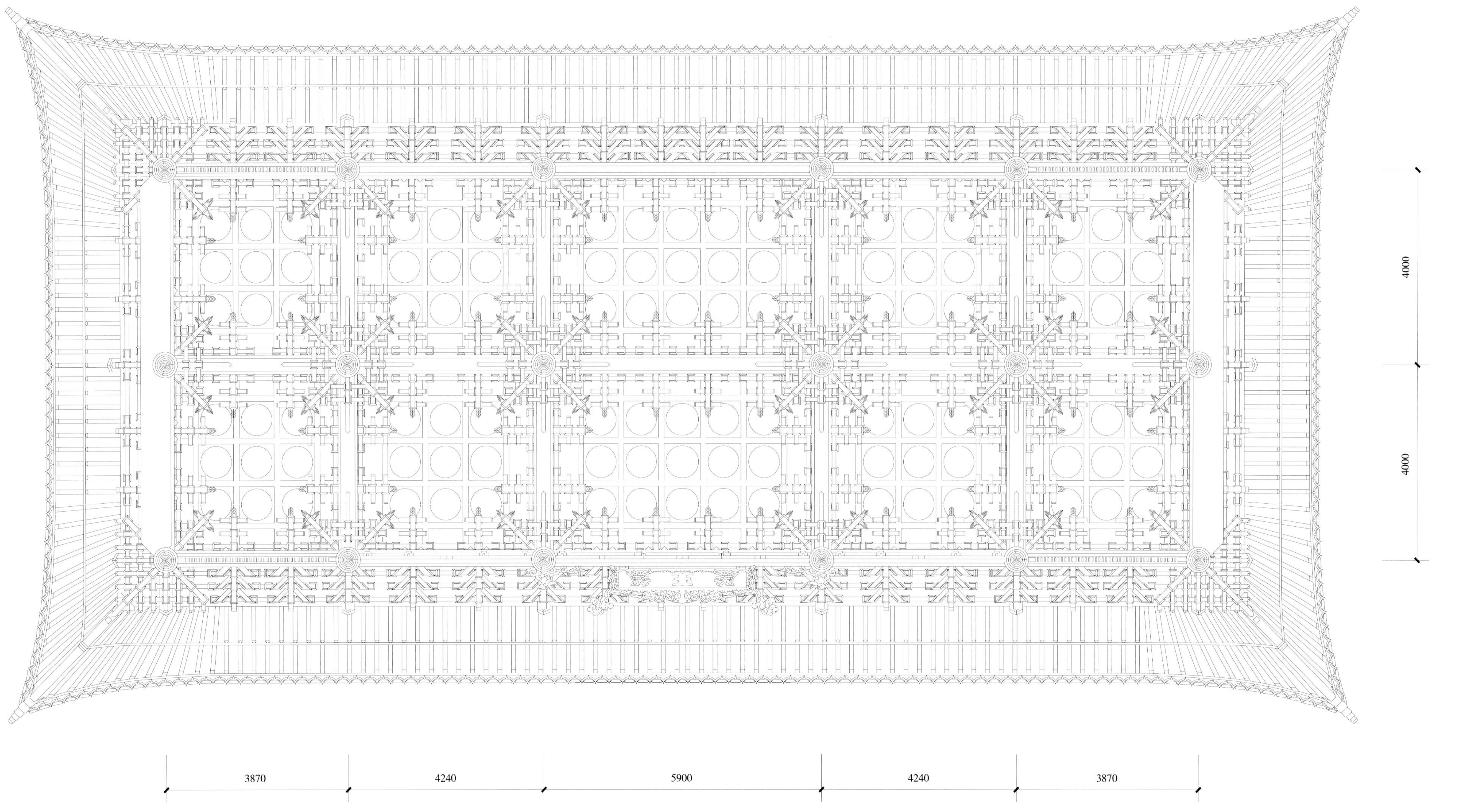
4000
4000
3870
4240
5900
4240
3870

平武报恩寺天王殿屋顶仰视图
Roof Framing of Tianwang Hall, Bao'en Temple, Pingwu

0 1 2 3 4 5m

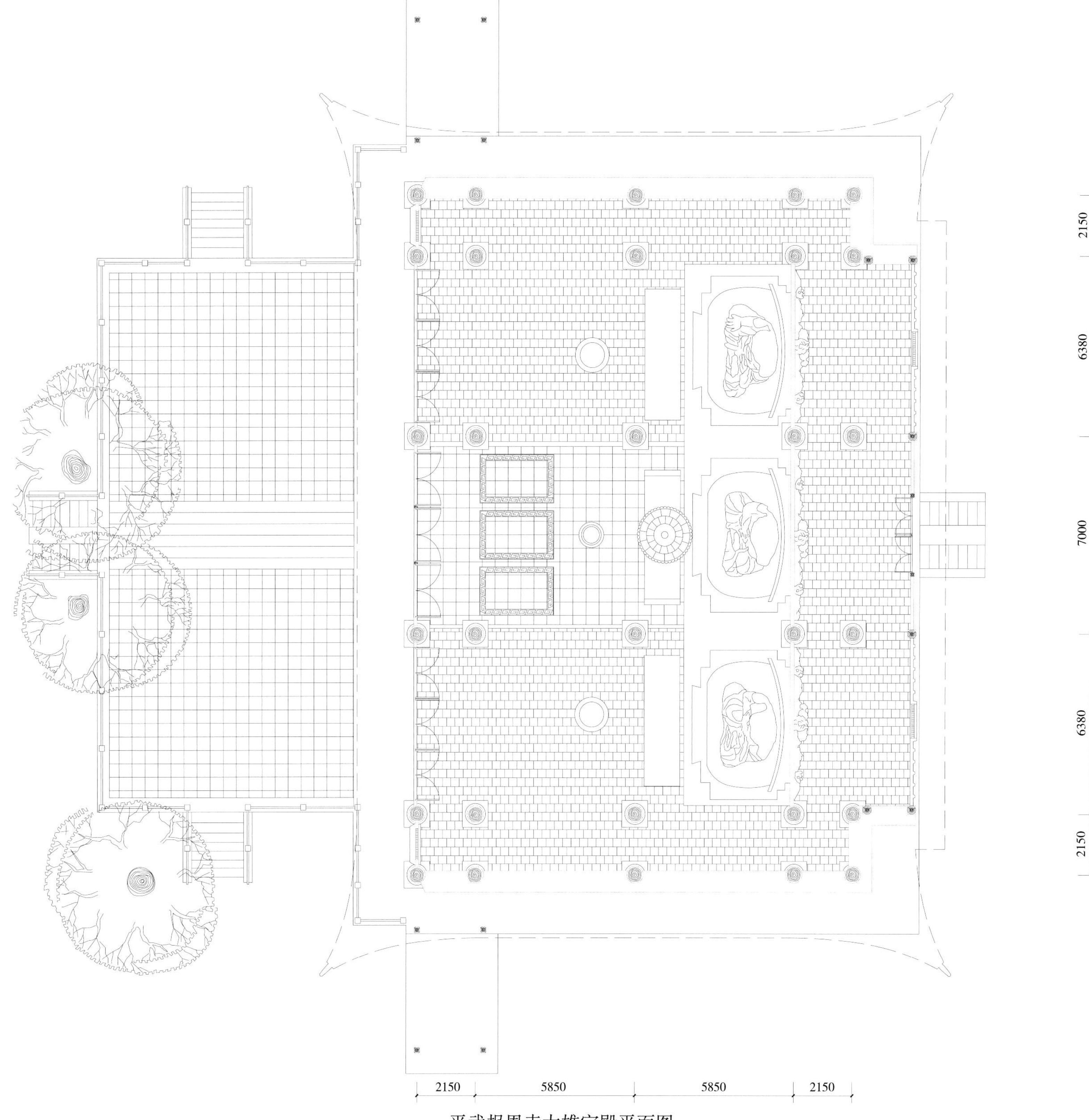
2150
6380
7000
6380
2150
2150
5850
5850
2150
0 1 2 3 4 5m

平武报恩寺大雄宝殿平面图
Ground Floor Plan of Mahavira Hall, Bao'en Temple, Pingwu

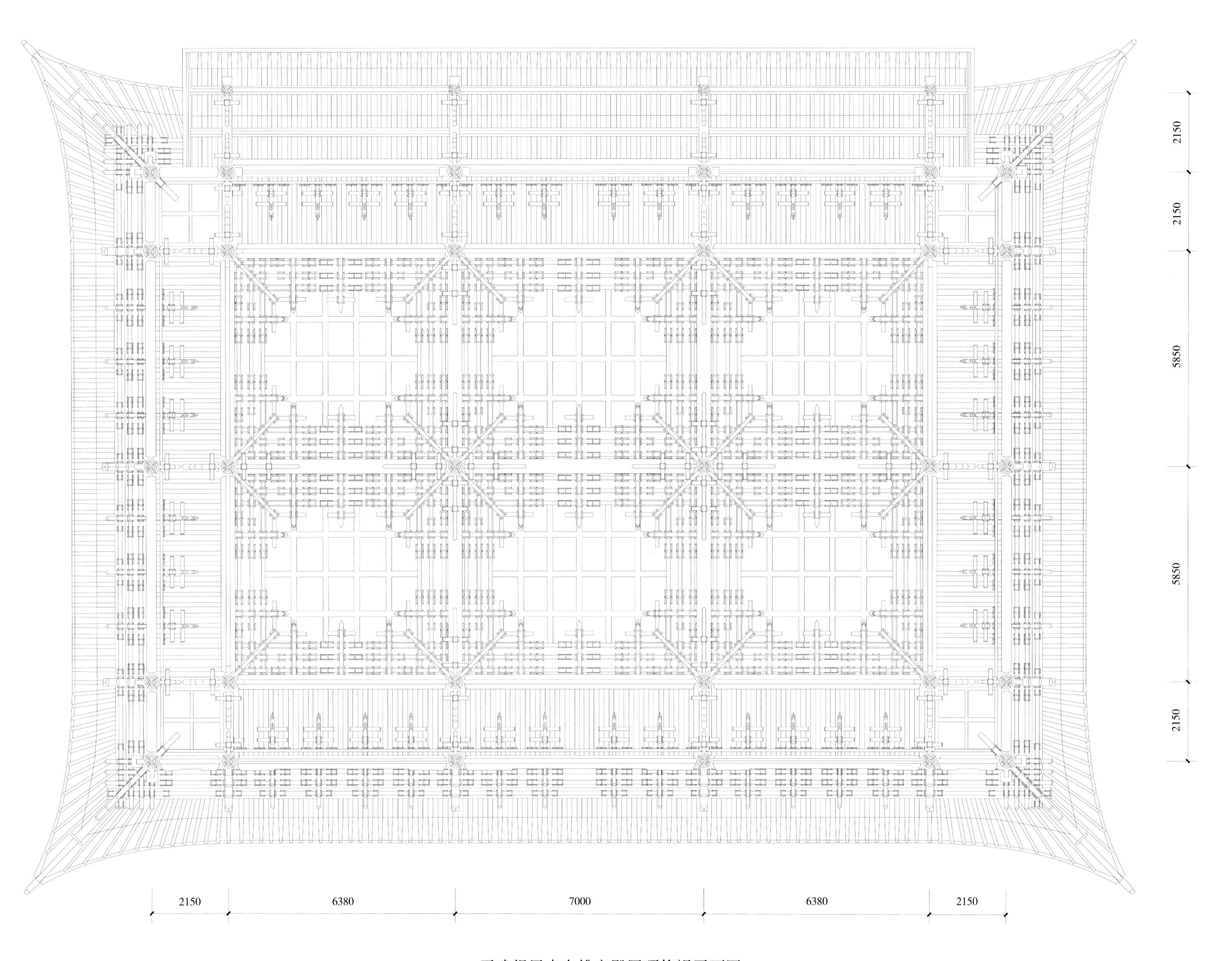
2150
2150
5850
5850
2150
2150
6380
7000
6380
2150
0 1 2 3 4 5m

平武报恩寺大雄宝殿屋顶仰视平面图
Roof Framing of Mahavira Hall, Bao'en Temple, Pingwu

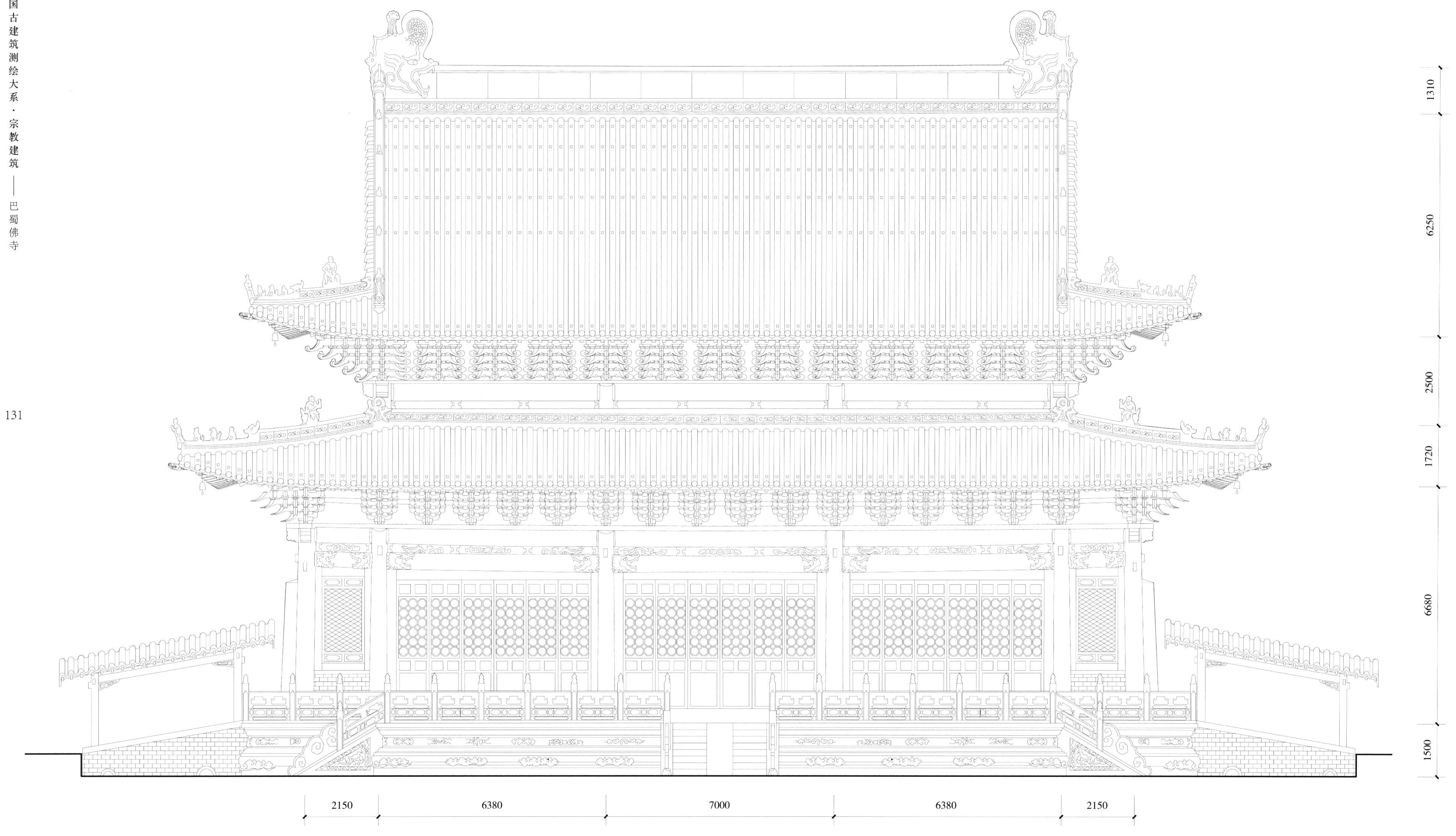
1310
6250
2500
1720
6680
1500
2150
6380
7000
6380
2150
0
1
2
3
4
5m

平武报恩寺大雄宝殿正立面图
Front Elevation of Mahavira Hall, Bao’en Temple, Pingwu

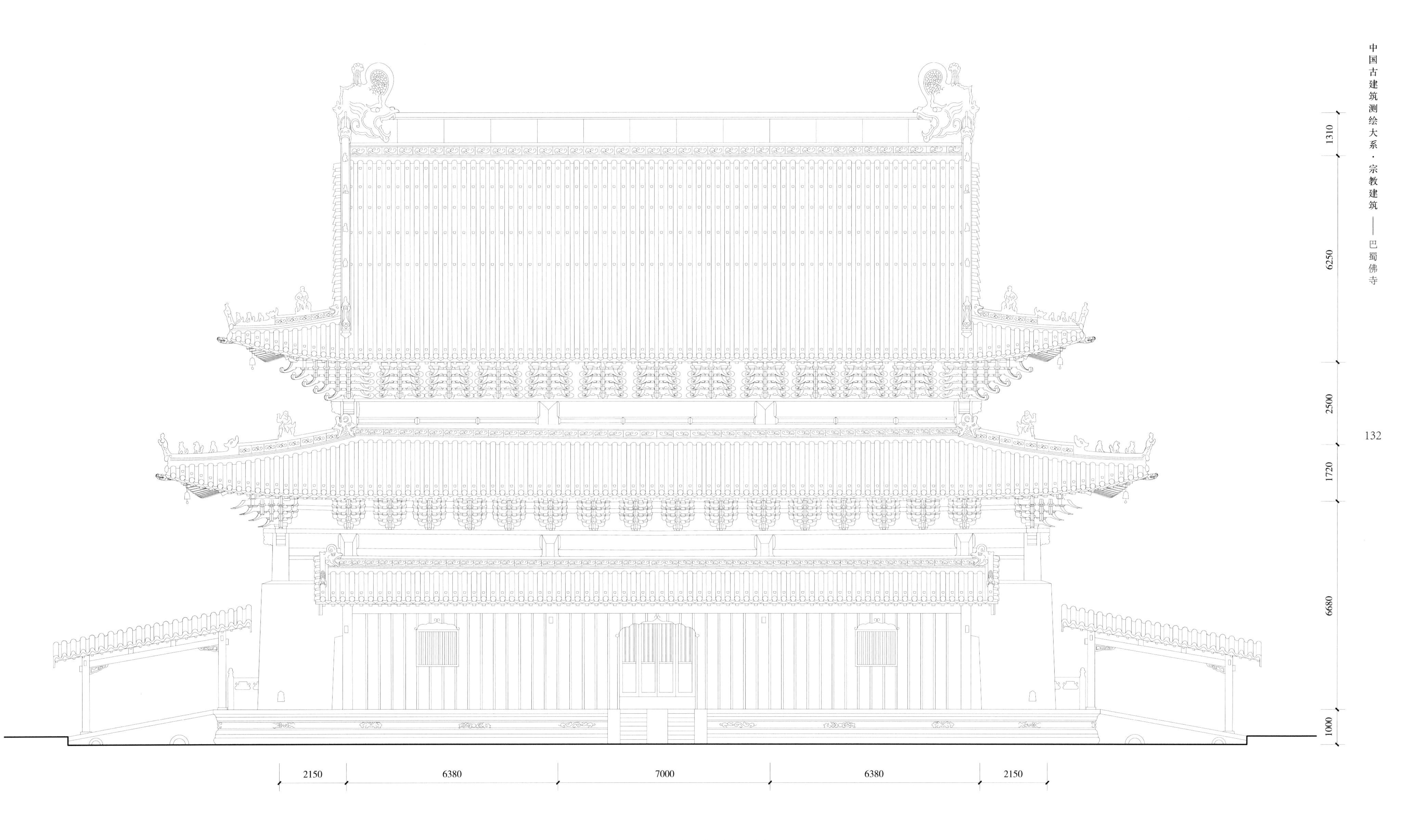
1310
6250
2500
1720
6680
1000
2150
6380
7000
6380
2150
0 1 2 3 4 5m

平武报恩寺大雄宝殿背立面图

Back Elevation of Mahavira Hall, Bao'en Temple, Pingwu

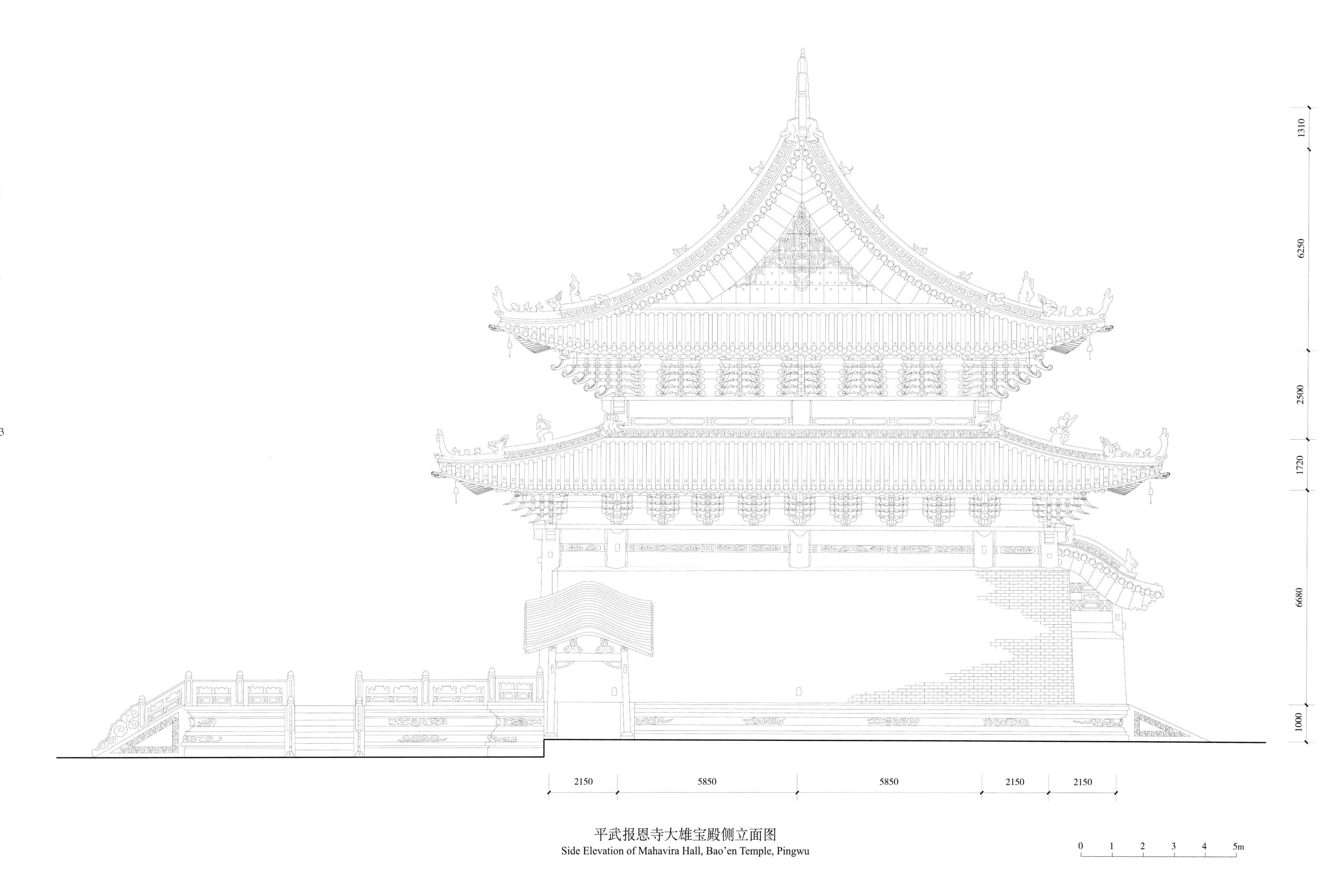
1310
6250
2500
1720
6680
1000
2150
5850
5850
2150
2150
0 1 2 3 4 5m

平武报恩寺大雄宝殿侧立面图
Side Elevation of Mahavira Hall, Bao'en Temple, Pingwu

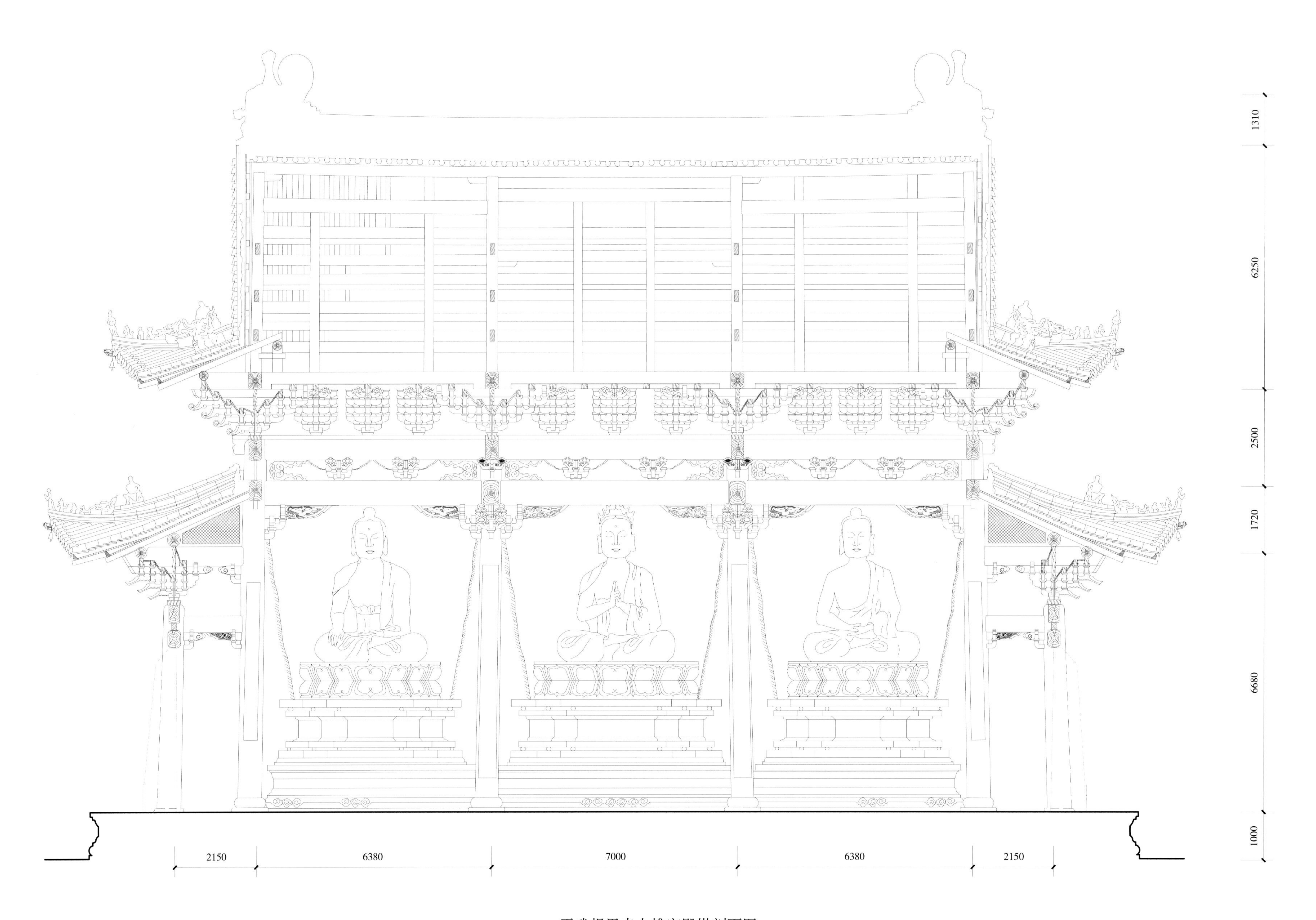
1310
6250
2500
1720
6680
1000
2150
6380
7000
6380
2150
0
1
2
3
4
5m

平武报恩寺大雄宝殿纵剖面图
Longitudinal Section of Mahavira Hall, Bao'en Temple, Pingwu

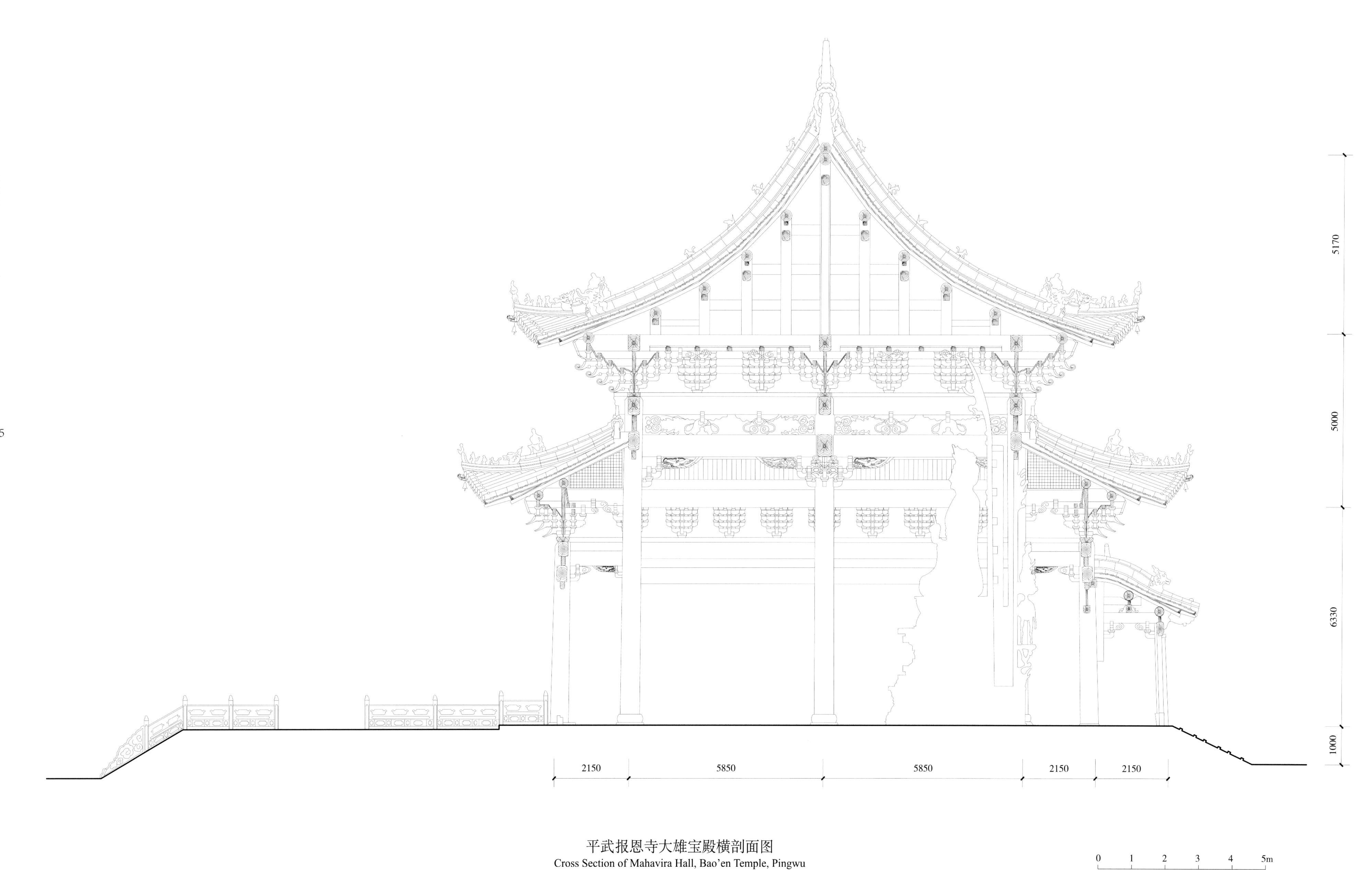
5170
5000
6330
1000
2150
5850
5850
2150
2150
0 1 2 3 4 5m

平武报恩寺大雄宝殿横剖面图

Cross Section of Mahavira Hall, Bao'en Temple, Pingwu

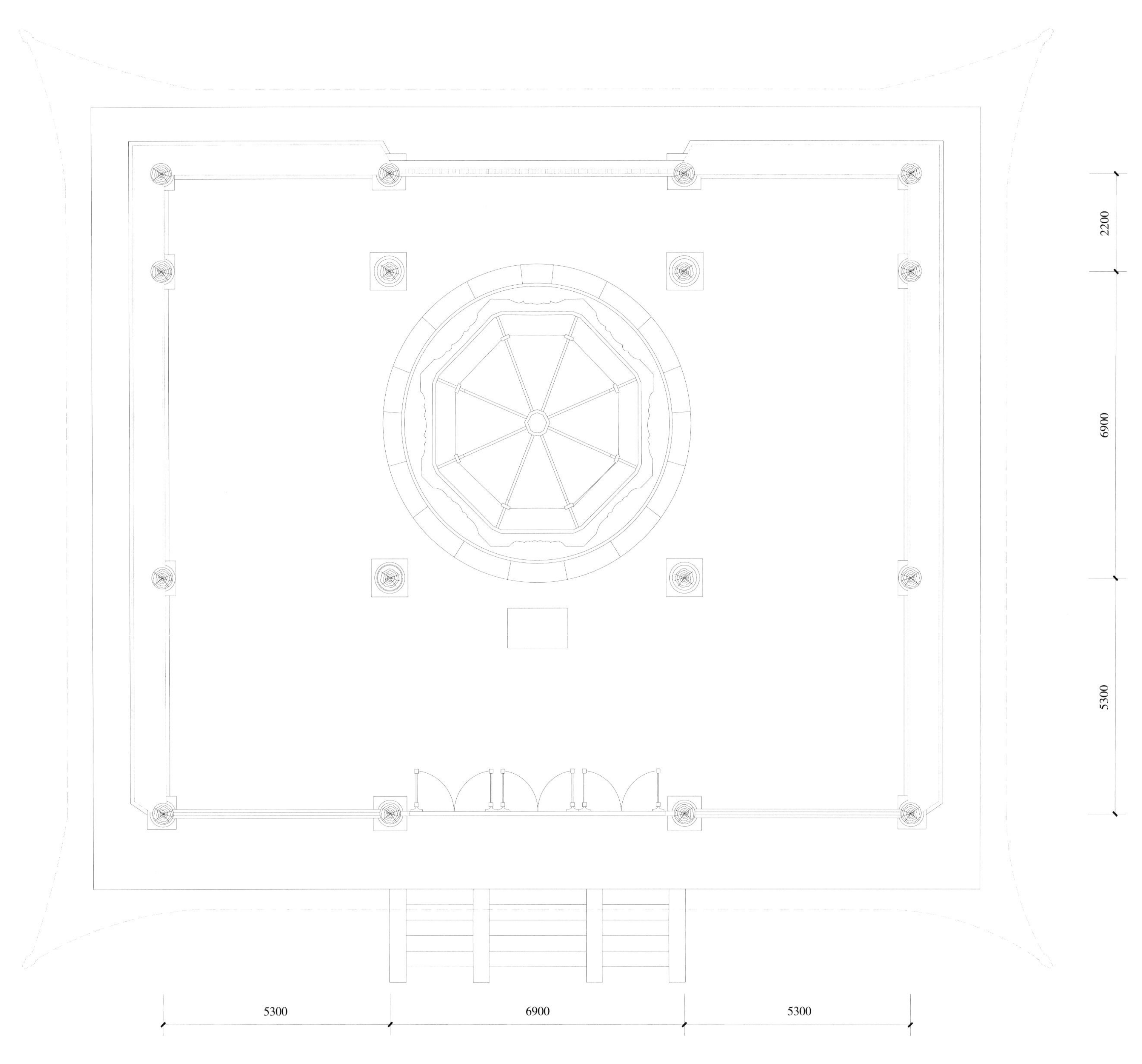
2200
6900
5300
5300
6900
5300
0 1 2 3m

平武报恩寺华严殿平面图
Ground Floor Plan of Huayan Hall, Bao’en Temple, Pingwu

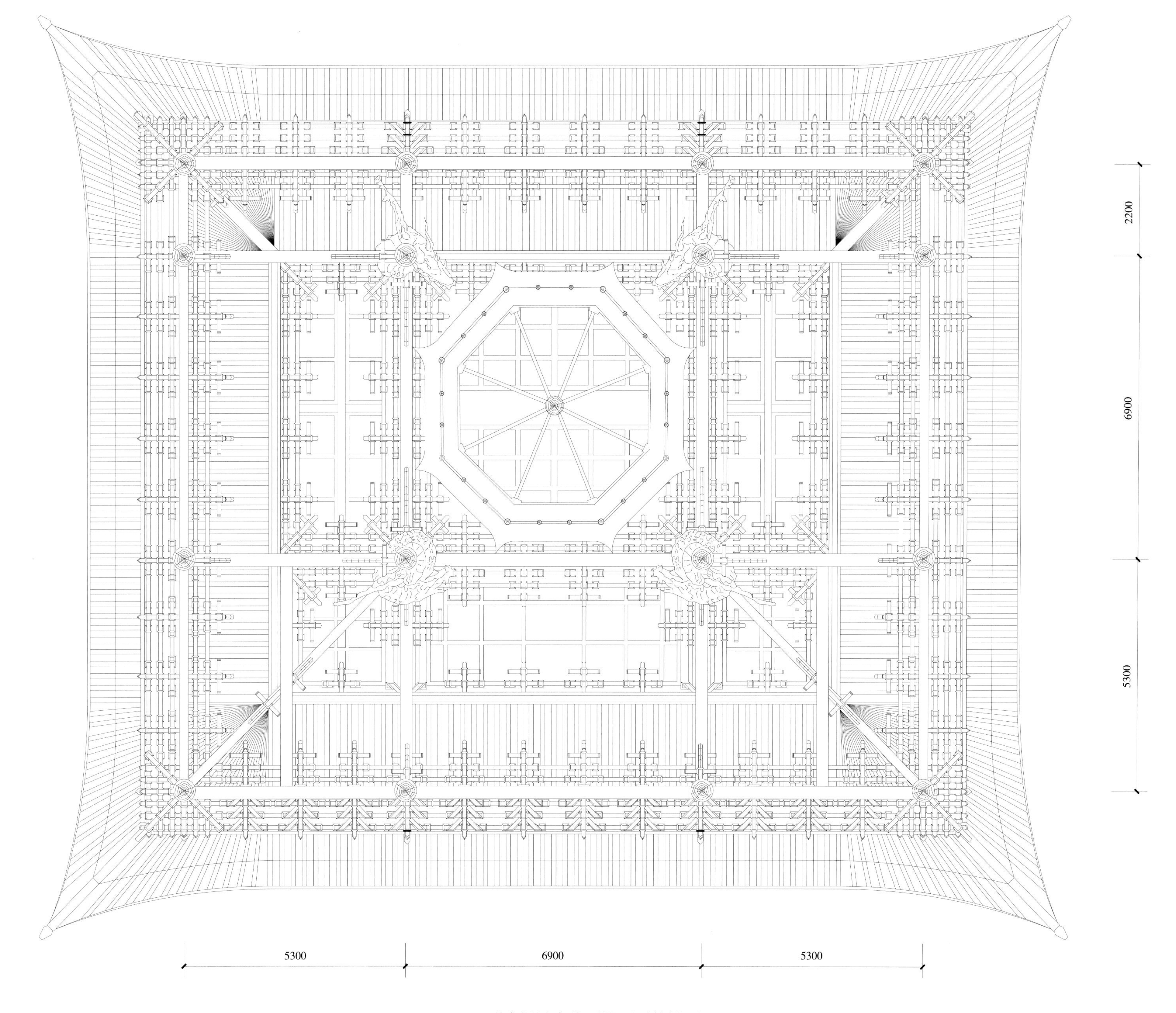
2200
6900
5300
5300
6900
5300
0
1
2
3m

平武报恩寺华严殿屋顶仰视图

Roof Framing of Huayan Hall, Bao'en Temple, Pingwu

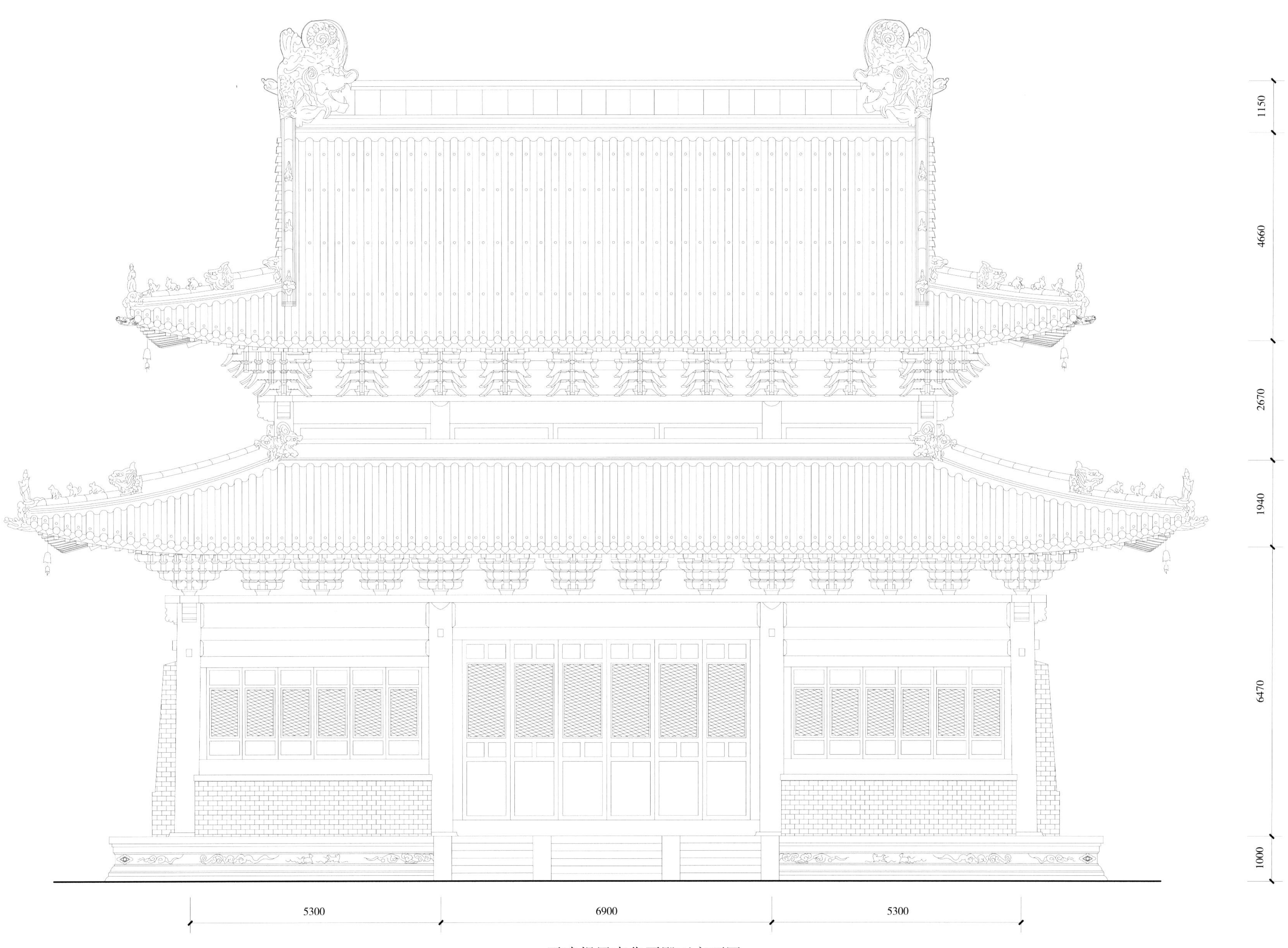
1150
4660
2670
1940
6470
1000
5300
6900
5300
0
1
2
3m

平武报恩寺华严殿正立面图
Front Elevation of Huayan Hall, Bao'en Temple, Pingwu

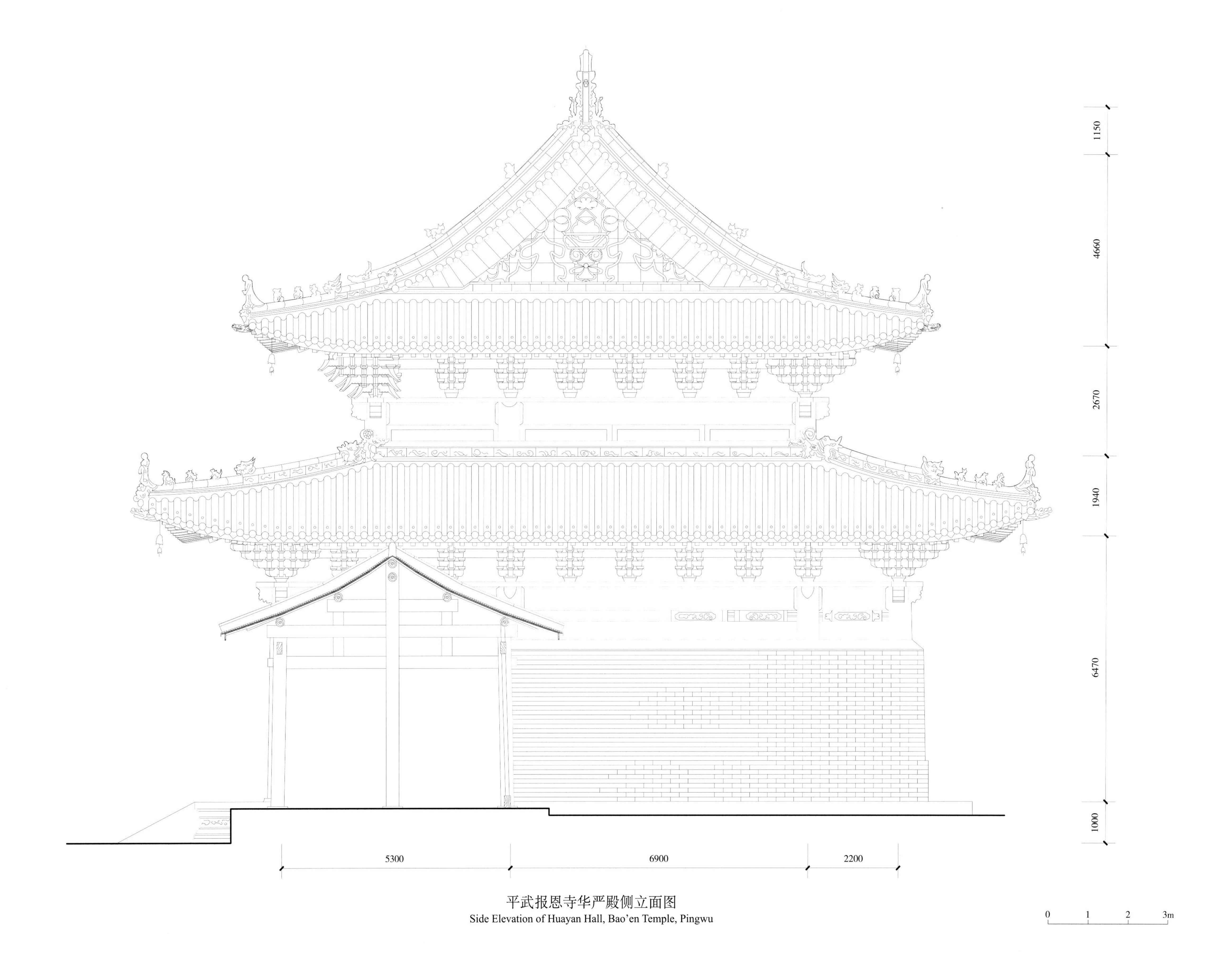
1150
4660
2670
1940
6470
1000
5300
6900
2200
0
1
2
3m

平武报恩寺华严殿侧立面图
Side Elevation of Huayan Hall, Bao'en Temple, Pingwu

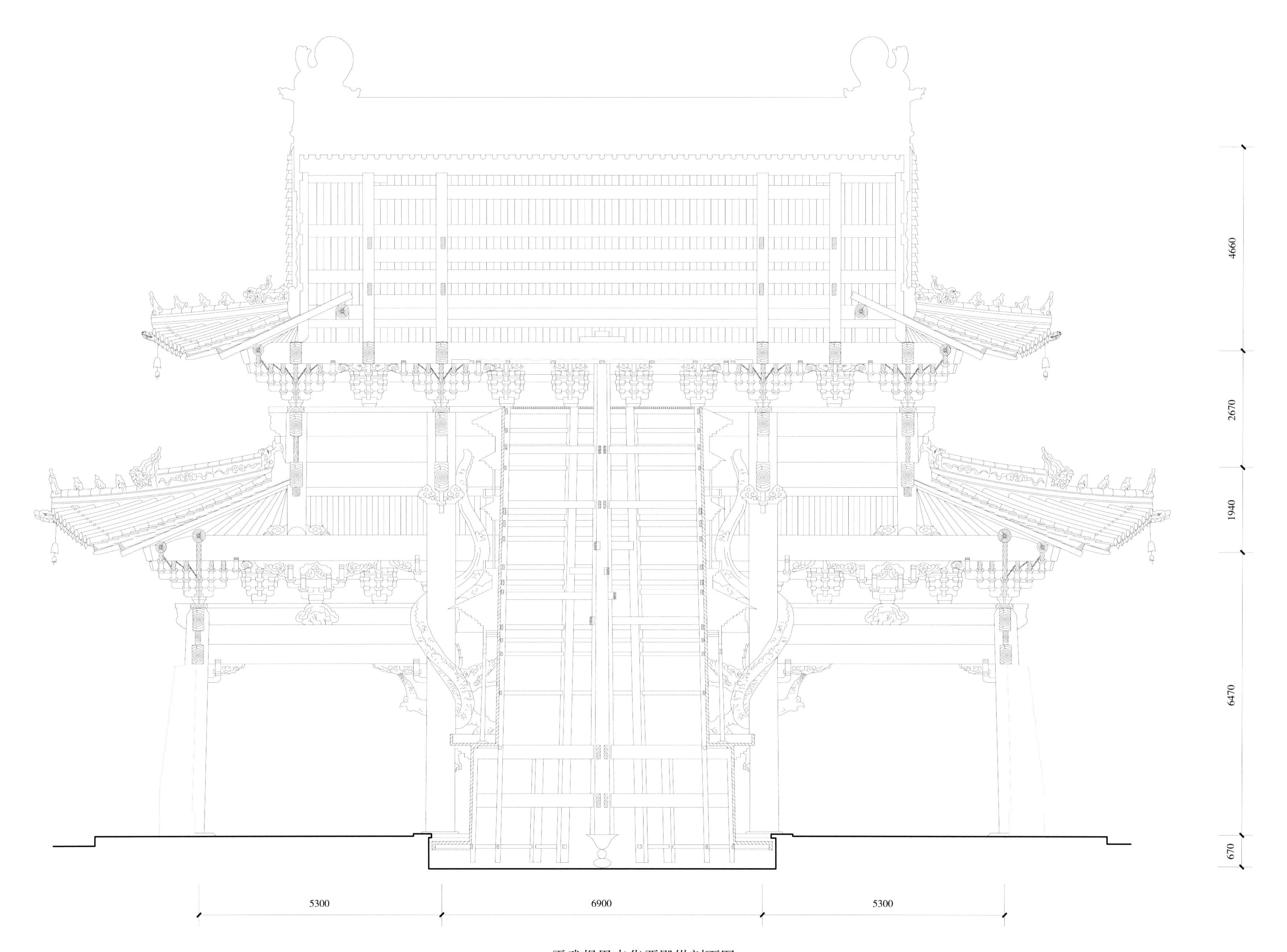
4660
2670
1940
6470
670
5300
6900
5300
0
1
2
3m

平武报恩寺华严殿纵剖面图

Longitudinal Section of Huayan Hall, Bao'en Temple, Pingwu

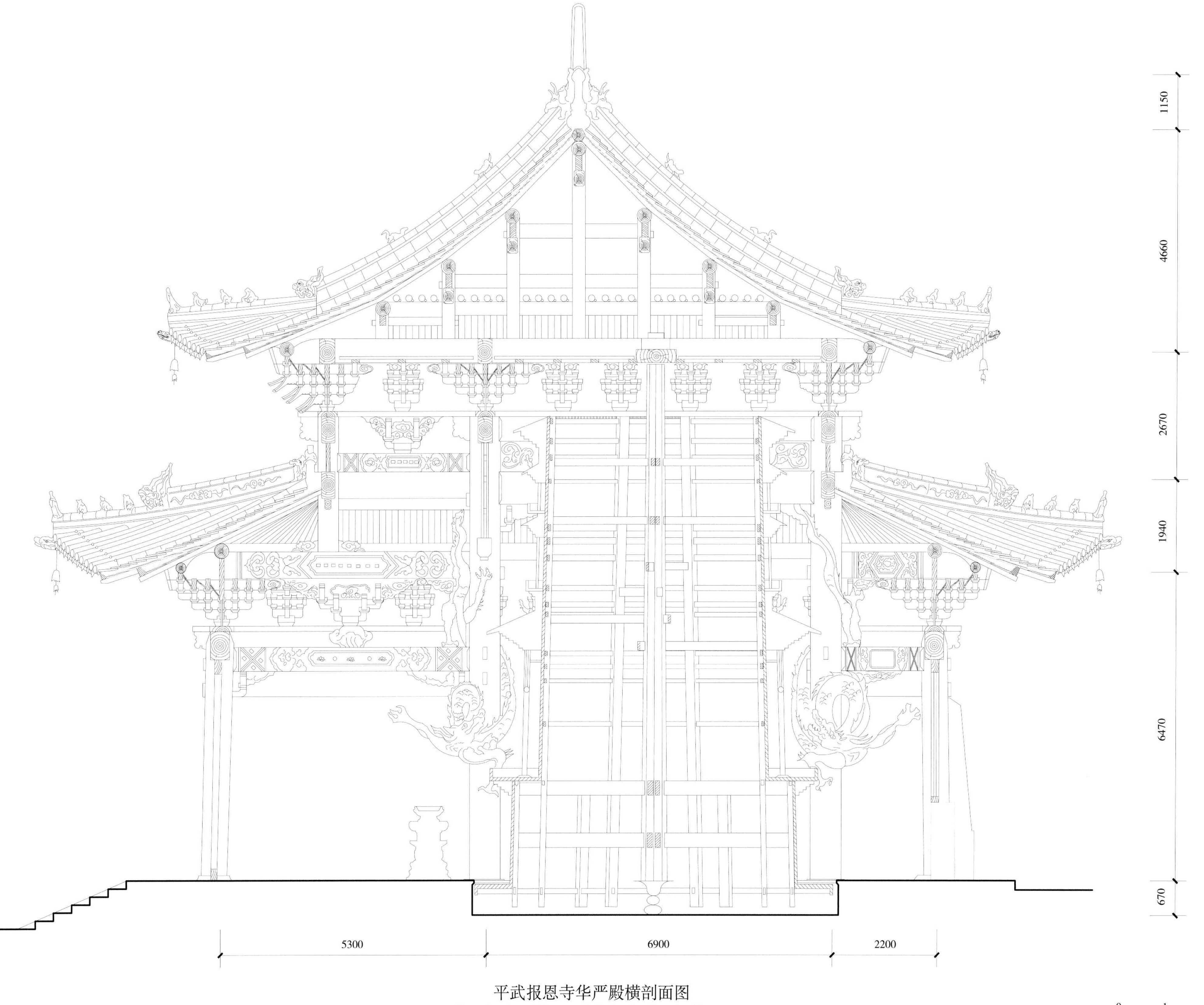
1150
4660
2670
1940
6470
670
5300
6900
2200
0
1
2
3m

平武报恩寺华严殿横剖面图
Cross Section of Huayan Hall, Bao'en Temple, Pingwu

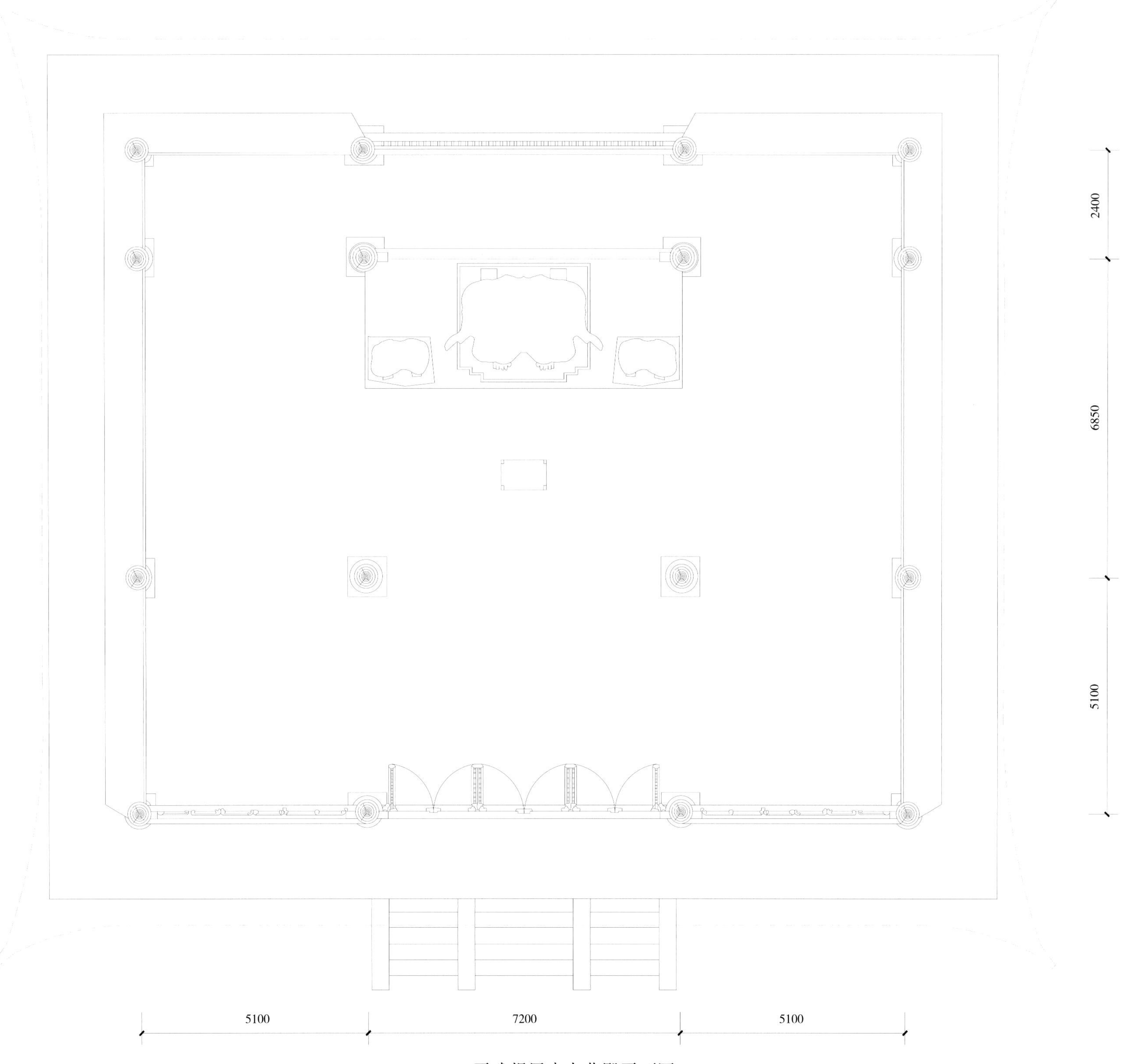

2400
6850
5100
5100
7200
5100
0 1 2 3m

平武报恩寺大悲殿平面图
Ground Floor Plan of Dabei Hall, Bao’en Temple, Pingwu

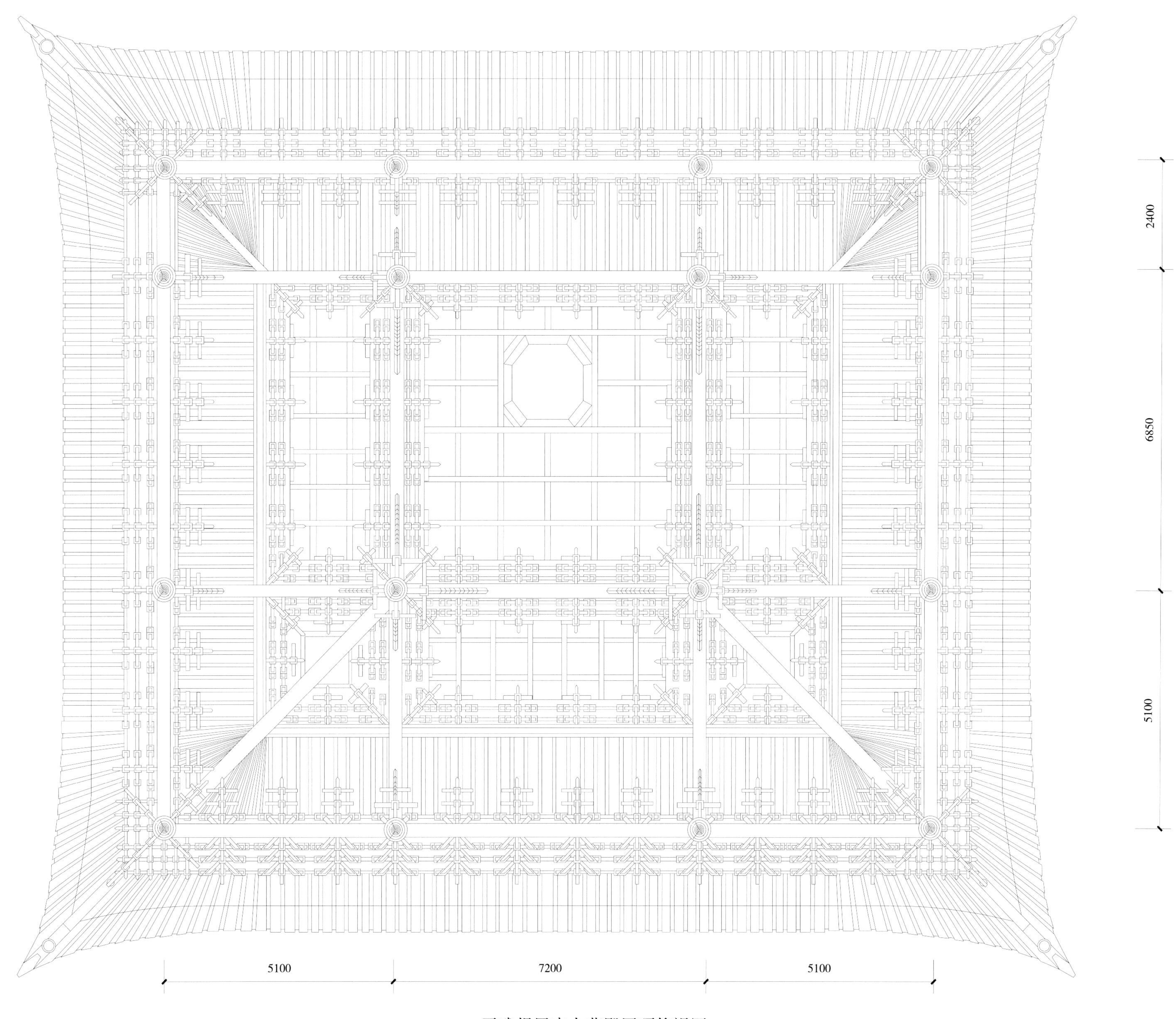
2400
6850
5100
5100
7200
5100
0
1
2
3m

平武报恩寺大悲殿屋顶仰视图
Roof Framing of Dabei Hall, Bao'en Temple, Pingwu

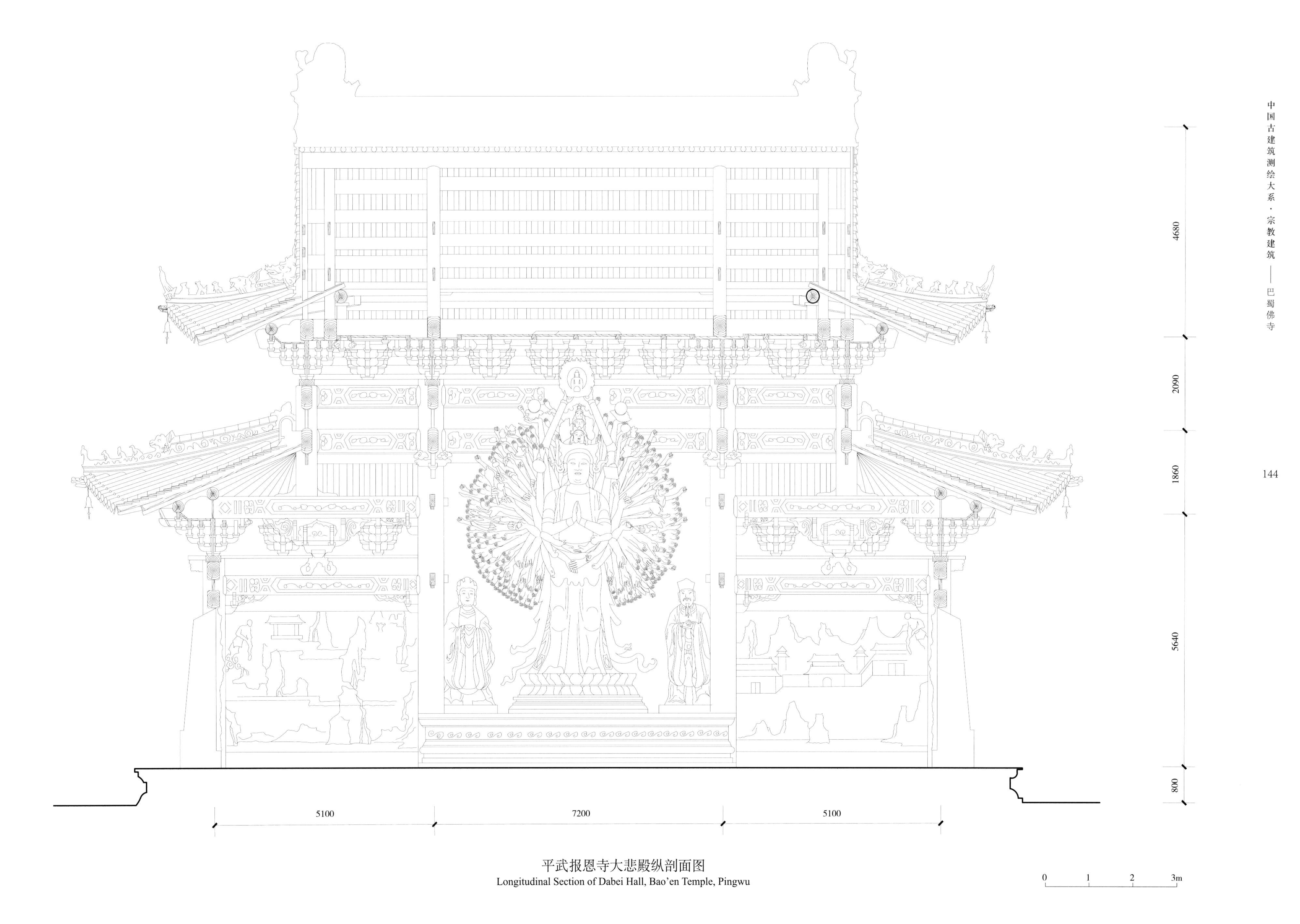
4680
2090
1860
5640
800
5100
7200
5100
0 1 2 3m

平武报恩寺大悲殿纵剖面图
Longitudinal Section of Dabei Hall, Bao'en Temple, Pingwu

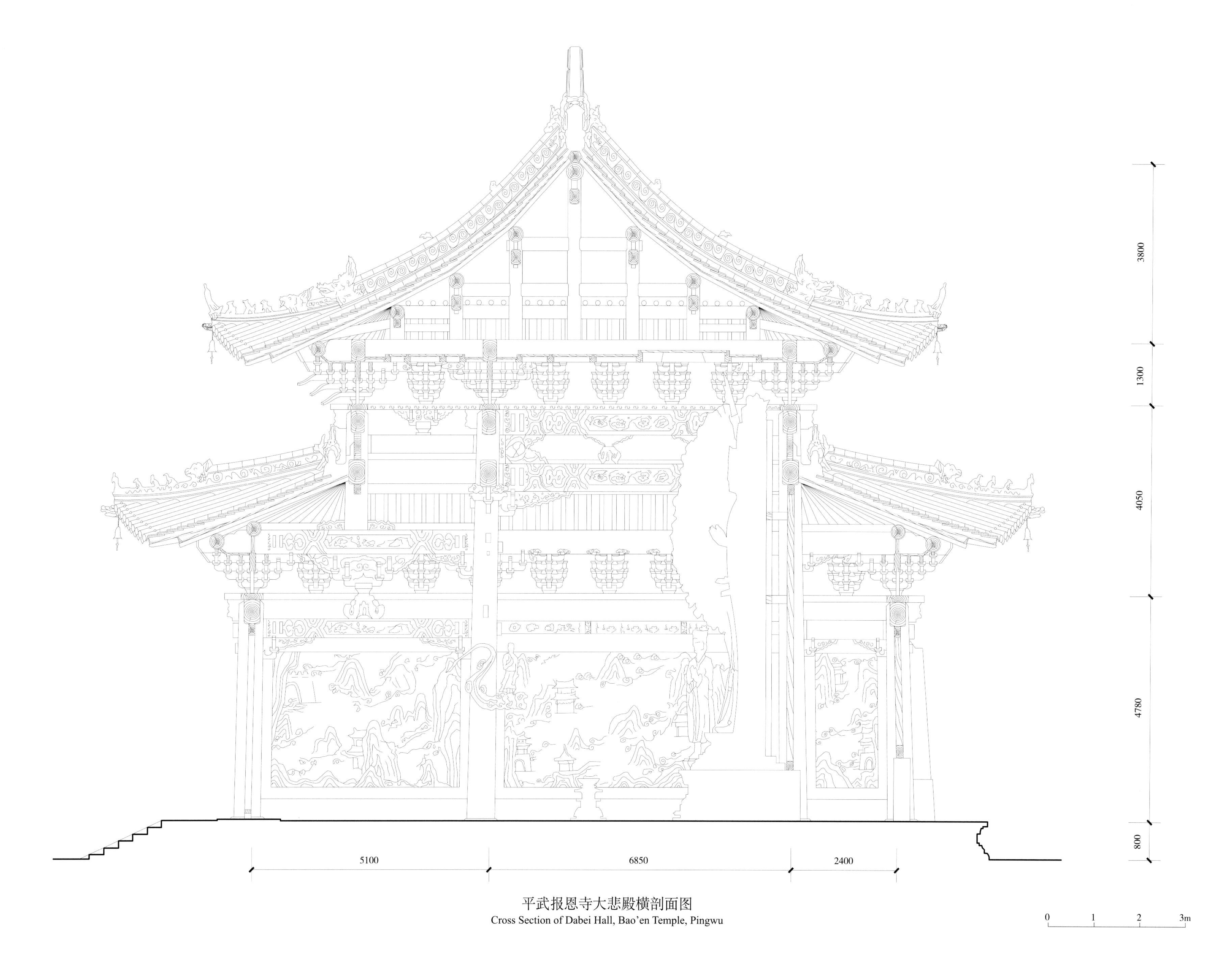
3800
1300
4050
4780
800
5100
6850
2400
0
1
2
3m

平武报恩寺大悲殿横剖面图
Cross Section of Dabei Hall, Bao'en Temple, Pingwu

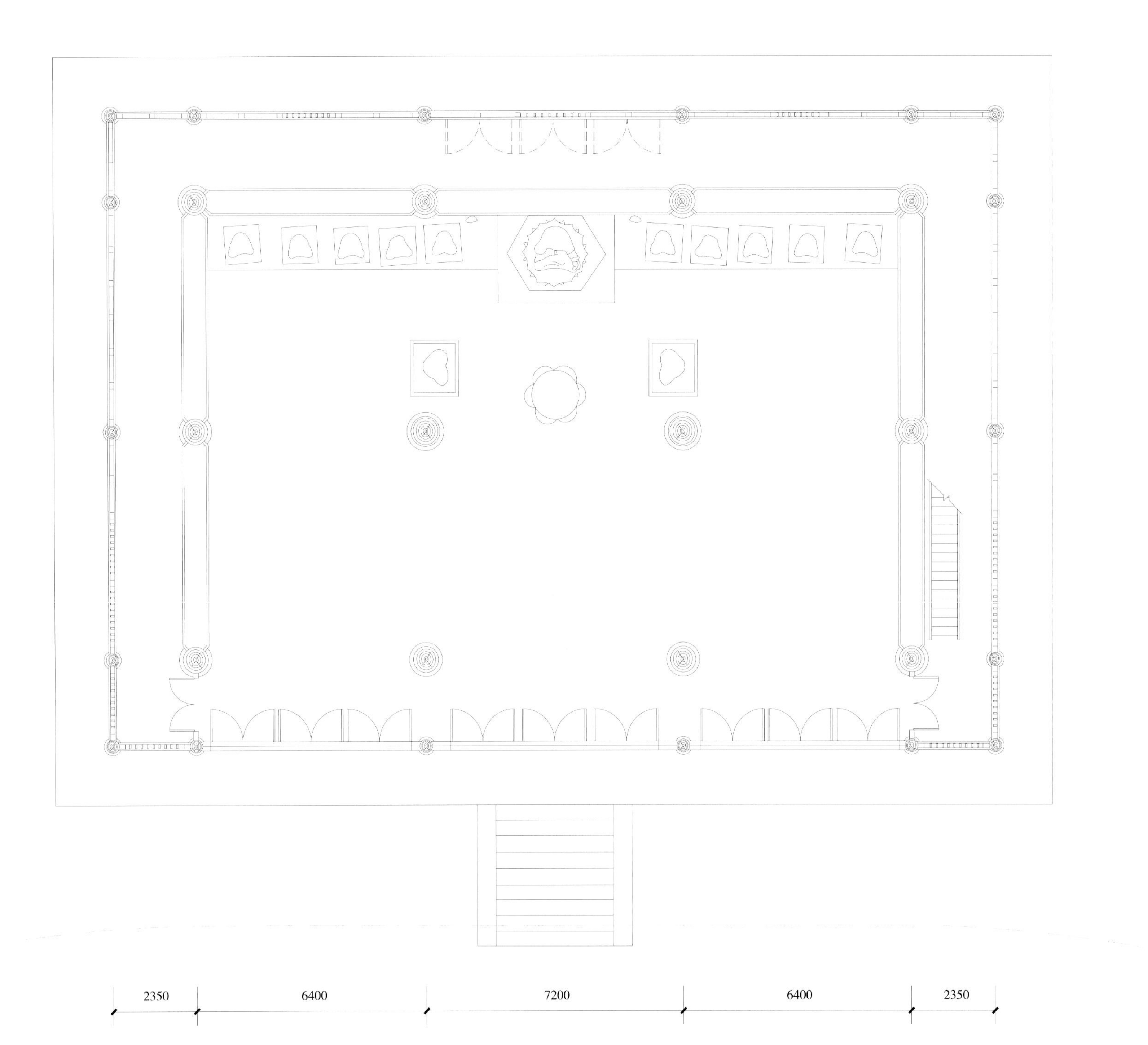
2350
6200
6200
2350
2350
6400
7200
6400
2350

平武报恩寺万佛阁一层平面图

Ground Floor Plan of Wanfo Pavilion, Bao'en Temple, Pingwu

0 1 2 3m

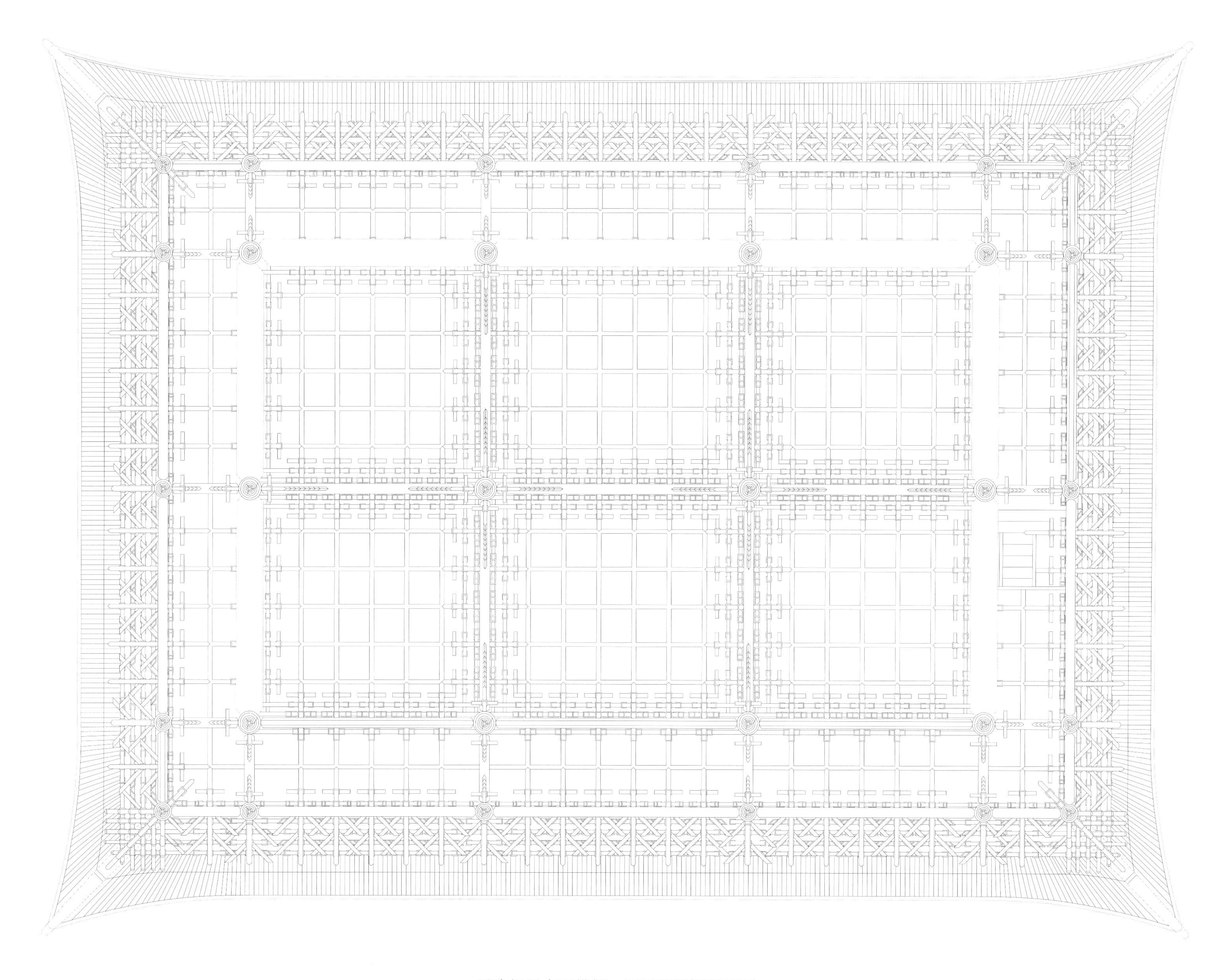
0 1 2 3m

平武报恩寺万佛阁一层屋顶仰视平面图
First Floor Roof Framing of Wanfo Pavilion, Bao'en Temple, Pingwu

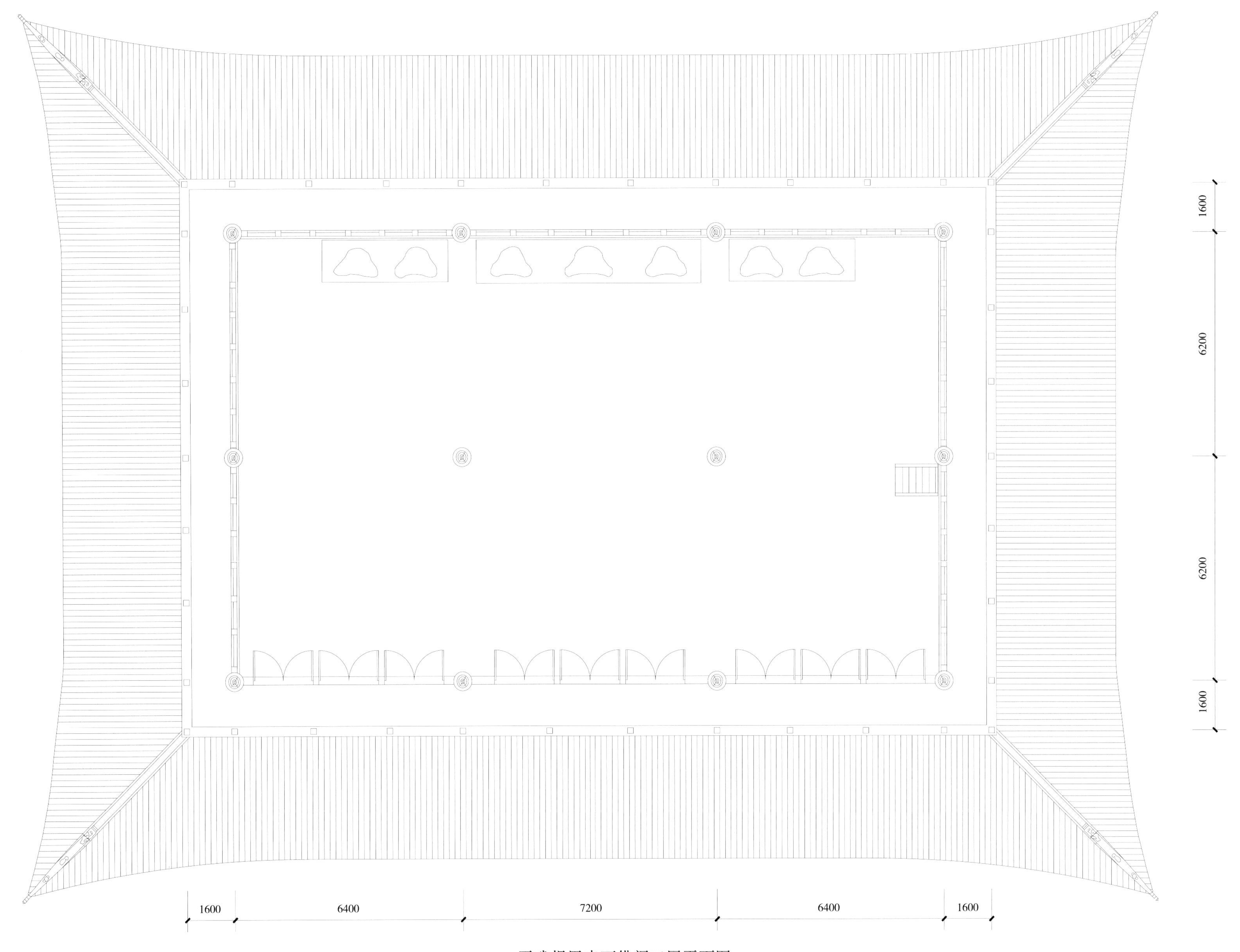
1600
6200
6200
1600
1600
6400
7200
6400
1600

平武报恩寺万佛阁二层平面图
First Floor Plan of Wanfo Pavilion, Bao'en Temple, Pingwu

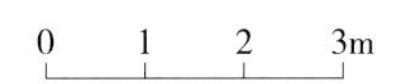
0 1 2 3m

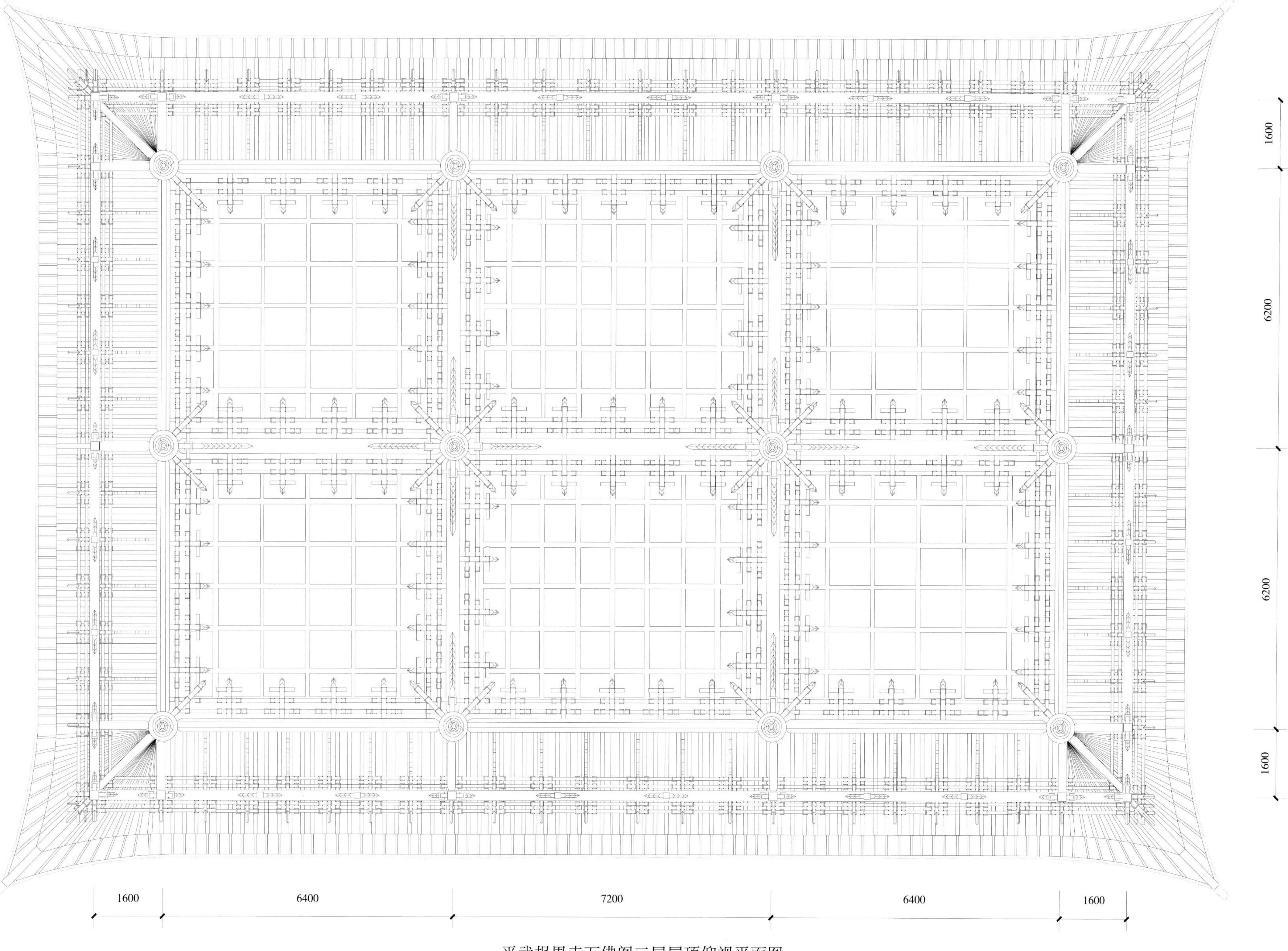
1600
6200
6200
1600
1600
6400
7200
6400
1600
0
1
2
3m

平武报恩寺万佛阁二层屋顶仰视平面图
Roof Framing of Wanfo Pavilion, Bao'en Temple, Pingwu

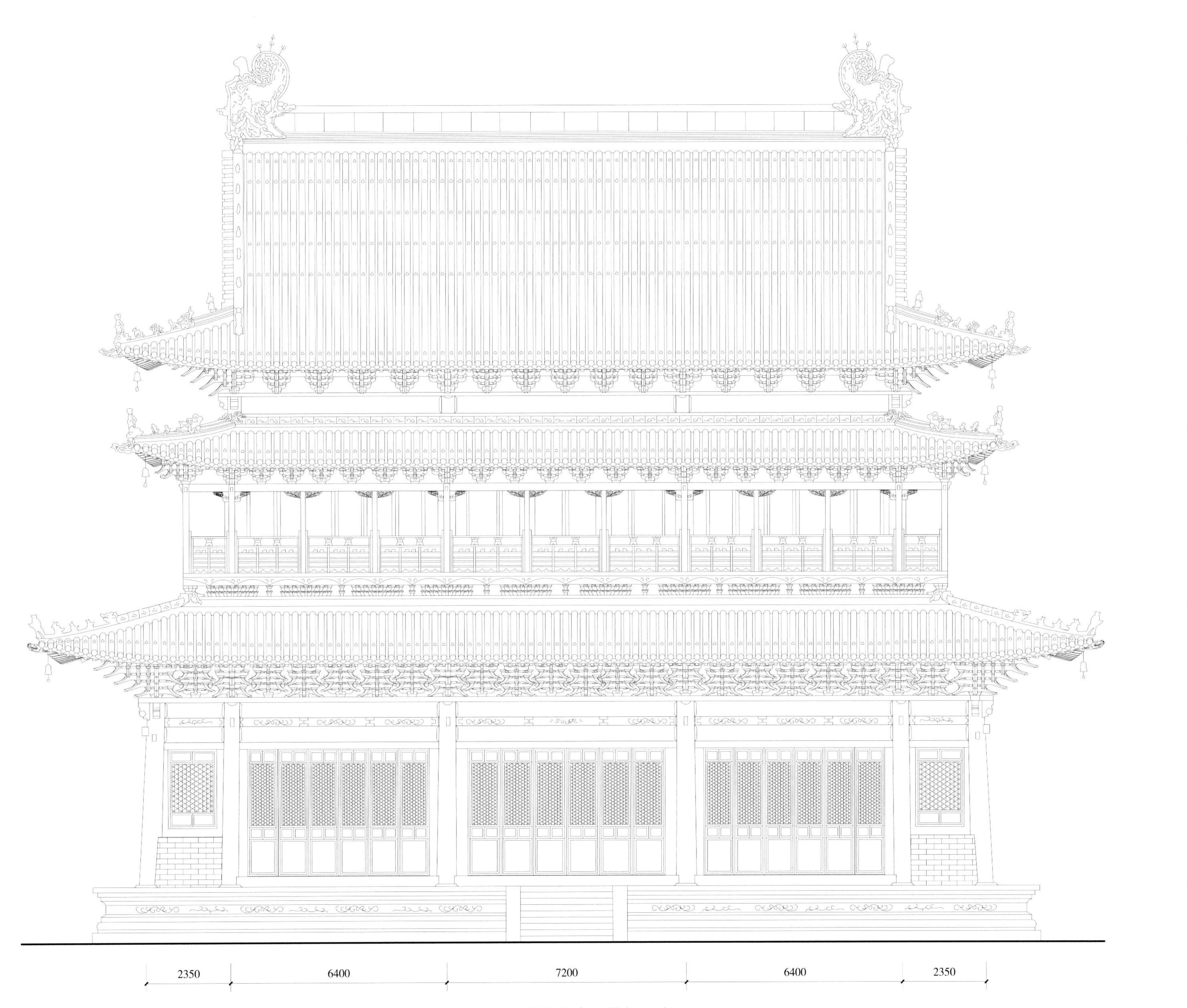
6570
1800
1100
4200
1590
6700
1600
2350
6400
7200
6400
2350
0 1 2 3m

平武报恩寺万佛阁正立面图

Front Elevation of Wanfo Pavilion, Bao'en Temple, Pingwu

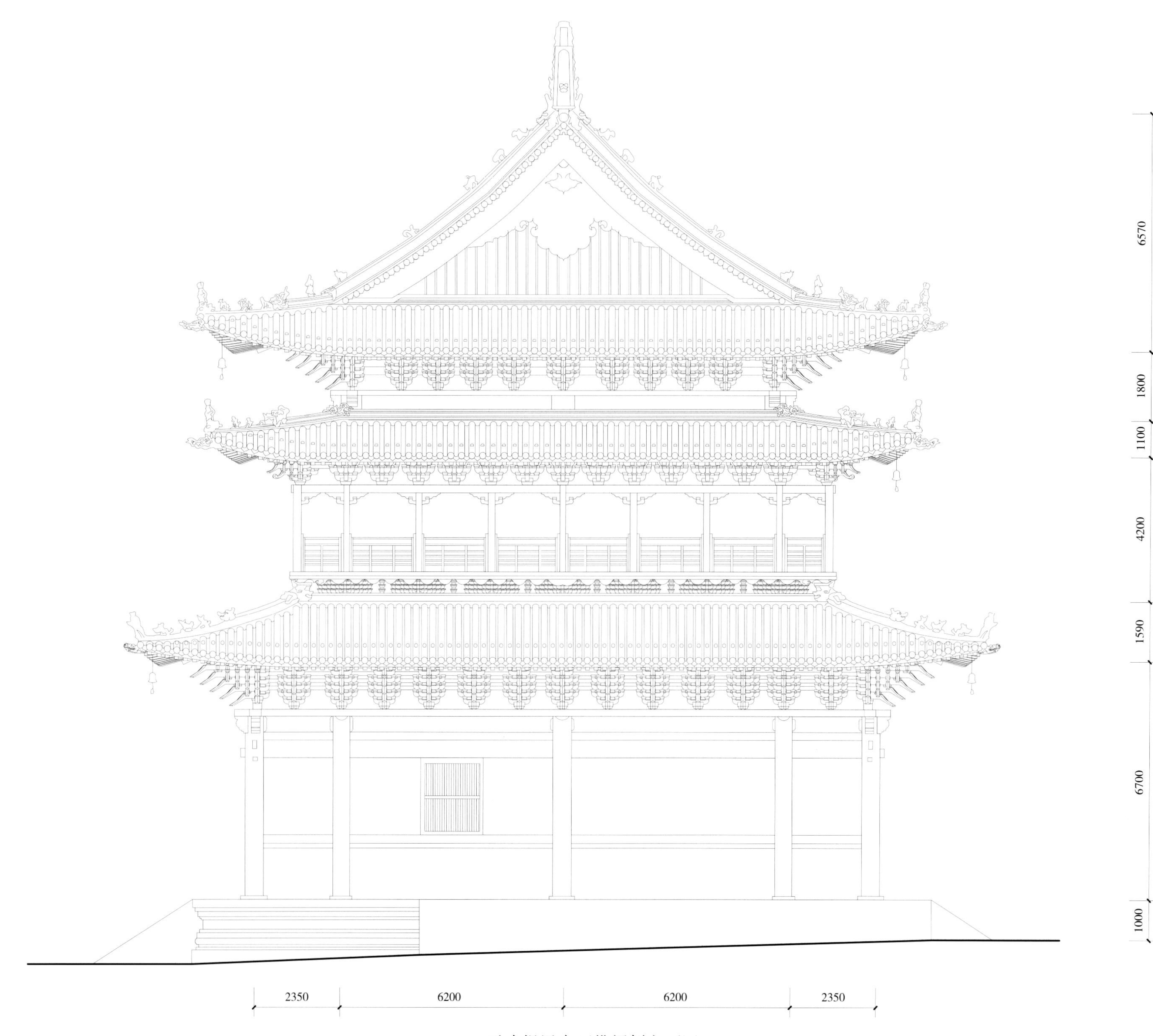
6570
1800
1100
4200
1590
6700
1000
2350
6200
6200
2350
0 1 2 3m

平武报恩寺万佛阁侧立面图
Side Elevation of Wanfo Pavilion, Bao'en Temple, Pingwu

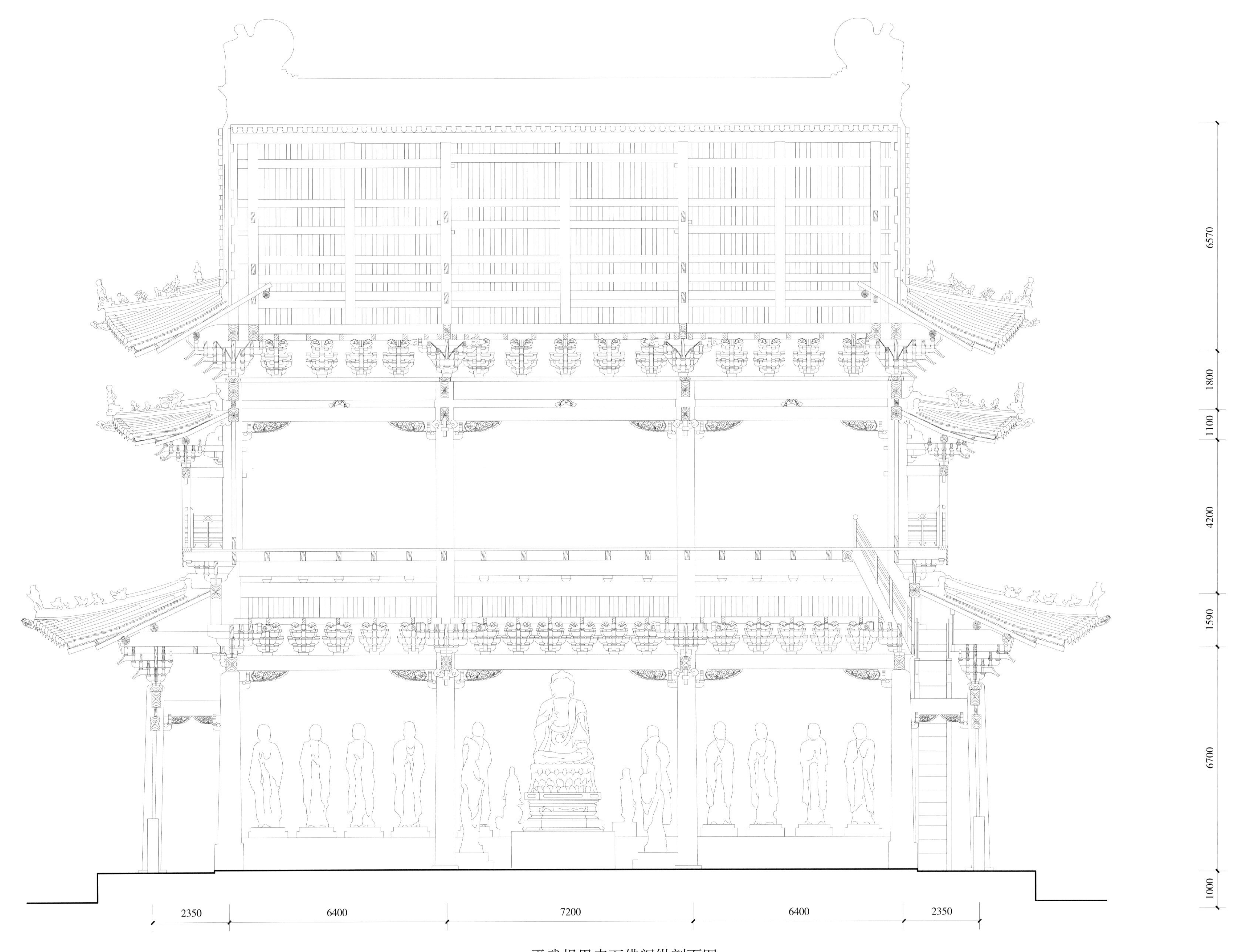
6570
1800
1100
4200
1590
6700
1000
2350
6400
7200
6400
2350
0 1 2 3m

平武报恩寺万佛阁纵剖面图

Longitudinal Section of Wanfo Pavilion, Bao'en Temple, Pingwu

6570
2900
5790
6700
2350
6200
6200
2350
0 1 2 3m

平武报恩寺万佛阁横剖面图

Cross Section of Wanfo Pavilion, Bao'en Temple, Pingwu

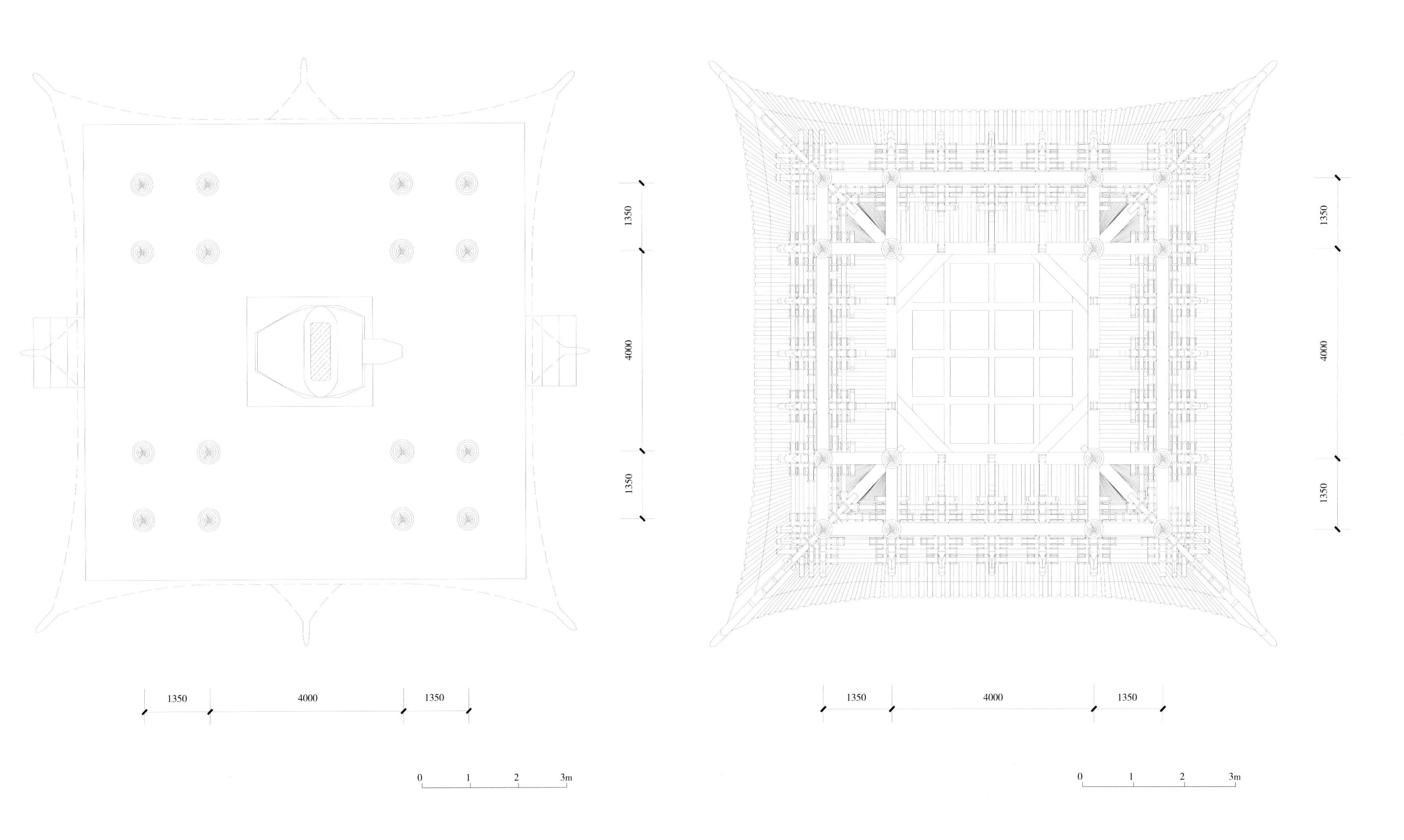
1350
4000
1350
1350
4000
1350
0
1
2
3m
1350
4000
1350
1350
4000
1350
0
1
2
3m

平武报恩寺碑亭平面图
Ground Floor Plan of Stele Pavilion, Bao'en Temple, Pingwu

平武报恩寺碑亭一层檐仰视图
Lower Roof Framing of Stele Pavilion, Bao'en Temple, Pingwu

1380
1120
1440

400 3840 460

0 1 2 3m

平武报恩寺碑亭重檐剖面图
Cross Section of Upper Roof Framing, Stele Pavilion, Bao'en Temple, Pingwu

3960 3840 3960

0 1 2 3m

平武报恩寺碑亭重檐仰视图
Upper Roof Framing of Stele Pavilion, Bao'en Temple, Pingwu

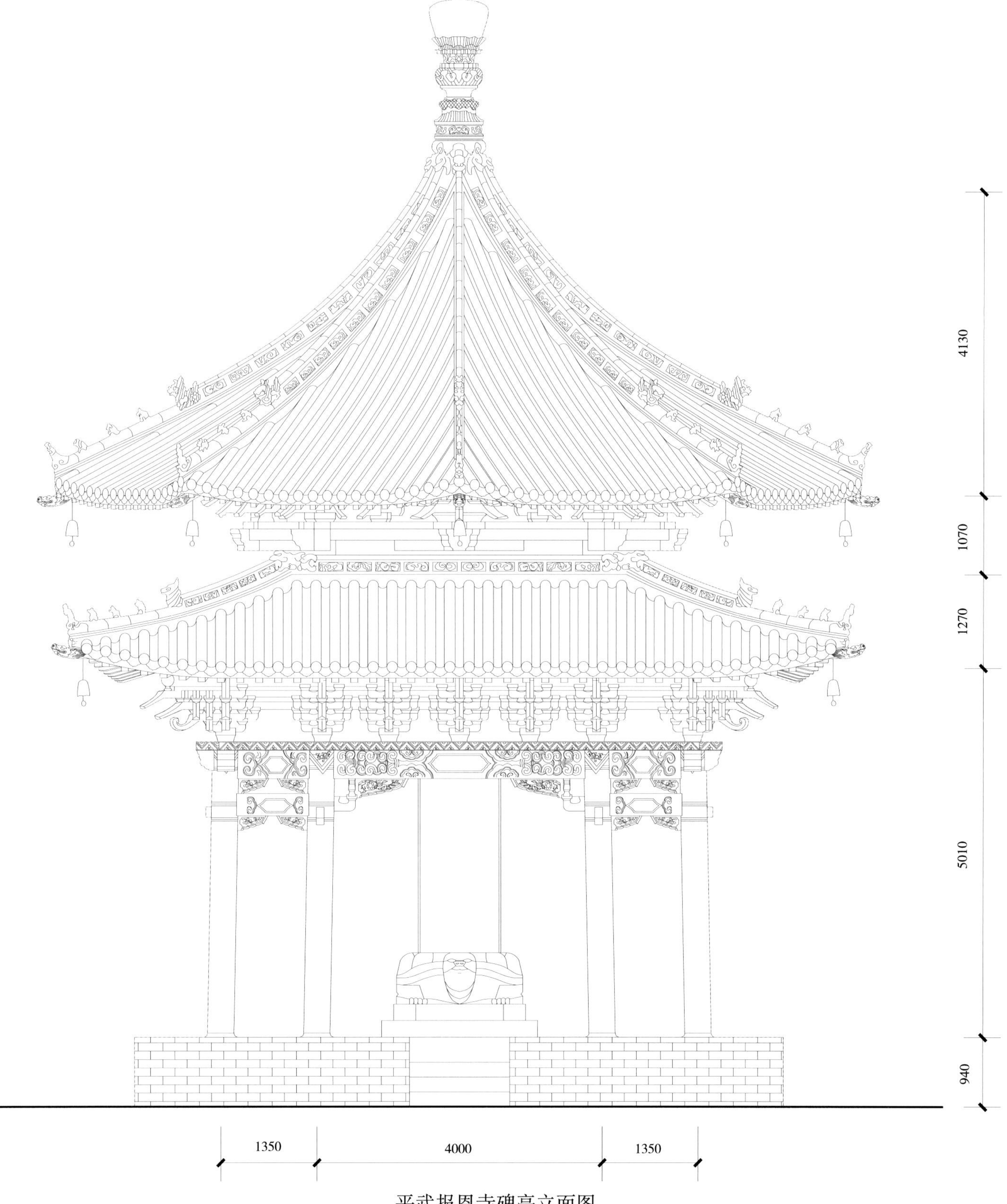
4130
1070
1270
5010
940
1350
4000
1350
0
1
2
3m

平武报恩寺碑亭立面图
Front Elevation of Stele Pavilion, Bao'en Temple, Pingwu

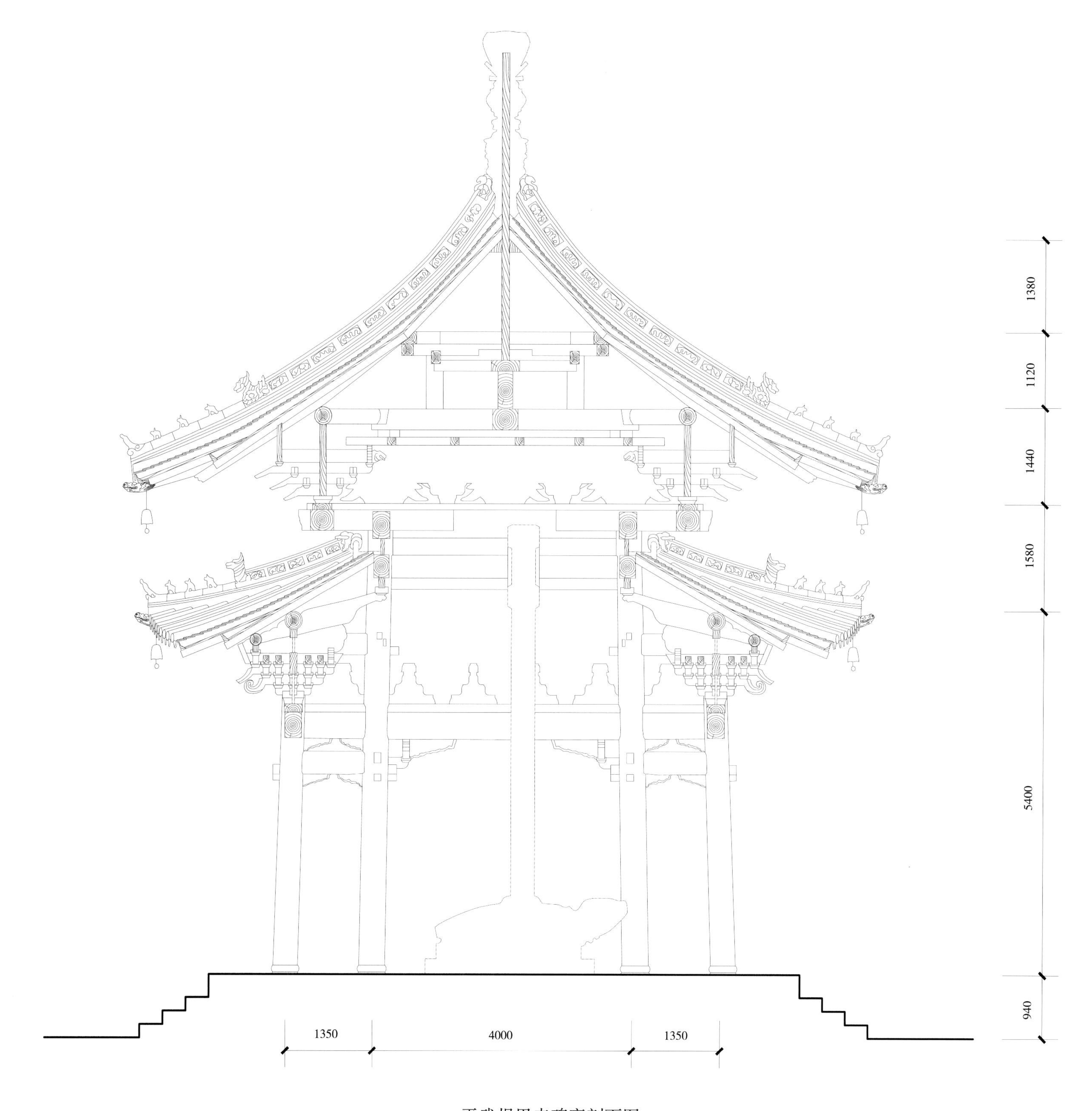
1380
1120
1440
1580
5400
940
1350
4000
1350
0
1
2
3m

平武报恩寺碑亭剖面图
Cross Section of Stele Pavilion, Bao'en Temple, Pingwu

0 20 40 60cm

天王殿梁枋彩画详图

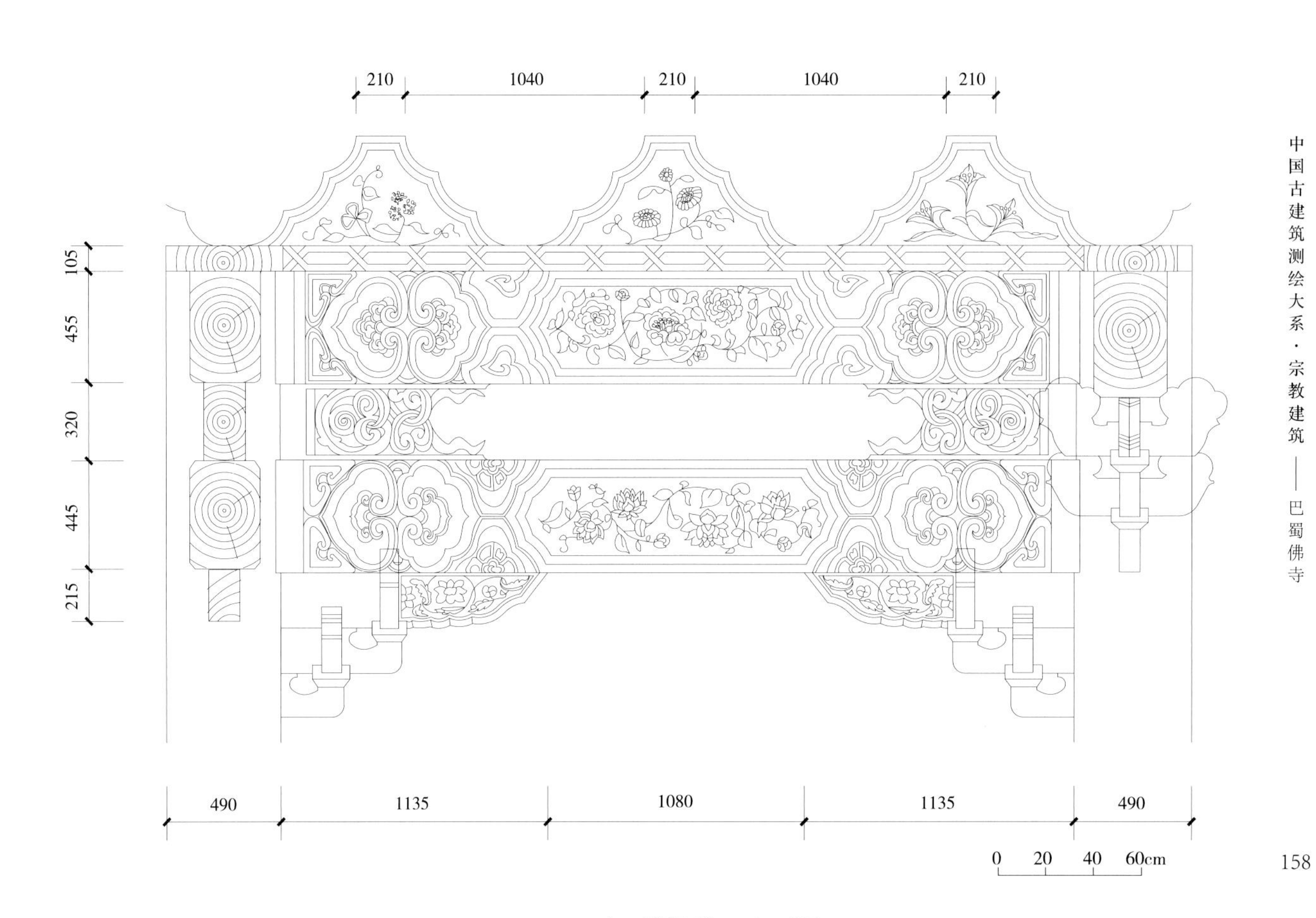
210
1040
210
1040
210
105
455
320
445
215
490
1135
1080
1135
490
0 20 40 60cm

天王殿梁枋正立面图

0 20 40 60cm

大雄宝殿梁枋彩画详图

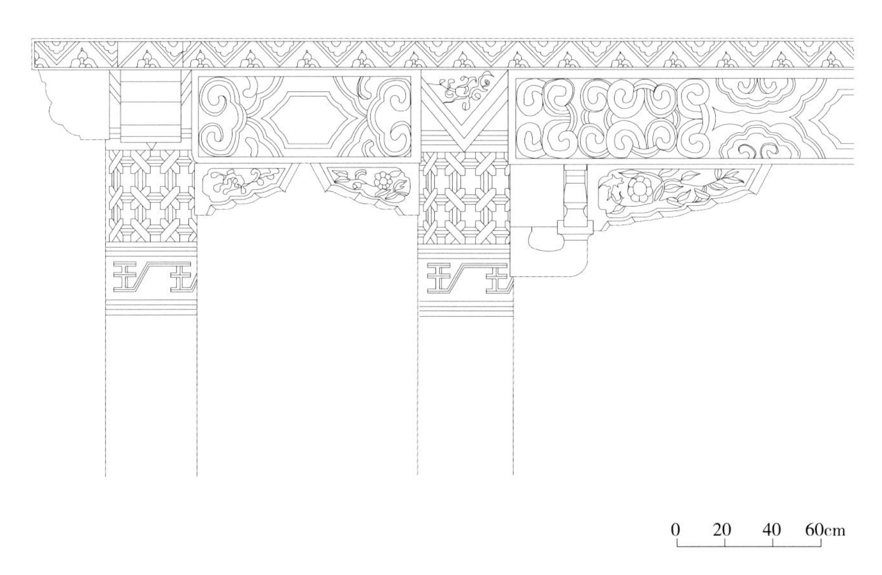
0 20 40 60cm

碑亭梁枋彩画式样图

平武报恩寺梁枋彩画式样图

Color Patterns on Beams and Lintels, Bao'en Temple, Pingwu

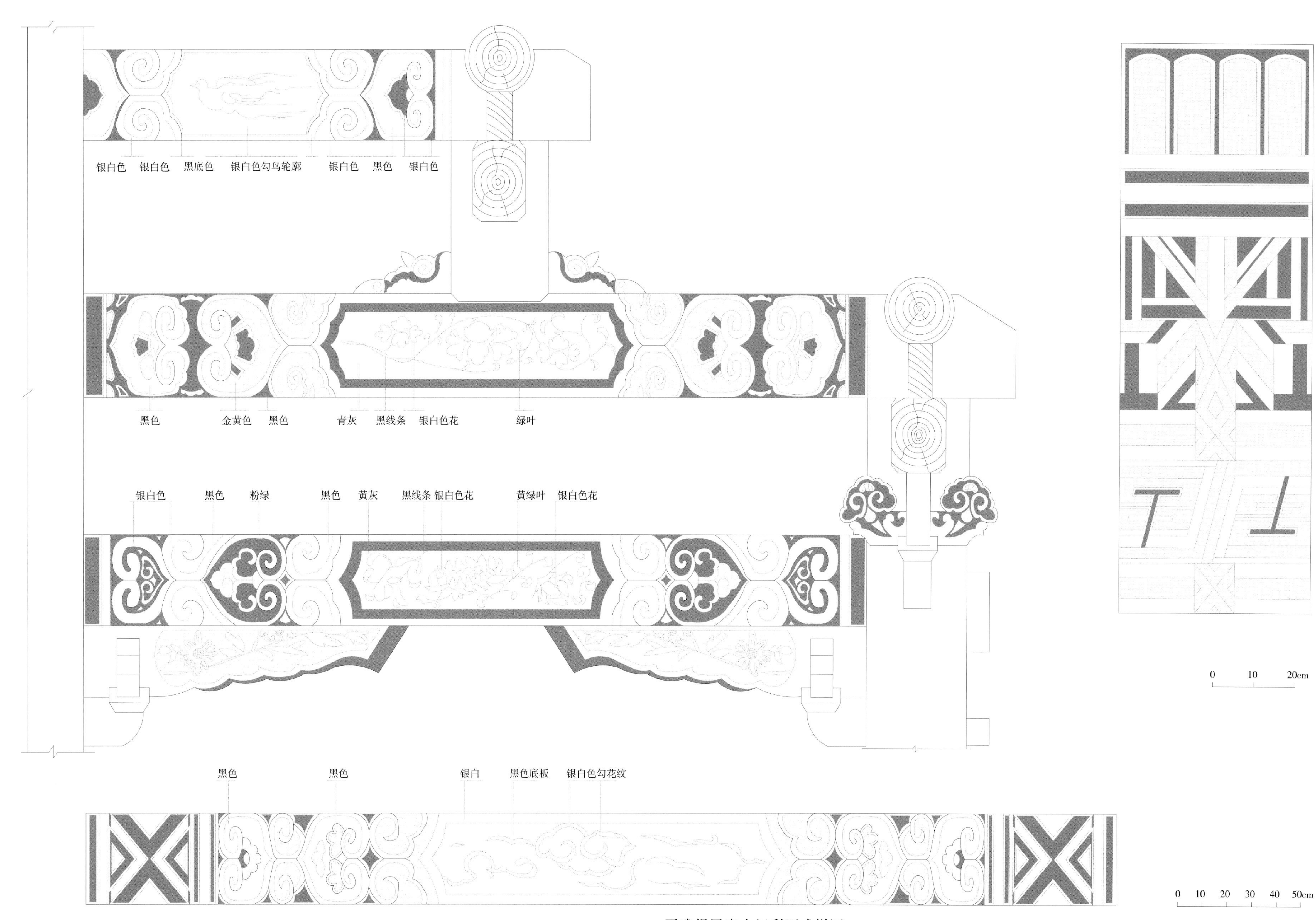
银白色
银白色
黑底色
银白色勾鸟轮廓
银白色
黑色
银白色
黑色
黑色
金黄色
黑色
青灰
黑线条
银白色花
绿叶
银白色
黑色
粉绿
黑色
黄灰
黑线条
银白色花
黄绿叶
银白色花
黑色
黑色
银白
黑色底板
银白色勾花纹
0 10 20cm
0 10 20 30 40 50cm

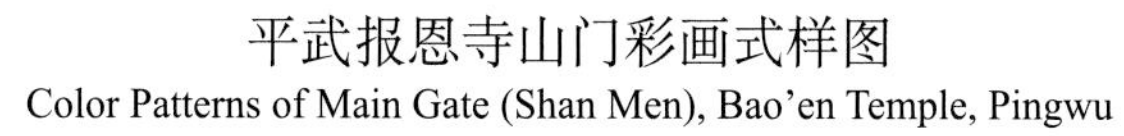
平武报恩寺山门彩画式样图

Color Patterns of Main Gate (Shan Men), Bao'en Temple, Pingwu

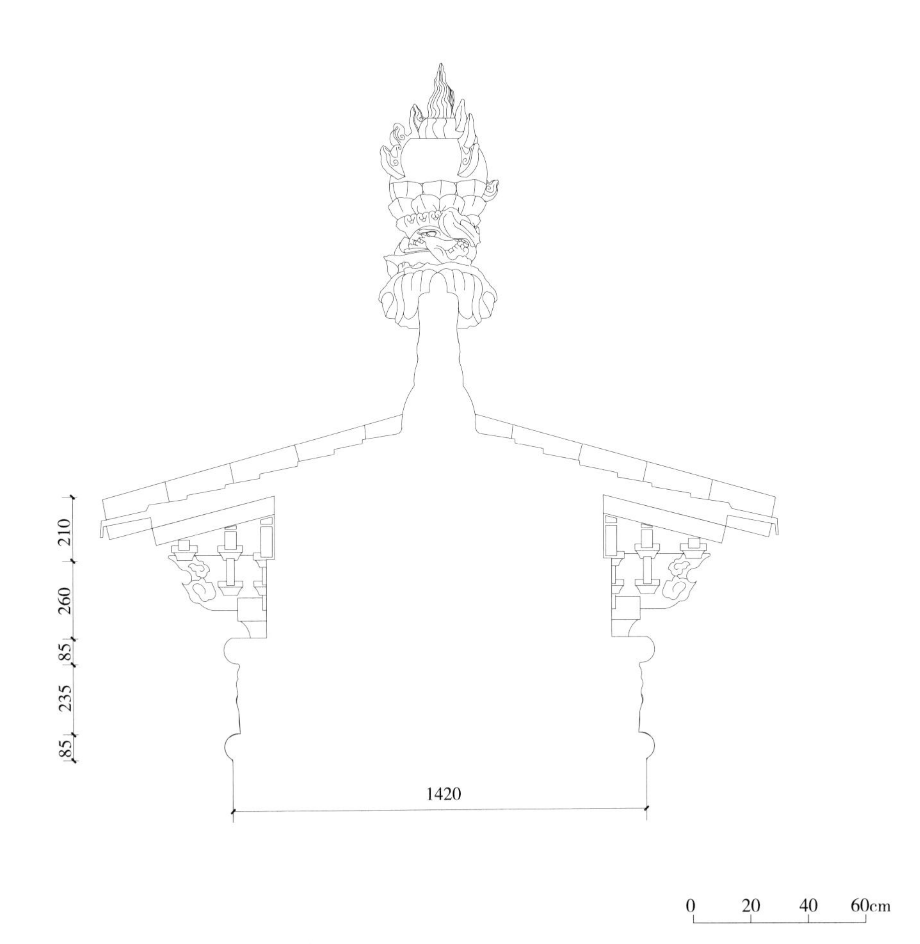
210
260
85
235
85
1420
0 20 40 60cm

平武报恩寺山门八字墙墙身剖面图
Detail, Section of Splayed Walls, Main Gate (Shan Men), Bao'en Temple, Pingwu

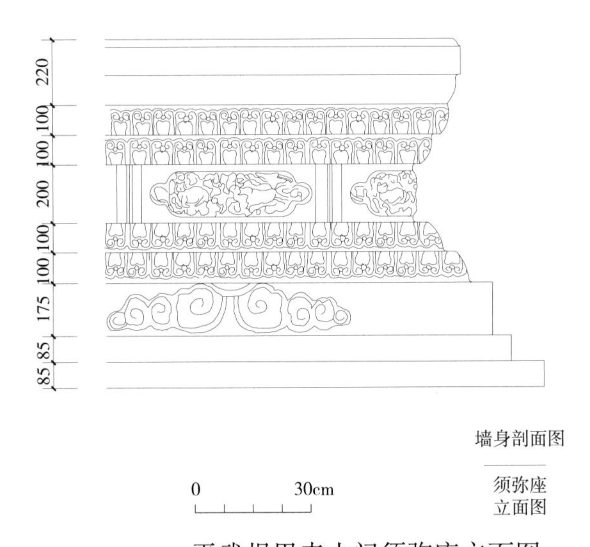
220
100
100
200
100
100
175
85
85
0 30cm
墙身剖面图
须弥座
立面图

平武报恩寺山门须弥座立面图
Detail, Elevation of Xumizuo, Main Gate (Shan Men), Bao'en Temple, Pingwu

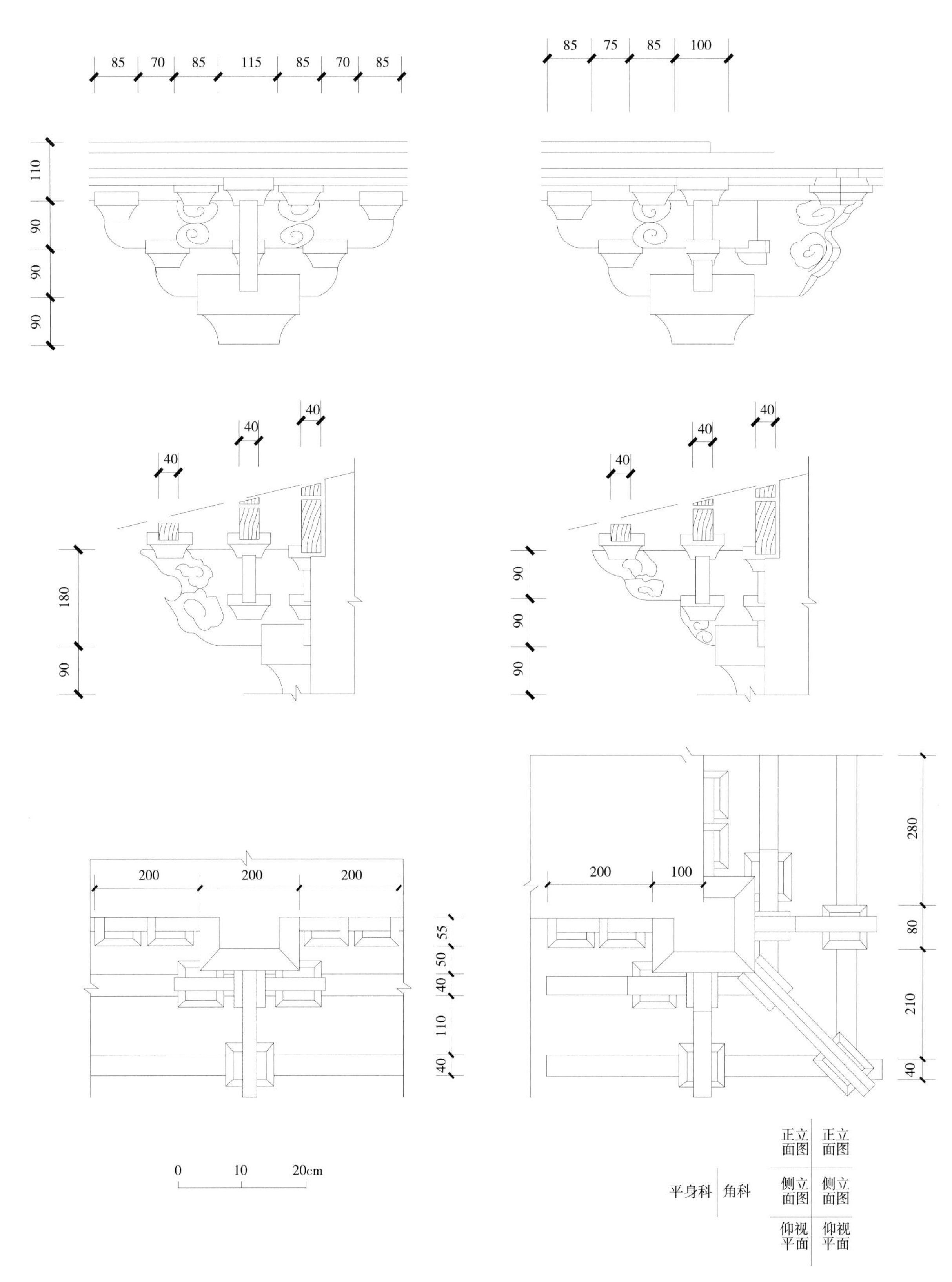
85 70 85 115 85 70 85
85 75 85 100
110
90
90
90
40
40
40
40
40
40
180
90
90
90
90
200 200 200
200 100
55
50
40
110
40
280
80
210
40
0 10 20cm
正立
面图
正立
面图
平身科
角科
侧立
面图
侧立
面图
仰视
平面
仰视
平面

平武报恩寺山门斗栱大样图
Detail, Dougong Sets, Main Gate (Shan Men), Bao'en Temple, Pingwu

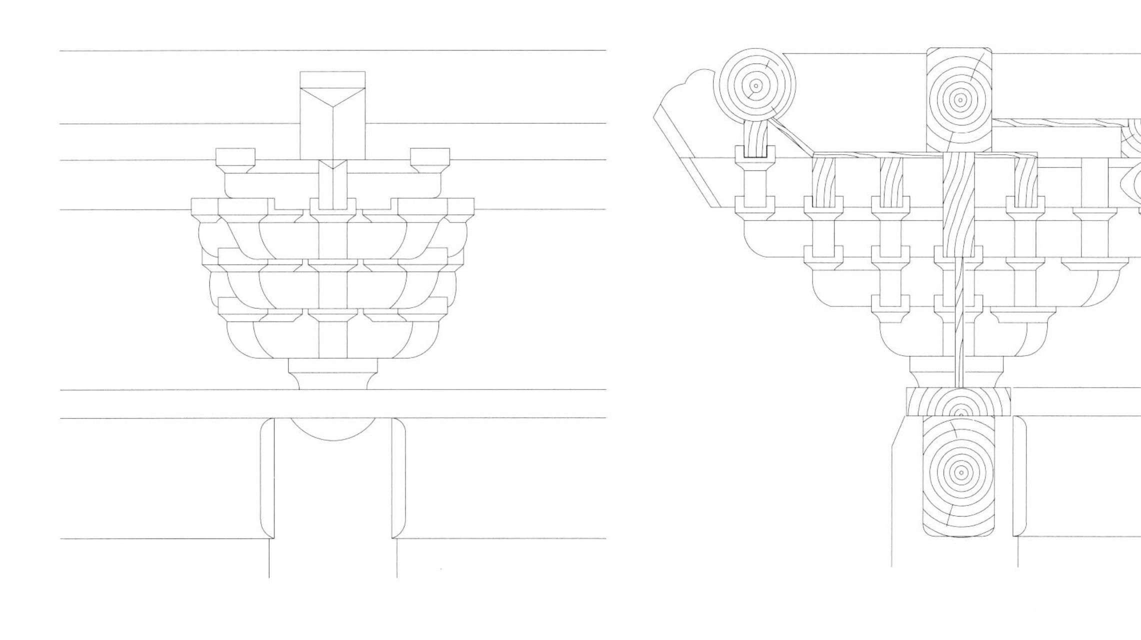

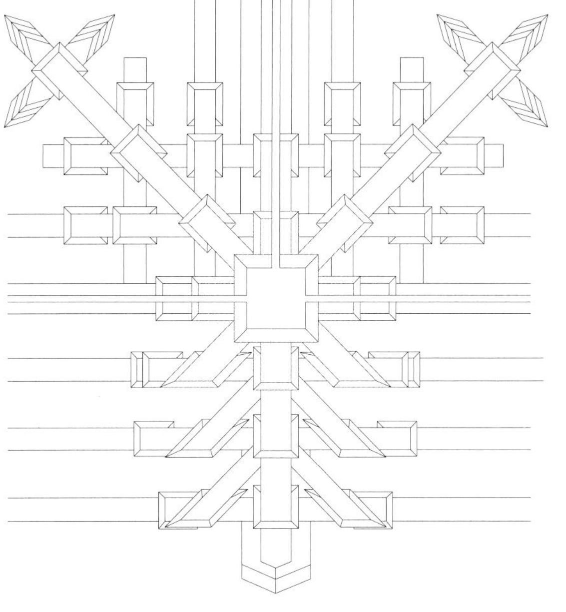

正立面图 侧立面图

仰视平面

0 20 40 60cm

天王殿外檐柱头科斗栱详图

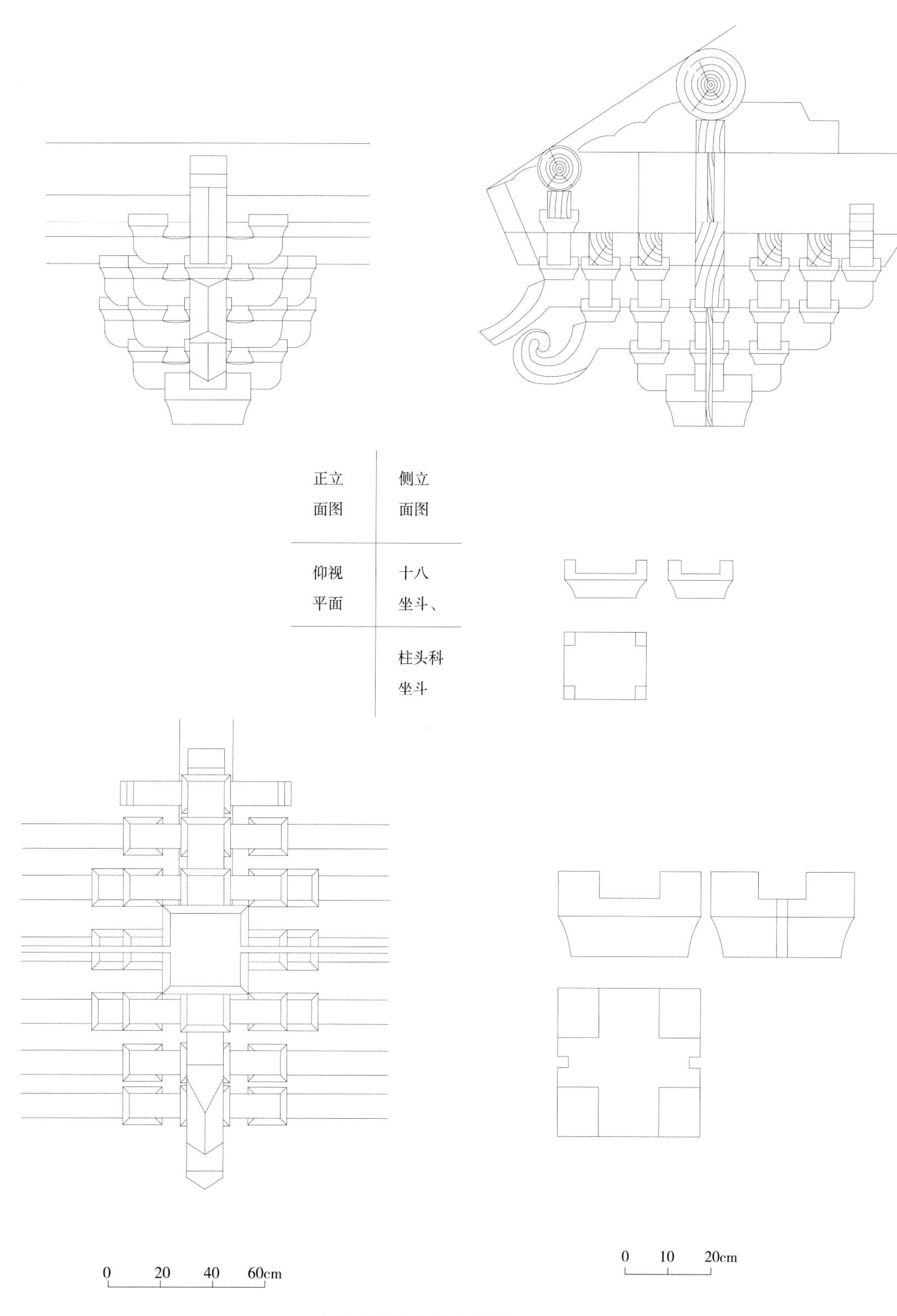
正立面图
侧立面图
仰视平面
十八坐斗、
柱头科坐斗
0 20 40 60cm
0 10 20cm

碑亭下檐柱头科斗栱详图

平武报恩寺柱头科斗栱大样图

Detail, Column Dougong Sets, Bao’en Temple, Pingwu

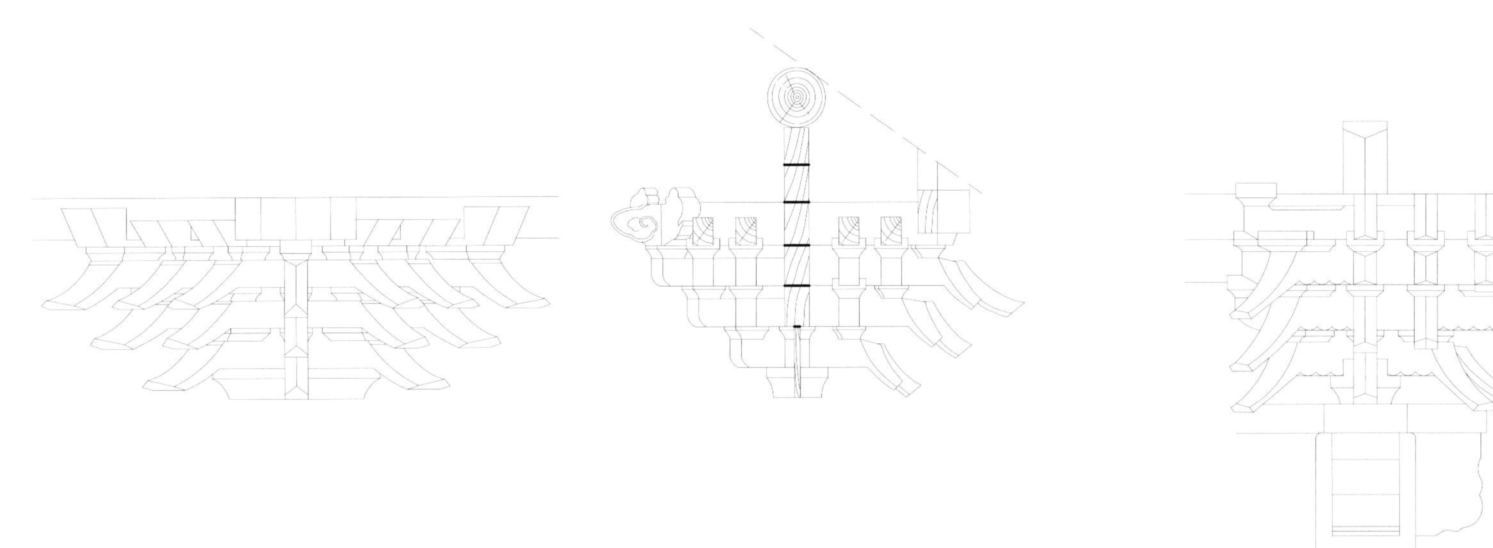

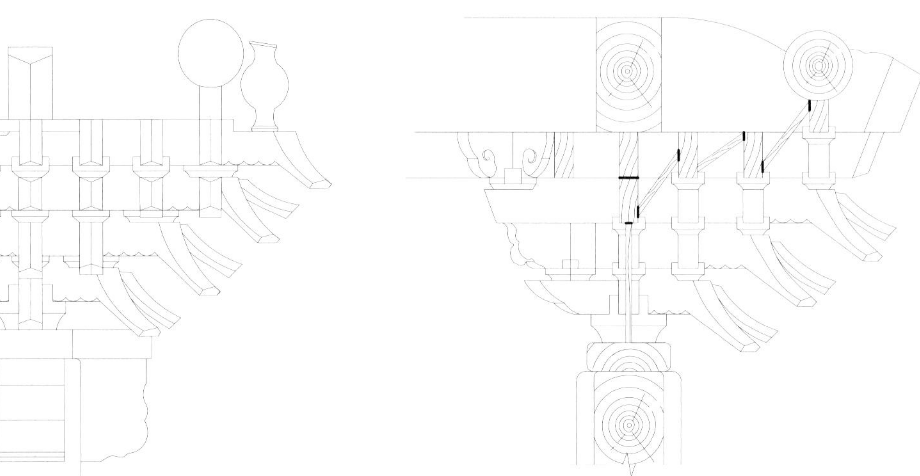

正立面图
侧立面图
仰视平面
大斗
菱形平盘斗

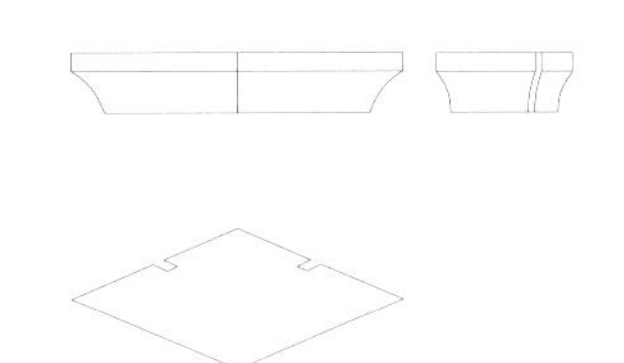

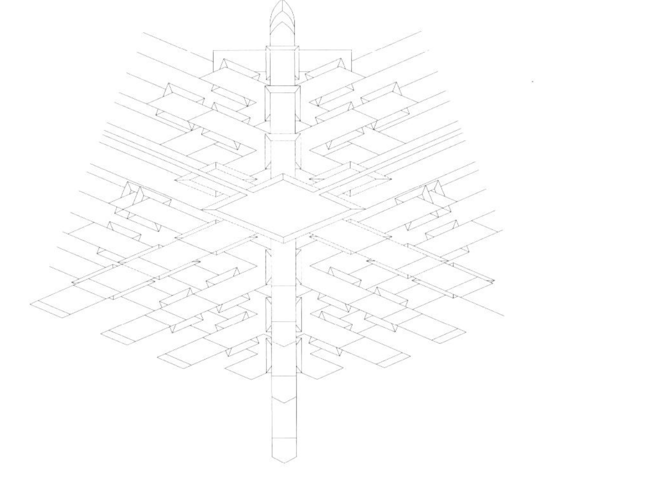

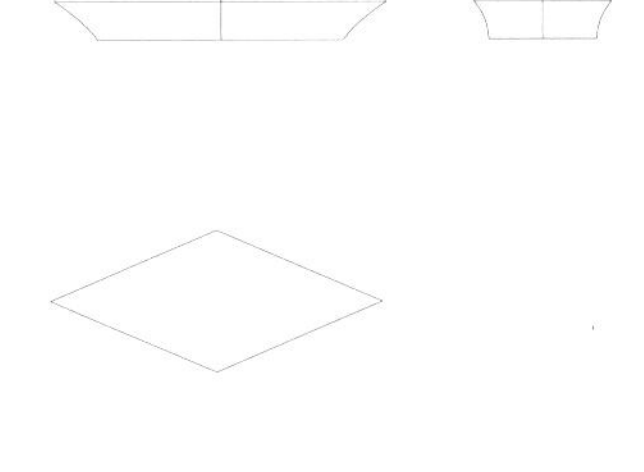

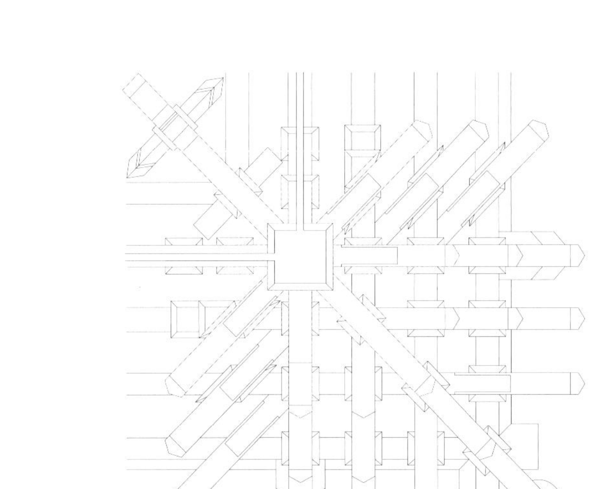

正立面图
侧立面图
仰视平面

0 20 40 60cm

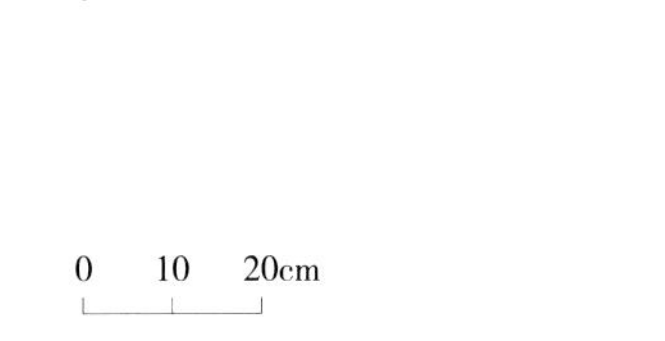
0 10 20cm

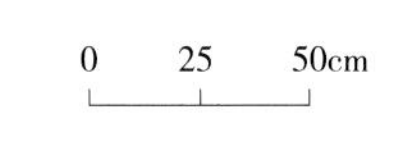
0 25 50cm

碑亭上檐角科斗栱详图

华严殿上檐角科斗栱详图

平武报恩寺角科斗栱大样图

Detail, Corner Dougong Sets, Bao’en Temple, Pingwu

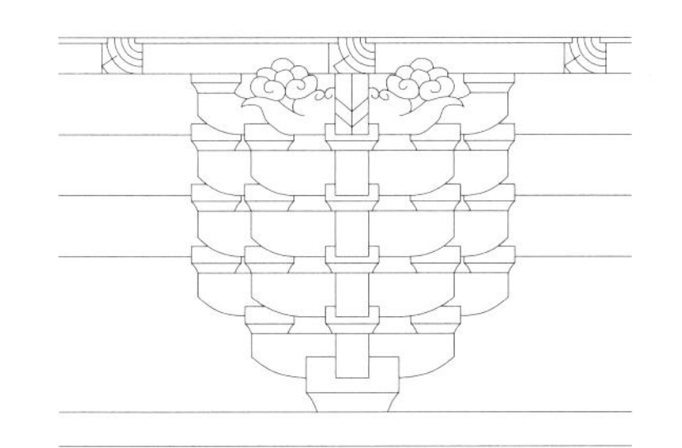
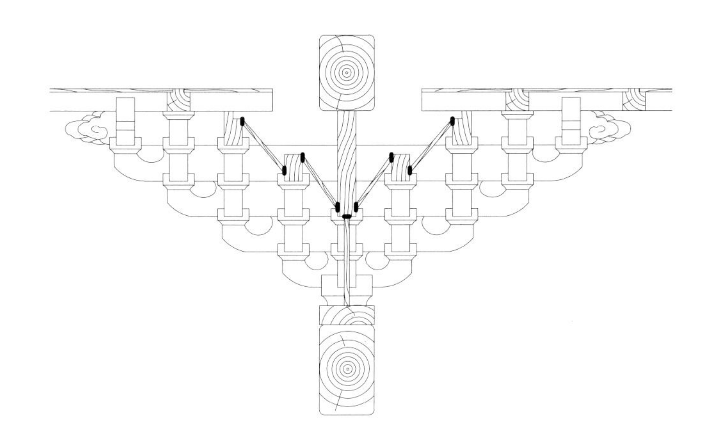
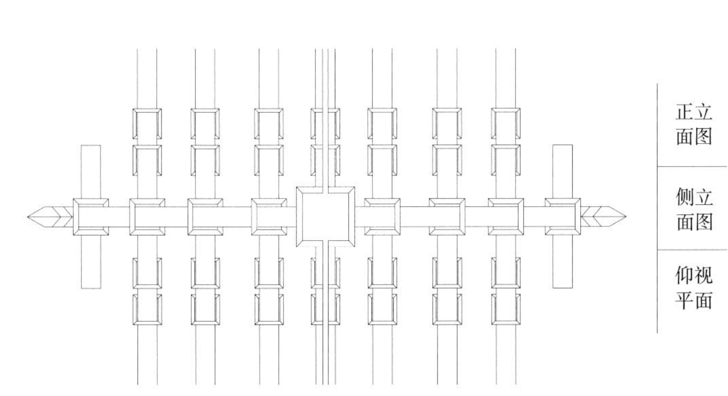
正立面图
侧立面图
仰视平面
0 25 50cm

大雄宝殿内檐斗栱详图

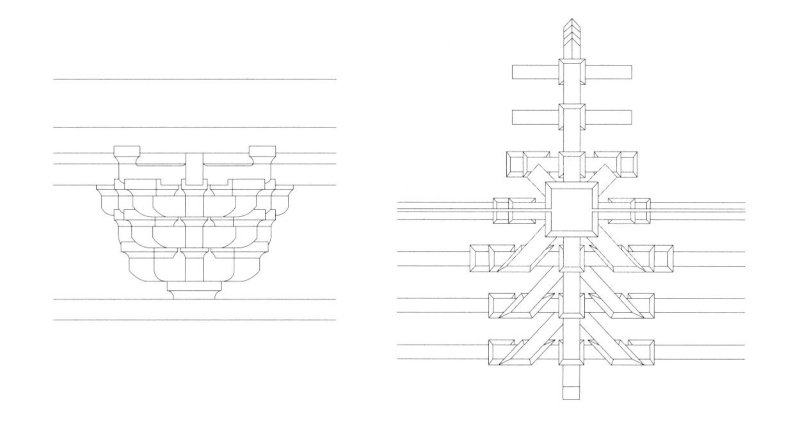
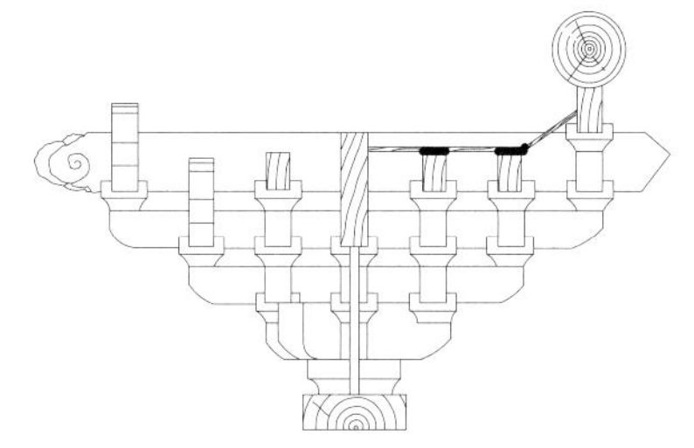
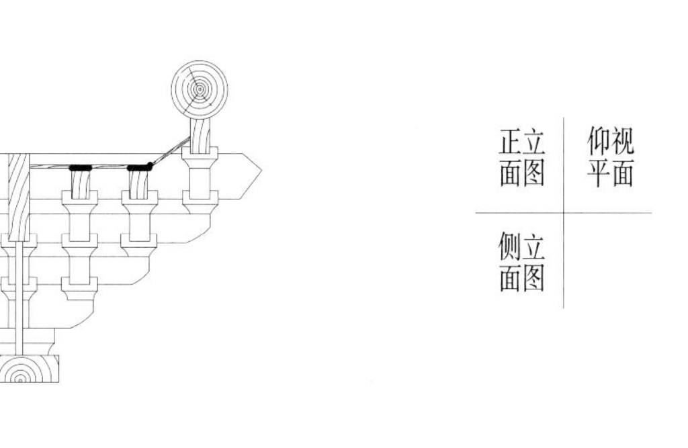
正立面图
仰视平面
侧立面图

大悲殿下檐平身科斗栱详图

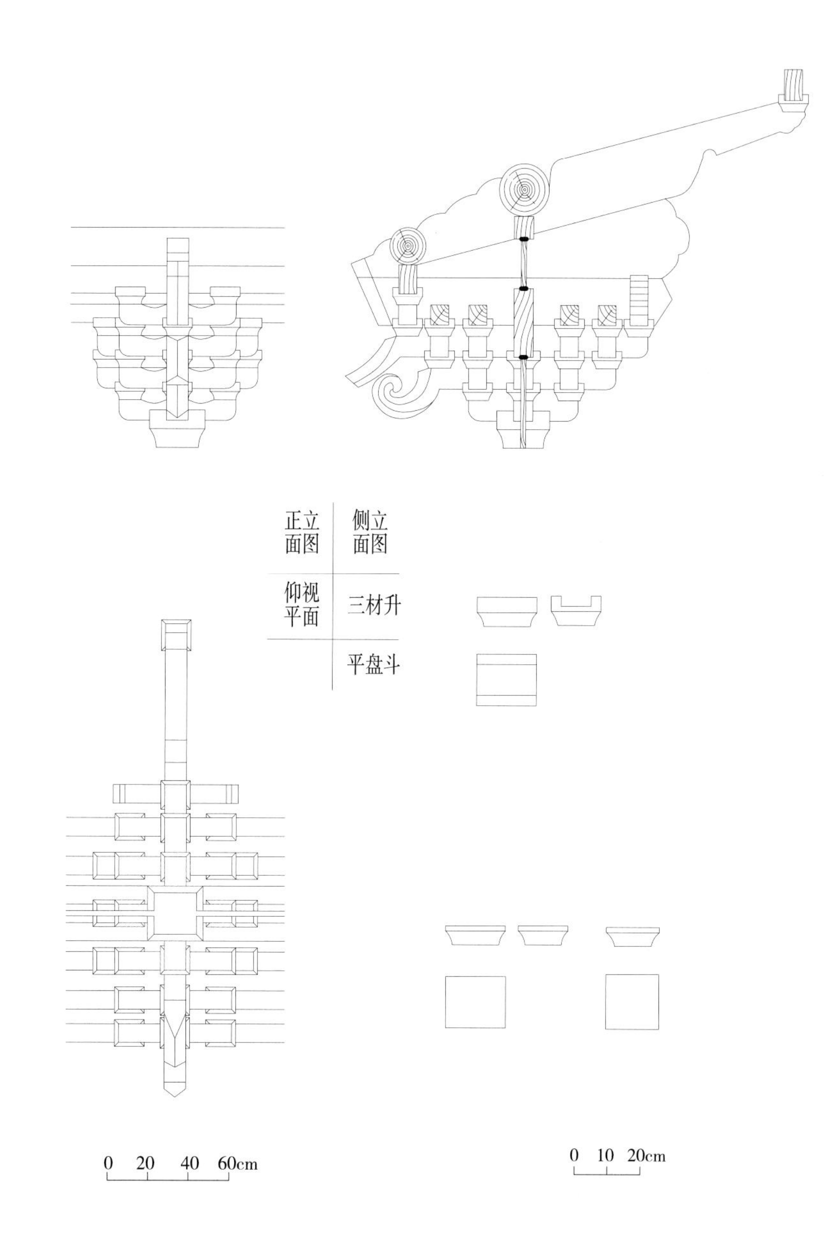
正立面图
侧立面图
仰视平面
三材升
平盘斗
0 20 40 60cm
0 10 20cm

碑亭下檐平身科斗栱详图

平武报恩寺平身科斗栱大样图

Detail, Intermediate Dougong Sets, Bao'en Temple, Pingwu

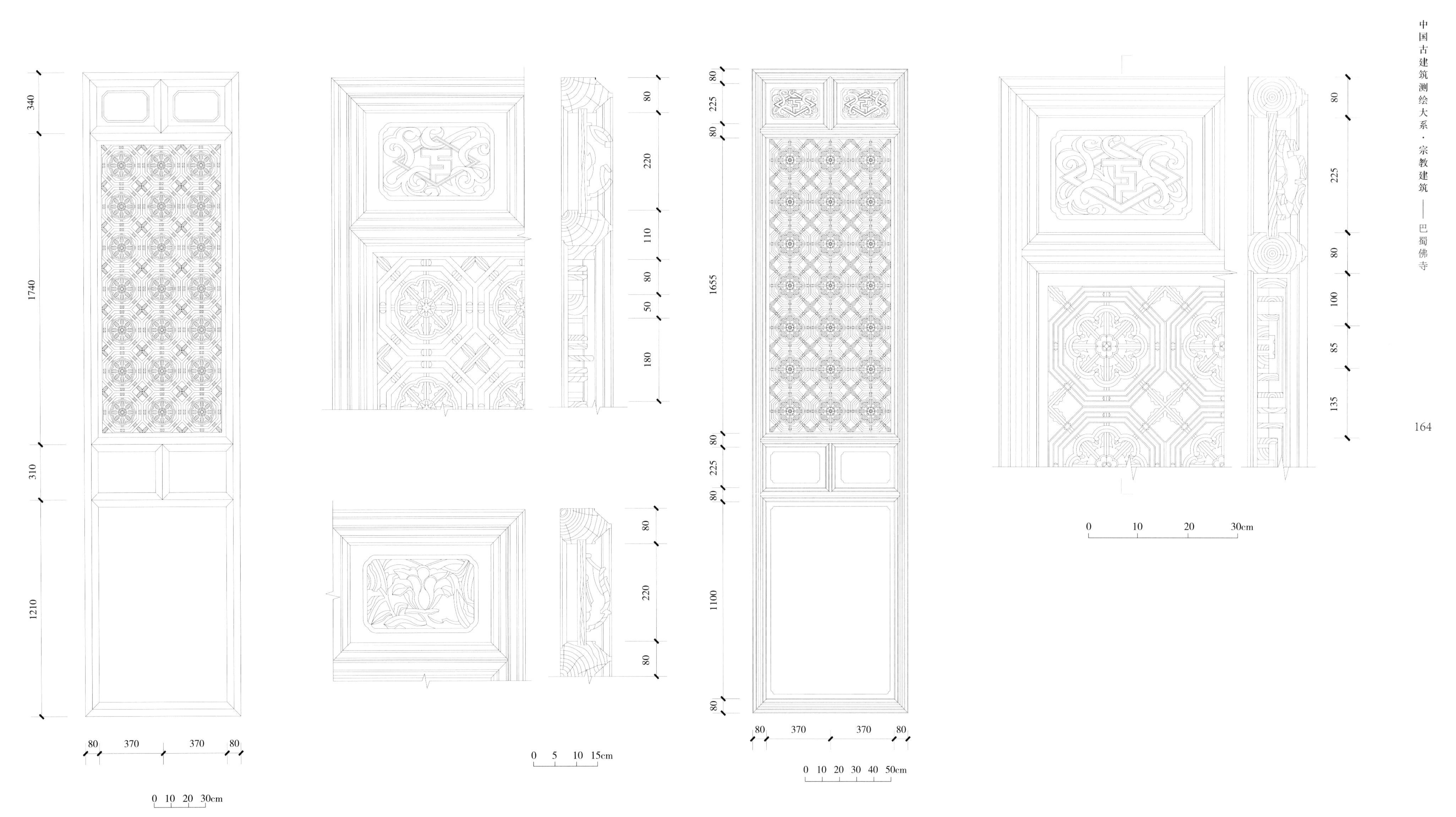
340
1740
310
1210
80 370 370 80
0 10 20 30cm
80
220
110
80
50
180
0 5 10 15cm
80
220
80
80
225
80
1655
80
225
80
1100
80
80 370 370 80
0 10 20 30 40 50cm
80
225
80
100
85
135
0 10 20 30cm

平武报恩寺大雄宝殿门扇式样图
Detail, Doors, Mahavira Hall, Bao’en Temple, Pingwu

平武报恩寺天王殿门扇式样图
Detail, Doors, Tianwang Hall, Bao’en Temple, Pingwu

0 50 100cm

平武报恩寺大雄宝殿天花彩画式样图（一）
Color Patterns on Ceiling (1), Mahavira Hall, Bao'en Temple, Pingwu

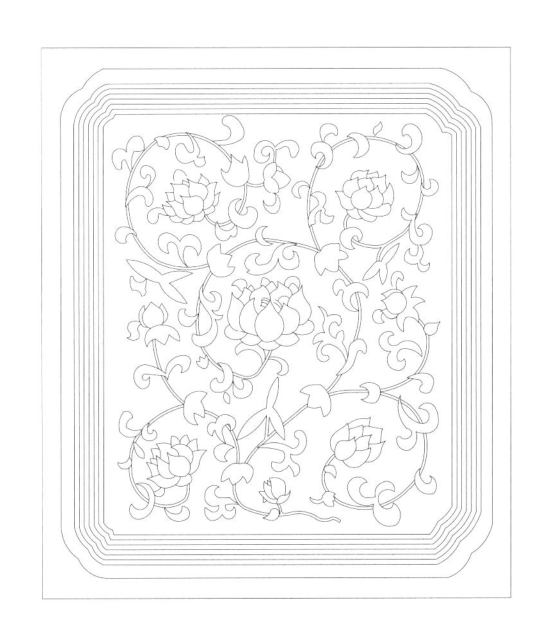

0 10 20cm

平武报恩寺大雄宝殿天花彩画式样图（二）
Color Patterns on Ceiling (2), Mahavira Hall, Bao'en Temple, Pingwu

Shengshou Temple at Dazu

Shengshou Temple perches on Baoding Mountain in Dazu District, Chongqing. It is said to have been first built in Tang Dynasty, and given its name by the Emperor during Xining period in Song Dynasty. The temple was destroyed by warfare in Yuan and Ming dynasties, and the existing buildings were rebuilt in the 23rd year of Kangxi's reign in Qing Dynasty (AD1684). In the 9th year of Tongzhi's reign (AD1870), it underwent a major restoration, when the whole complex was reconstructed on the site of Ming Dynasty remains. Major Buddhist structures on the central axis include: the entrance gate, the Sakra Hall, the Great Buddha Hall, the Trikaya Hall, the Dipamkara Hall, and the Vimalakirti Hall. Baoding Mountain is where the monk Zhao Zhifeng created the Vimalakirti Bodhimanda in the Southern Song Dynasty. On the cliffs of Great and Small Buddha Valleys that are closely related to the Shengshou Temple, there are massive carved Buddhist statues, which are celebrated as the final culmination of grotto statue creations in China. The Dabei Hall and Muniu Hanging Gallery built for the cliff carvings are distinctive architectural creations. Except for the Vimalakirti Hall that have bracket sets, buildings in Shengshou Temple share a simple technique of eave support like overhang or horizontal braces. The small grey roof tile, stuccoed roof ridge with porcelain decoration, and upturned roof-ridge have strong Ba-Shu regional characteristics.

大足圣寿寺

圣寿寺位于重庆大足的宝顶山，相传寺庙始建于唐，宋熙宁年间敕赐名圣寿寺。元代和明代都曾毁于兵火，现存建筑为清代康熙二十三年（1684年）重修，同治九年（1870年）又进行过大的修葺，整个建筑群在明代遗存基址上重建。中轴线上的主要佛教殿堂有：山门殿、帝释殿、大雄宝殿、三世佛殿、燃灯殿、维摩殿。宝顶山是南宋时期赵智凤创建的维摩祖师道场，与圣寿寺紧密相连的大佛湾和小佛湾崖壁上镌刻有规模巨大的摩崖造像群，成为中国石窟造像的最后一个高潮，结合摩崖石刻造像的大悲殿、牧牛摩崖悬挑廊亭极富建筑创造特色。圣寿寺的建筑除维摩殿有斗栱外，其余殿堂的屋檐出挑均采用撑栱和直接用水平挑枋支撑，小青瓦屋面，灰塑屋脊采用瓷片贴装饰，翼角起翘等都具有浓郁的巴蜀地域风格特色。

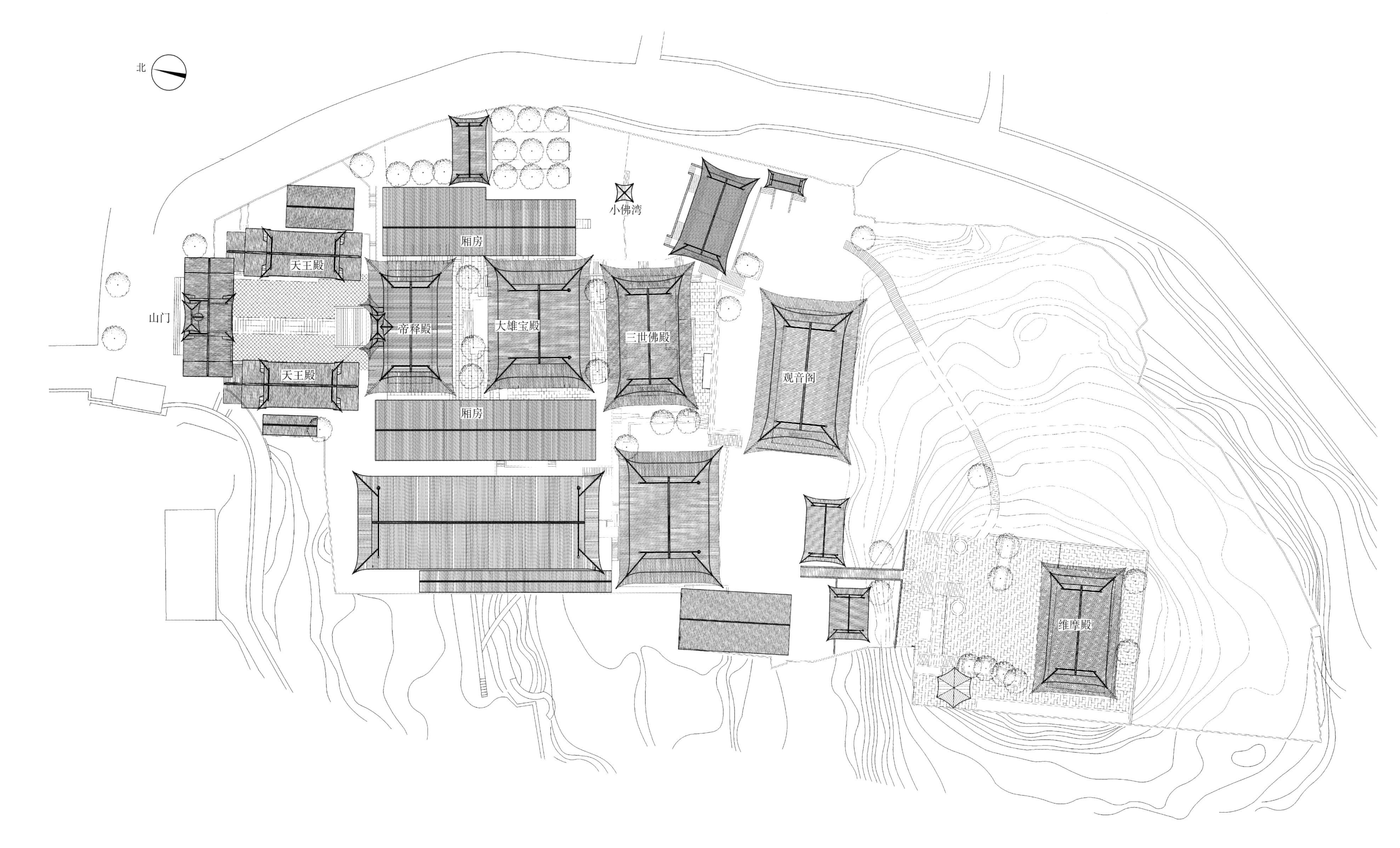

北
小佛湾
厢房
天王殿
山门
帝释殿
大雄宝殿
三世佛殿
观音阁
天王殿
厢房
维摩殿
0 5 10 15m

大足圣寿寺建筑群总平面图

Site Plan of Shengshou Temple, Dazu

图一 大足圣寿寺建筑群鸟瞰图

圣寿寺
售票处

图二 大足圣寿寺山门

图三 大足圣寿寺山门背立面

Fig.1 Bird's Eye View of Shengshou Temple, Dazu
Fig.2 Main Gate (Shan Men), Shengshou Temple, Dazu
Fig.3 Back Elevation of Main Gate (Shan Men), Shengshou Temple, Dazu

图六 大足圣寿寺大雄宝殿

图四 大足圣寿寺帝释殿

图五 大足圣寿寺帝释殿背立面

Fig.4 Śakra Hall, Shengshou Temple, Dazu
Fig.5 Back Elevation of Śakra Hall, Shengshou Temple, Dazu
Fig.6 Mahavira Hall, Shengshou Temple, Dazu

图七　大足圣寿寺三世佛殿

图八　大足圣寿寺三世佛殿当心间

图九　大足圣寿寺维摩殿

Fig.7 The Three Buddhas Hall, Shengshou Temple, Dazu
Fig.8 Central Bay (Dangxinjian) of The Three Buddhas Hall, Shengshou Temple, Dazu
Fig.9 Vimalakirti Hall, Shengshou Temple, Dazu

图十 大足圣寿寺万岁楼

图十一 大足圣寿寺大悲殿

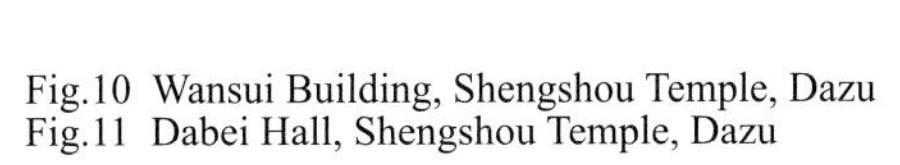

Fig.10 Wansui Building, Shengshou Temple, Dazu
Fig.11 Dabei Hall, Shengshou Temple, Dazu

图十二　大足圣寿寺大悲殿

图十三　大足圣寿寺牧牛亭

图十四　大足圣寿寺牧牛亭

Fig.12 Dabei Hall, Shengshou Temple, Dazu
Fig.13 Muniu Pavilion, Shengshou Temple, Dazu
Fig.14 Muniu Pavilion, Shengshou Temple, Dazu

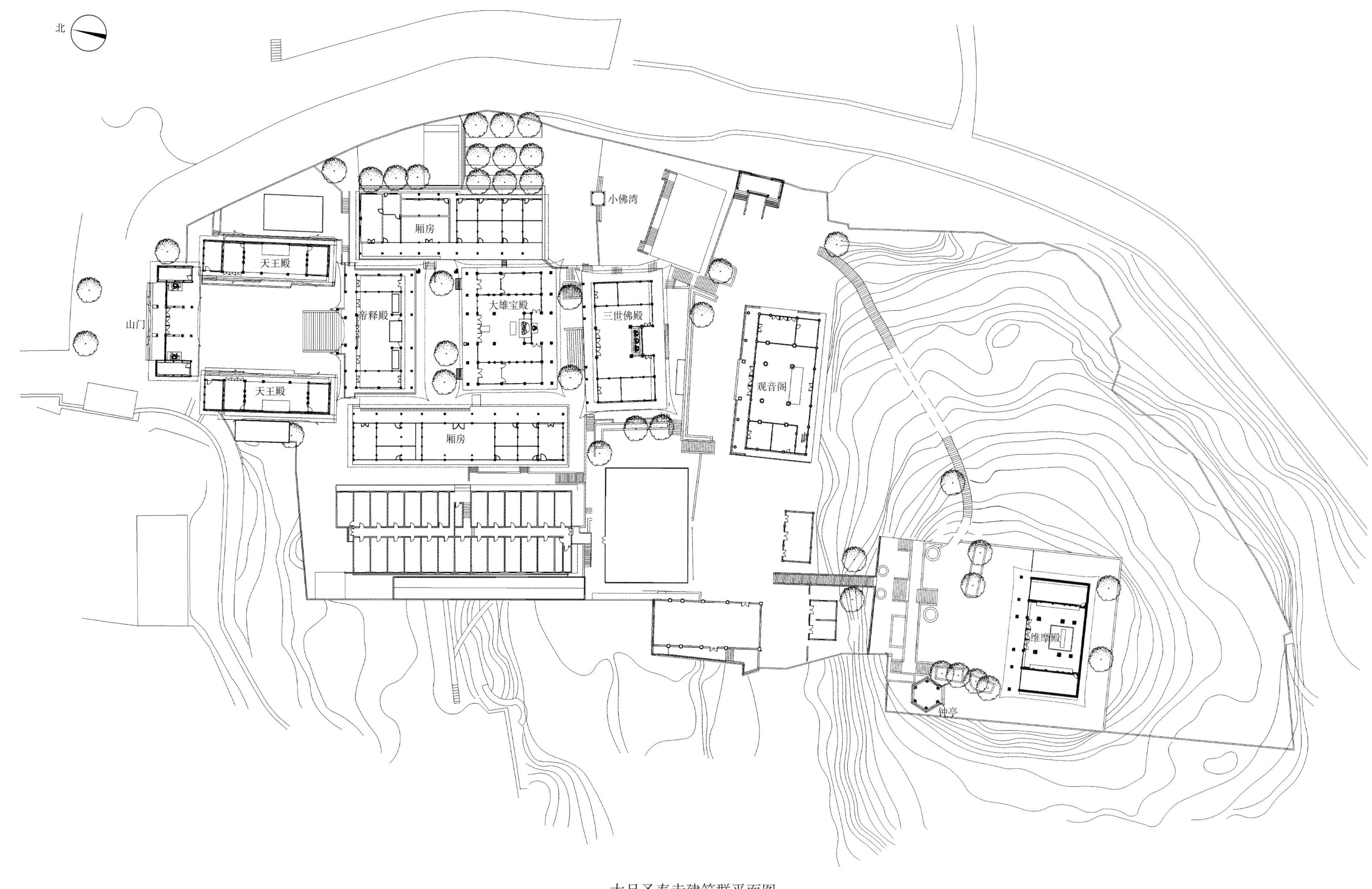
北
小佛湾
厢房
天王殿
大雄宝殿
三世佛殿
帝释殿
山门
观音阁
天王殿
厢房
维摩殿
钟亭
0 5 10 15m

大足圣寿寺建筑群平面图

Ground Floor Plan of Entire Compound, Shengshou Temple, Dazu

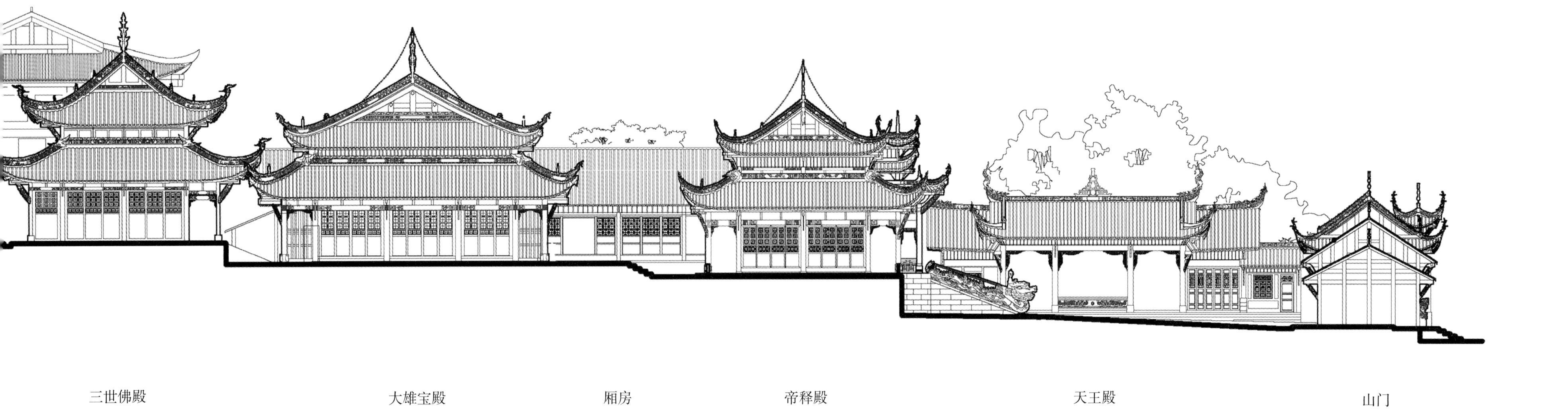
三世佛殿
大雄宝殿
厢房
帝释殿
天王殿
山门
0
5
10
15m

大足圣寿寺建筑群侧立面图
Side Elevation of Entire Compound, Shengshou Temple, Dazu

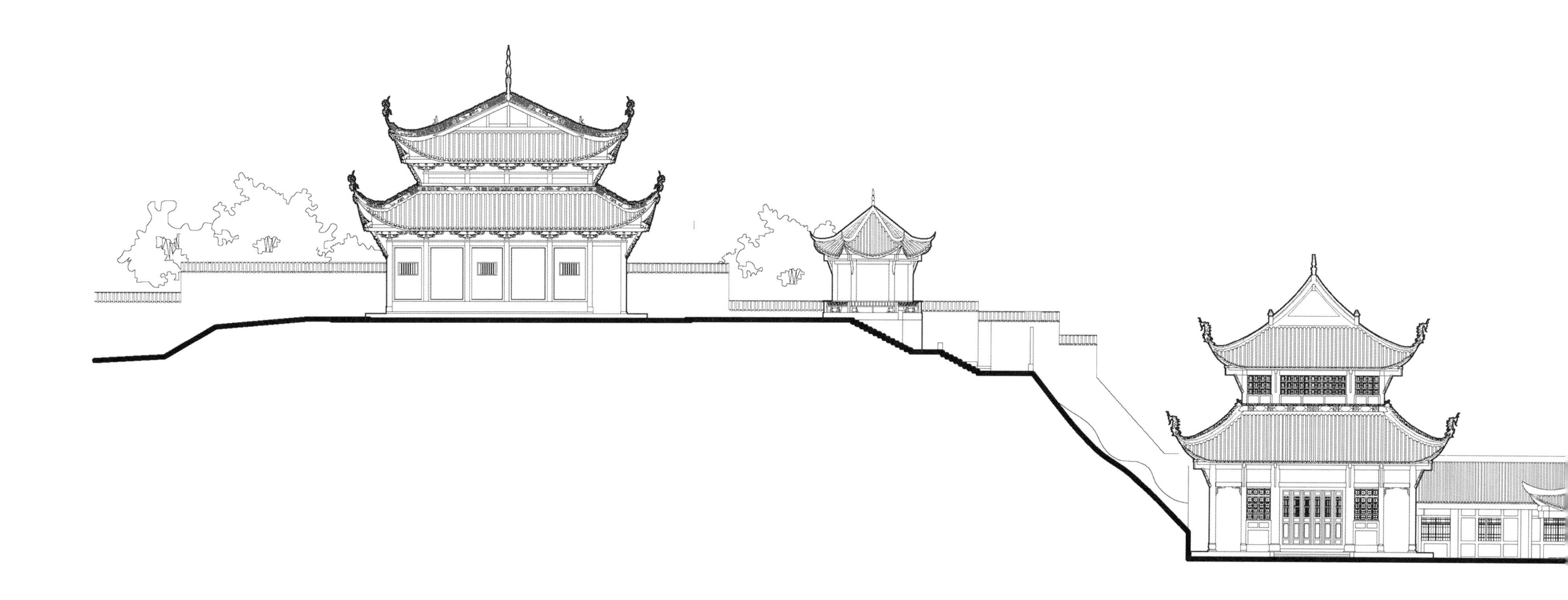

维摩殿

观音阁

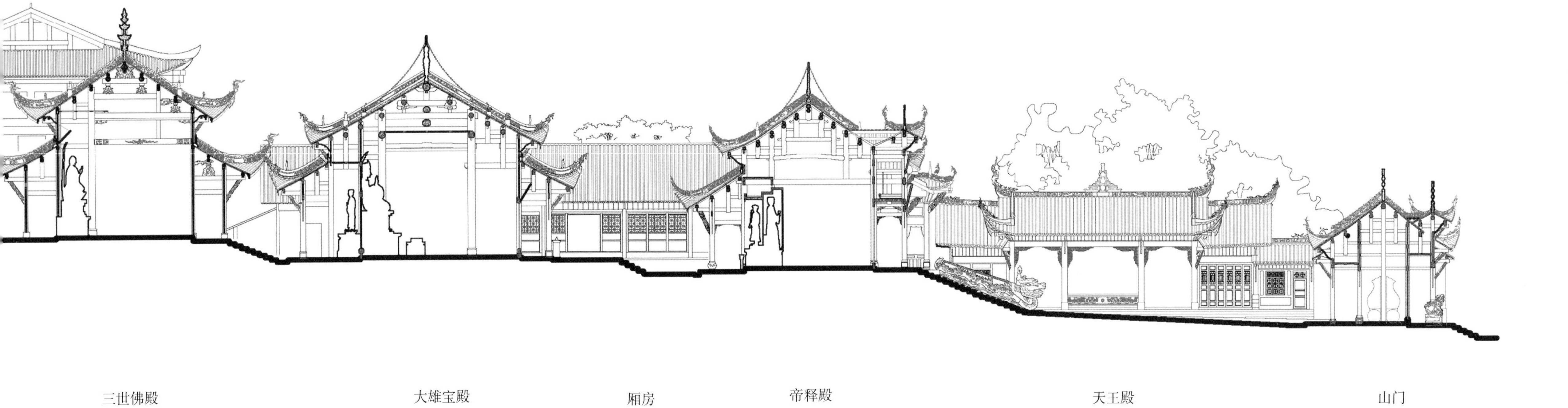
三世佛殿
大雄宝殿
厢房
帝释殿
天王殿
山门
0
5
10
15m

大足圣寿寺建筑群剖面图
Section of Entire Compound, Shengshou Temple, Dazu

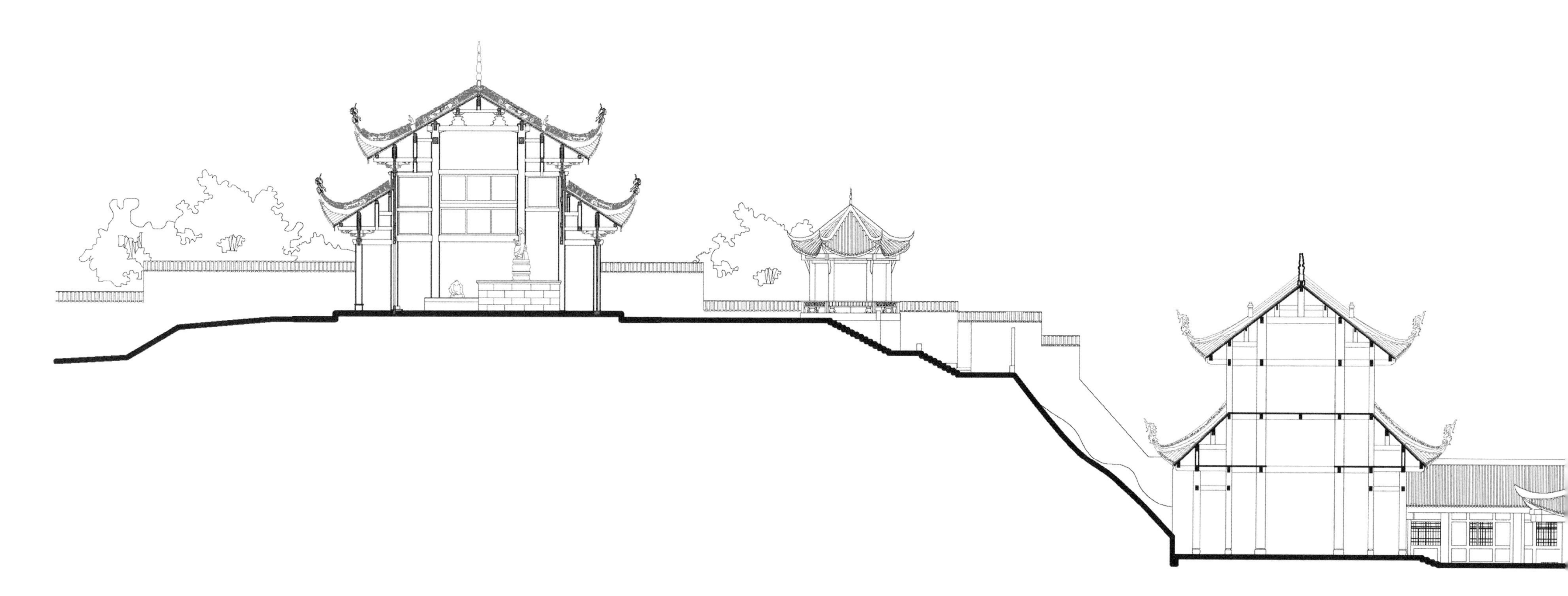

维摩殿

观音阁

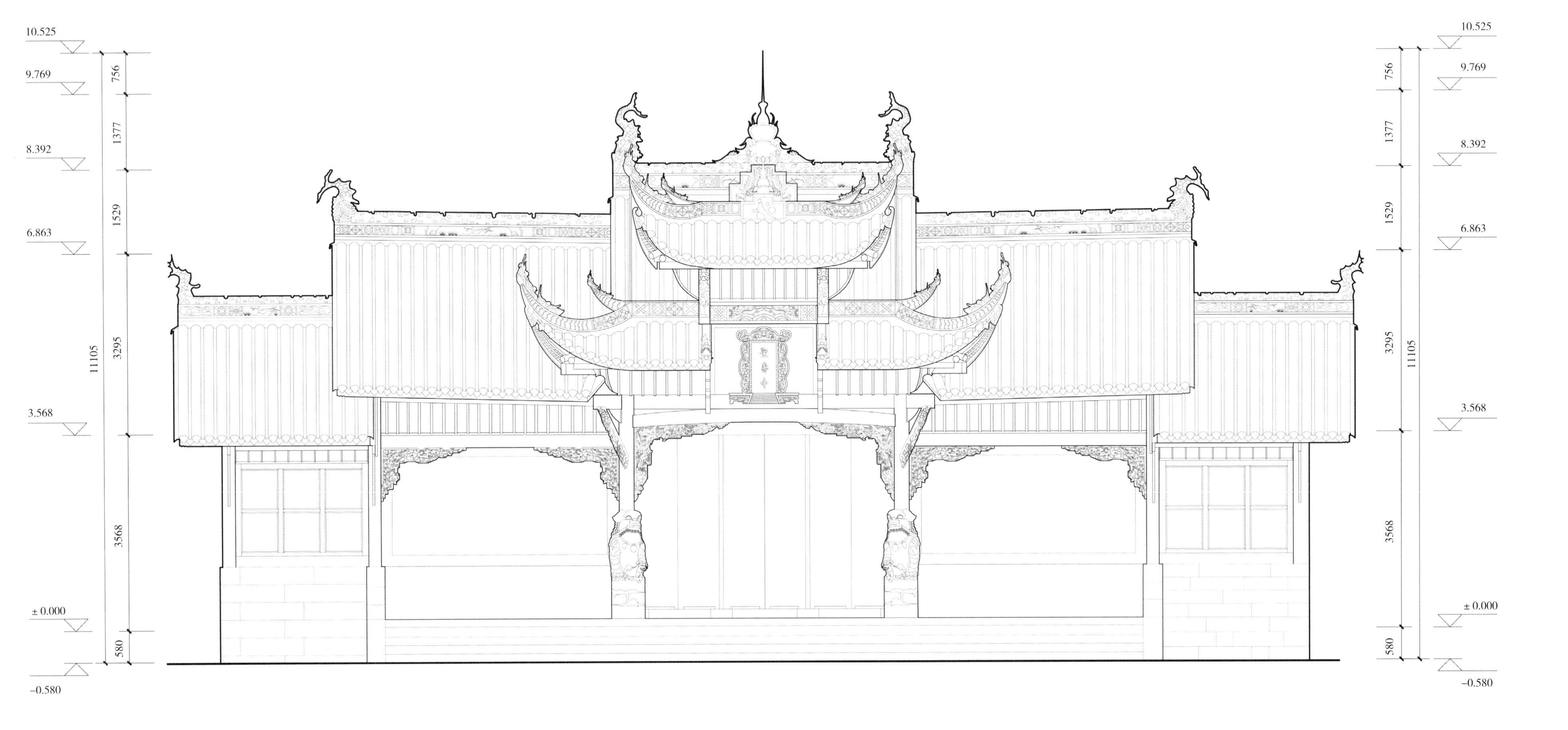
10.525
9.769
8.392
6.863
3.568
± 0.000
-0.580
756
1377
1529
3295
3568
580
11105

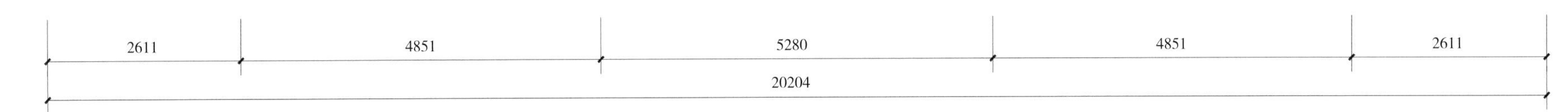
2611
4851
5280
4851
2611
20204

大足圣寿寺山门正立面图

Front Elevation of Main Gate (Shan Men), Shengshou Temple, Dazu

0 1 2m

9.567
8.205
4.742
3.300
−0.140
−0.790
1362
3463
1442
3440
650
10357
9.567
8.205
4.742
3.300
−0.140
−0.245
1362
3463
1442
3440
105
9812

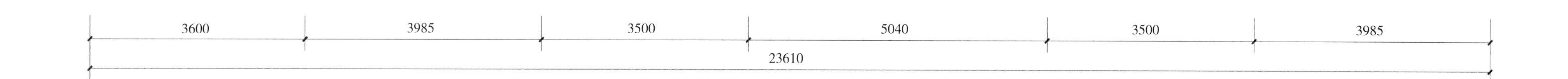
3600
3985
3500
5040
3500
3985
23610

大足圣寿寺天王殿左殿正立面图

Front Elevation, The Left Hall of Tianwang Hall, Shengshou Temple, Dazu

0 1 2 3m

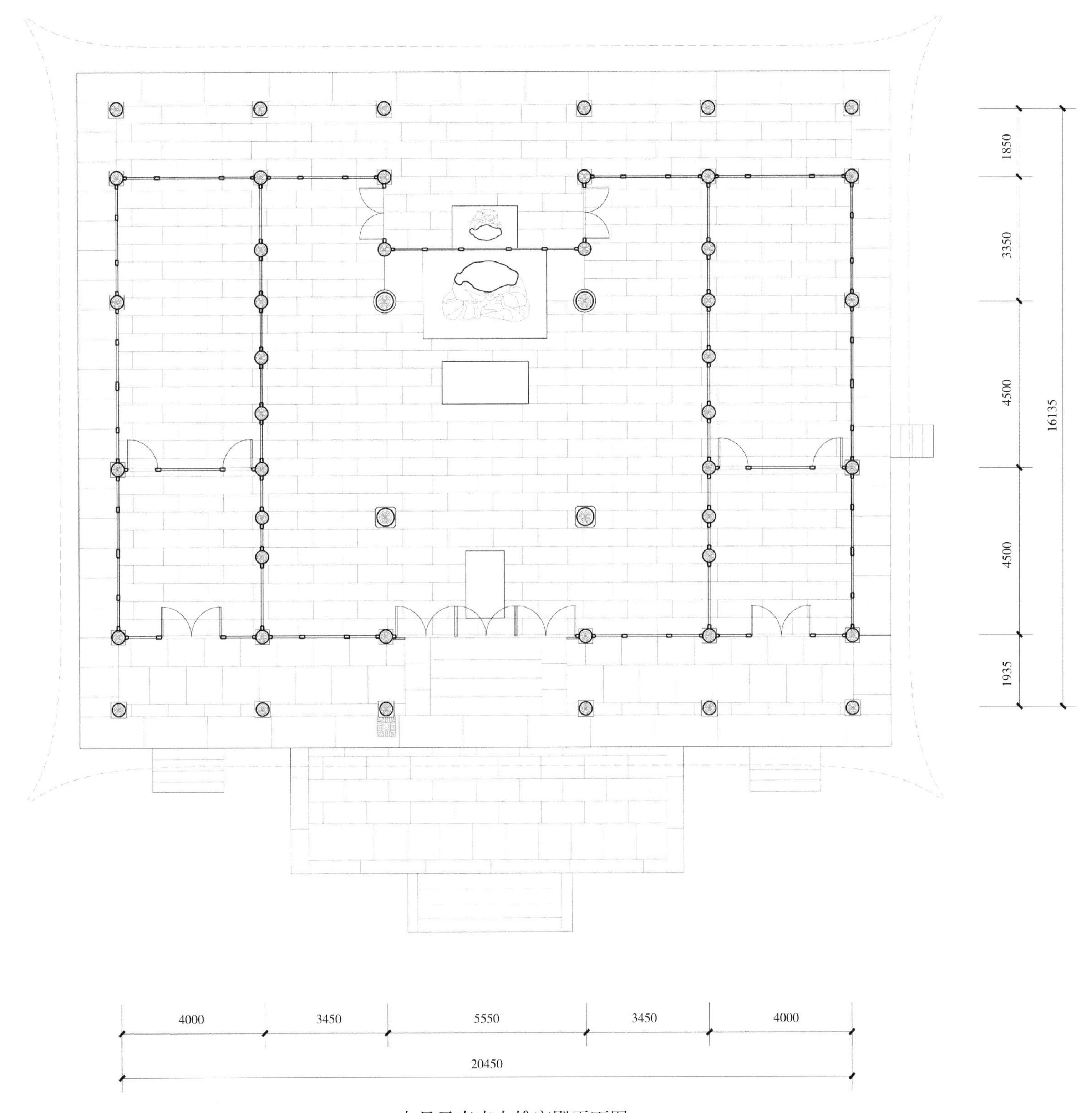
1850
3350
4500
4500
1935
16135
4000
3450
5550
3450
4000
20450

大足圣寿寺大雄宝殿平面图
Ground Floor Plan of Mahavira Hall, Shengshou Temple, Dazu

0 1 2 3m

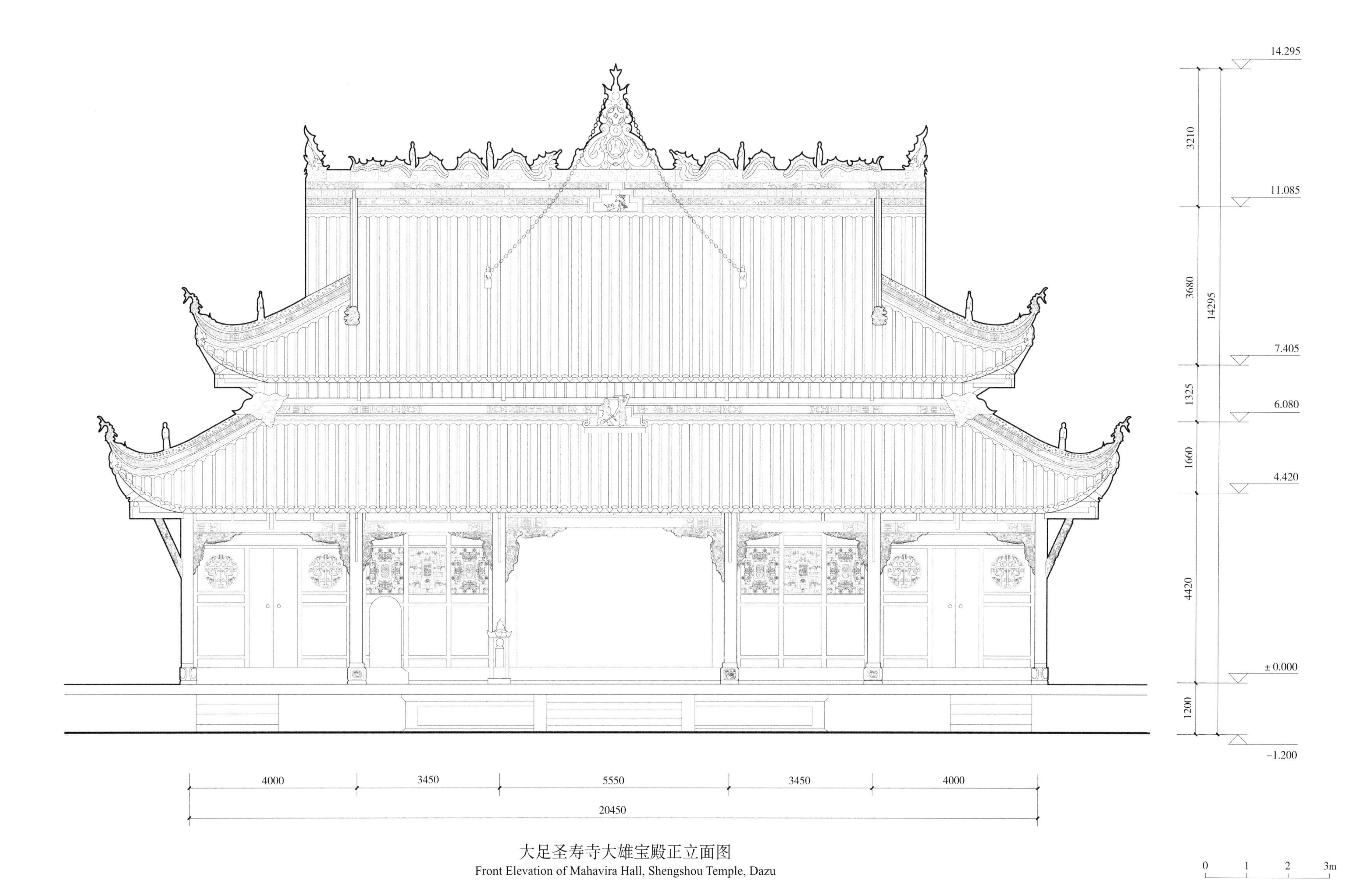
14.295
11.085
7.405
6.080
4.420
± 0.000
-1.200
3210
3680
1325
1660
4420
1200
14295
4000
3450
5550
3450
4000
20450
0
1
2
3m

大足圣寿寺大雄宝殿正立面图

Front Elevation of Mahavira Hall, Shengshou Temple, Dazu

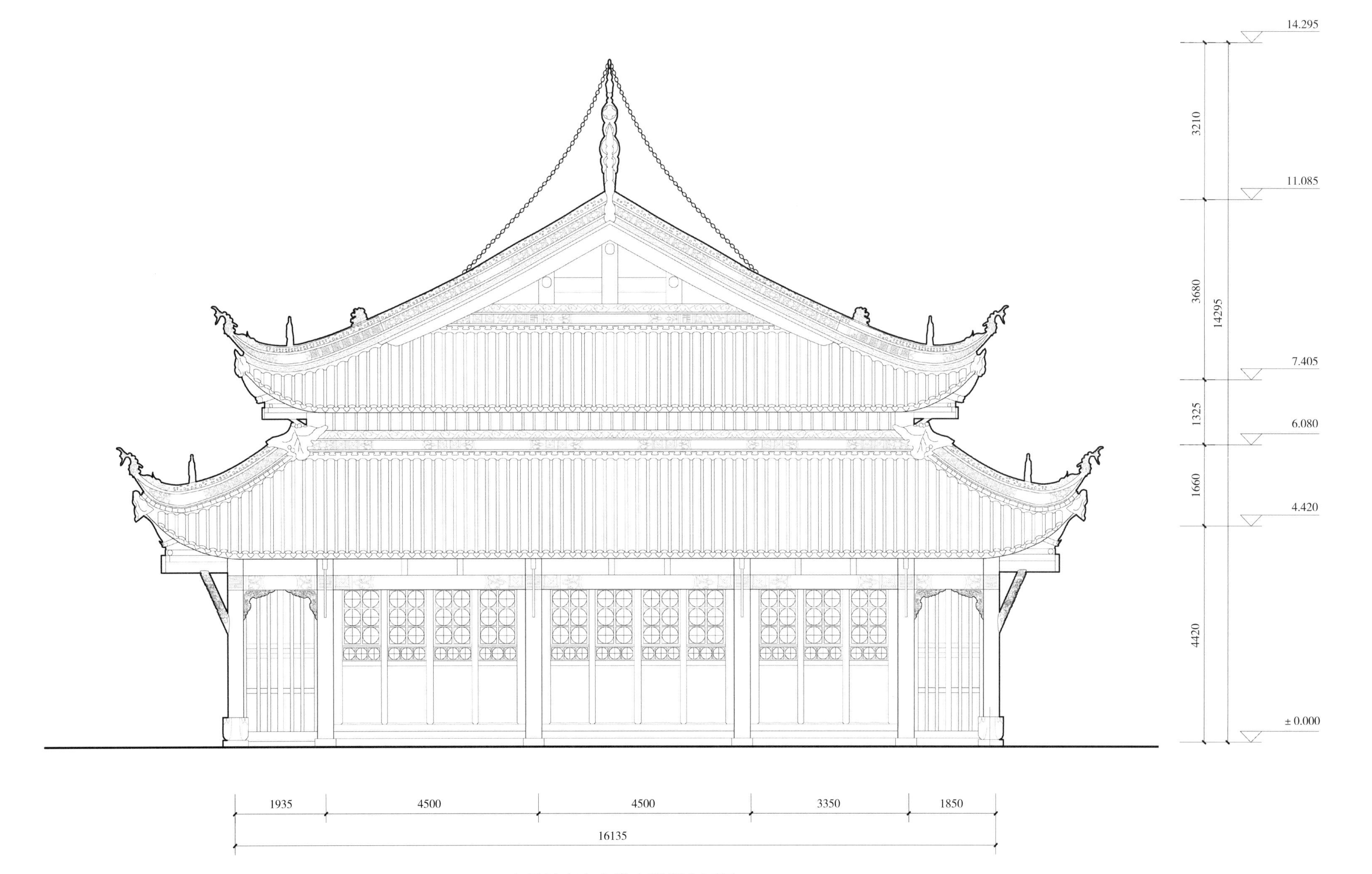
14.295
11.085
7.405
6.080
4.420
± 0.000
3210
3680
1325
1660
4420
14295
1935
4500
4500
3350
1850
16135
0
1
2
3m

大足圣寿寺大雄宝殿侧立面图
Side Elevation of Mahavira Hall, Shengshou Temple, Dazu

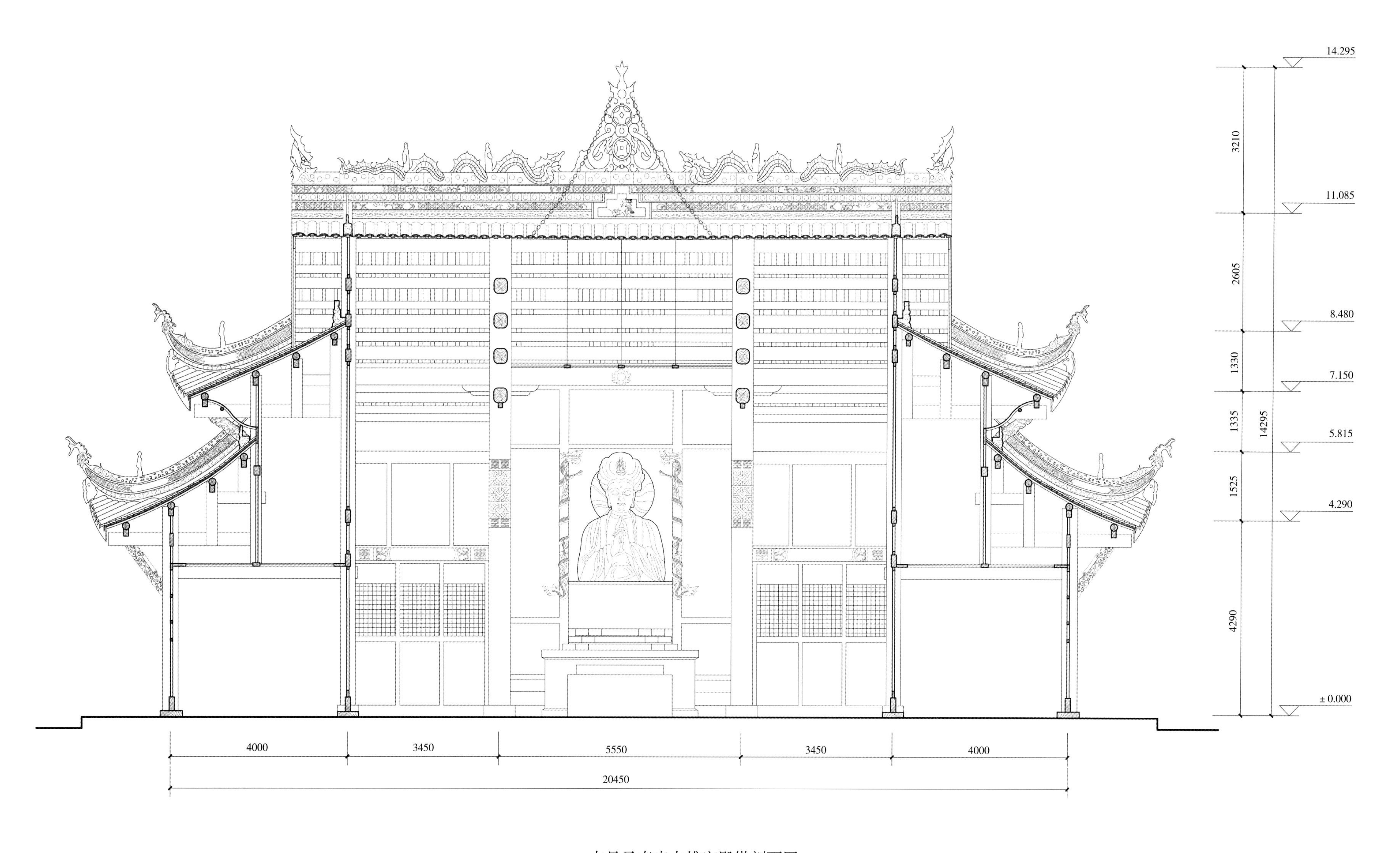
14.295
11.085
8.480
7.150
5.815
4.290
±0.000
3210
2605
1330
1335
1525
4290
14295
4000
3450
5550
3450
4000
20450
0
1
2
3m

大足圣寿寺大雄宝殿纵剖面图

Longitudinal Section of Mahavira Hall, Shengshou Temple, Dazu

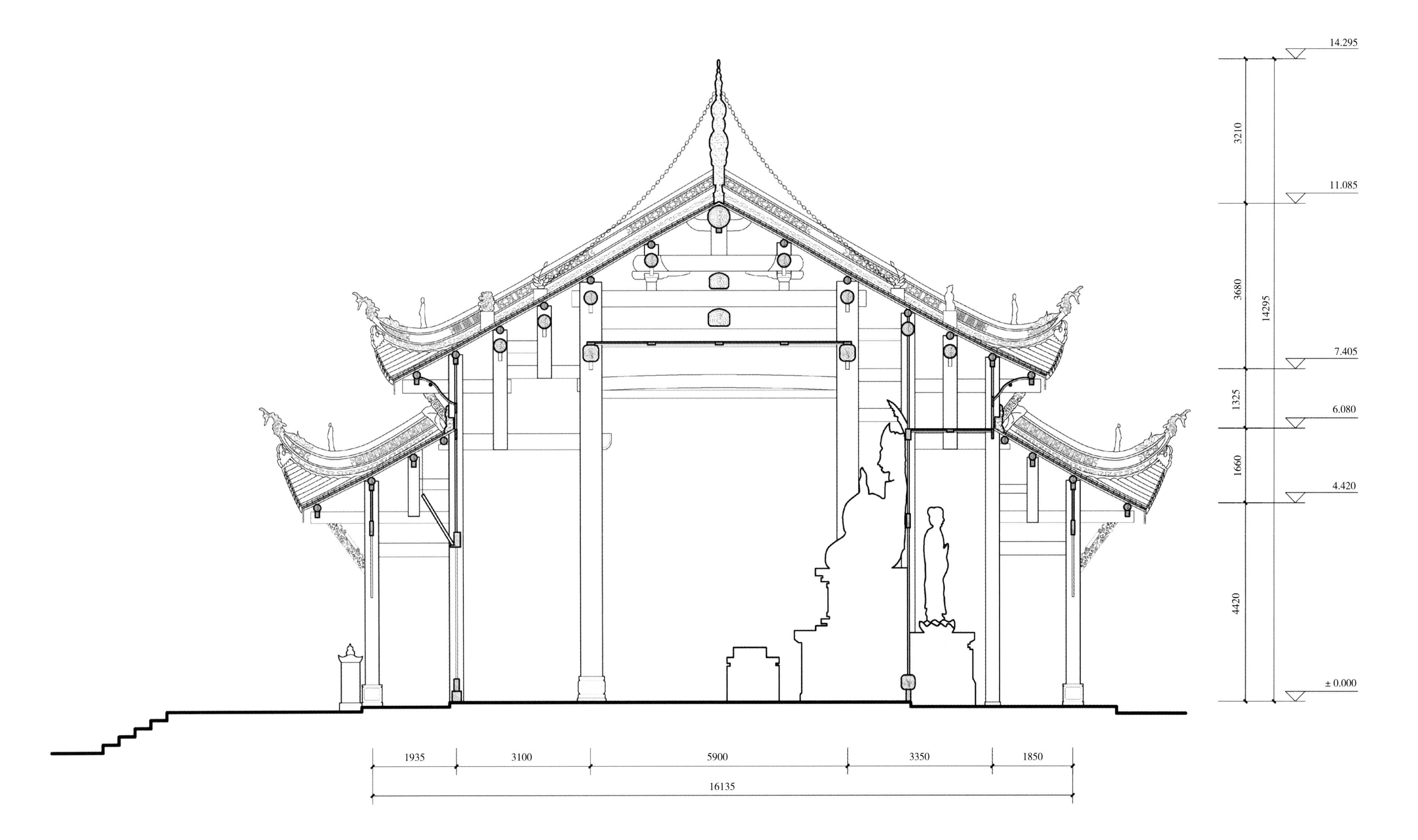
14.295
11.085
7.405
6.080
4.420
± 0.000
3210
3680
1325
1660
4420
14295
1935
3100
5900
3350
1850
16135

大足圣寿寺大雄宝殿明间横剖面图
Cross Section of Central Bay (Mingjian), Mahavira Hall, Shengshou Temple, Dazu

0 1 2 3m

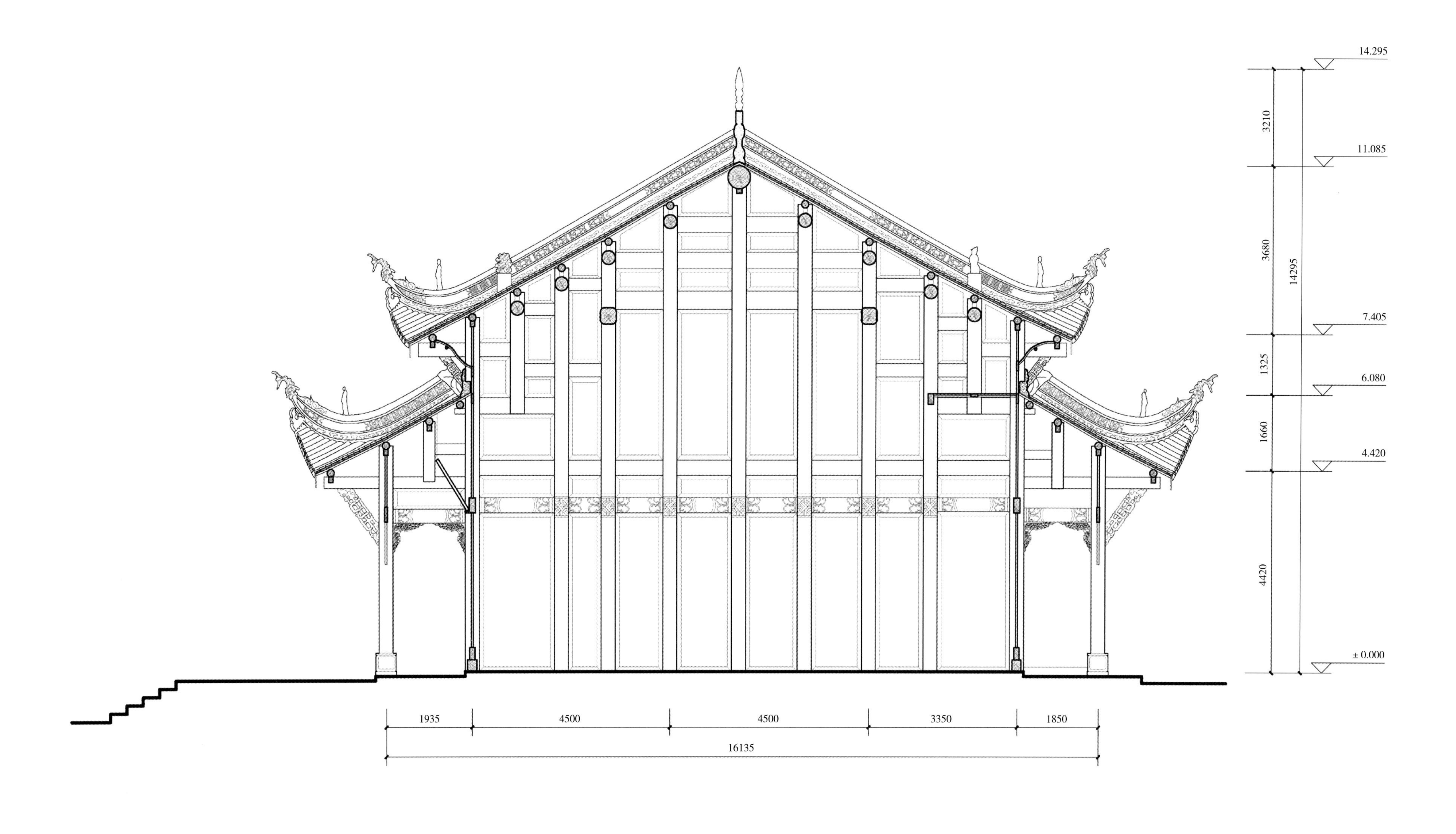
14.295
3210
11.085
3680
14295
7.405
1325
6.080
1660
4.420
4420
± 0.000
1935
4500
4500
3350
1850
16135

大足圣寿寺大雄宝殿次间横剖面图

Cross Section of In-between Bay (Cijian), Mahavira Hall, Shengshou Temple, Dazu

0 1 2 3m

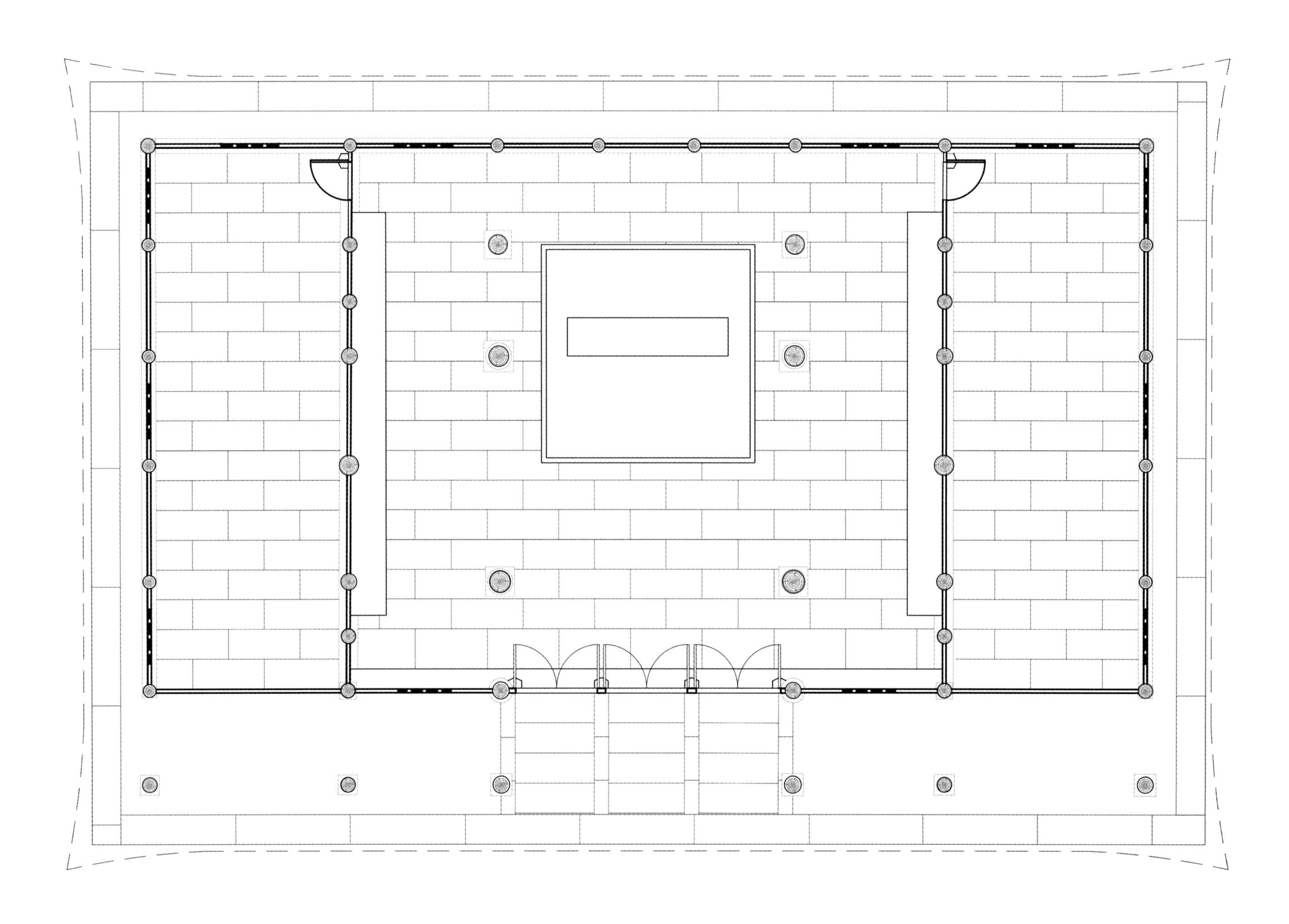

大足圣寿寺维摩殿平面图
Ground Floor Plan of Vimalakirti Hall, Shengshou Temple, Dazu

0 1 2 3m

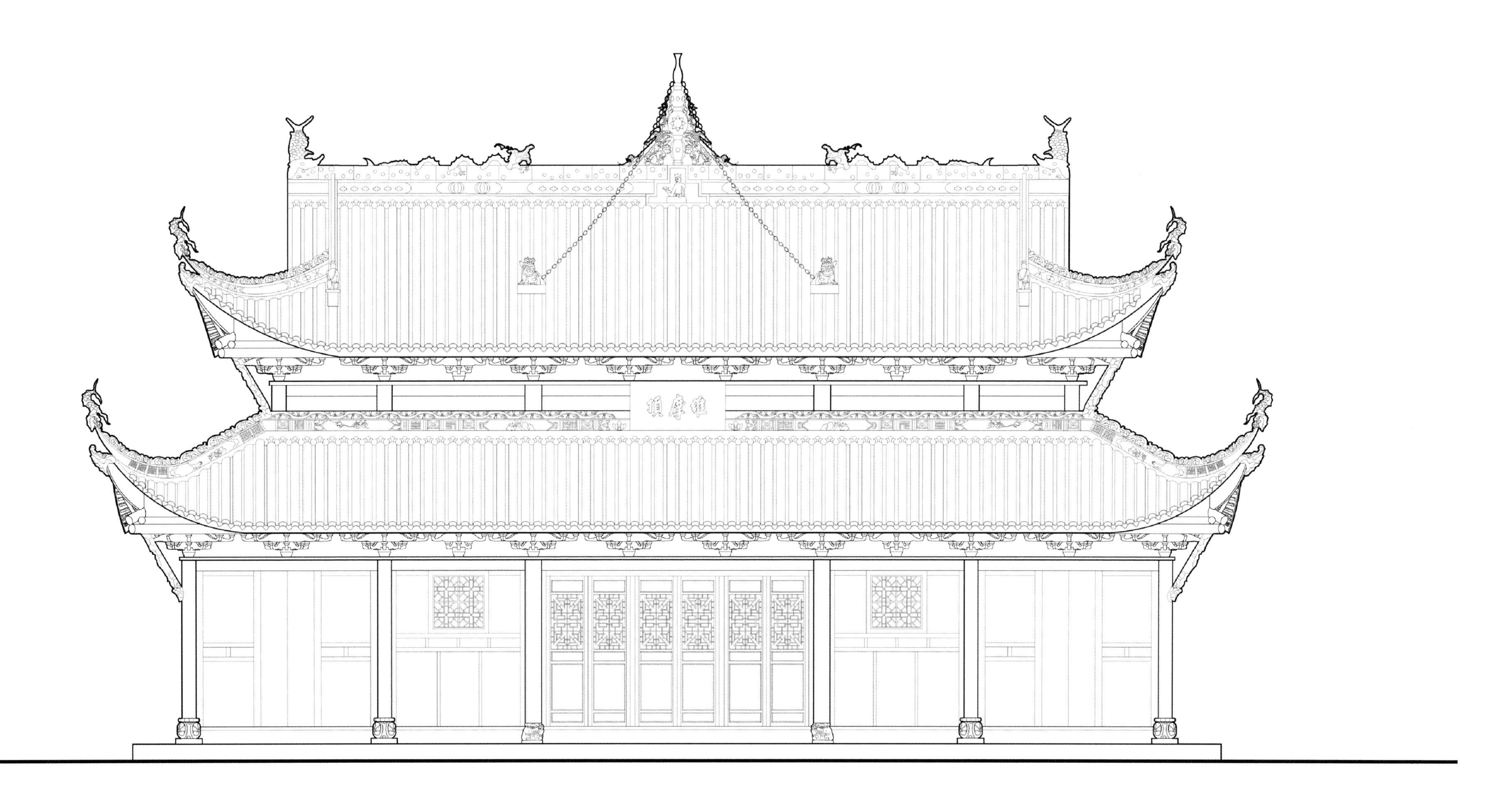
维摩殿
0 1 2 3m

大足圣寿寺维摩殿正立面图
Front Elevation of Vimalakirti Hall, Shengshou Temple, Dazu

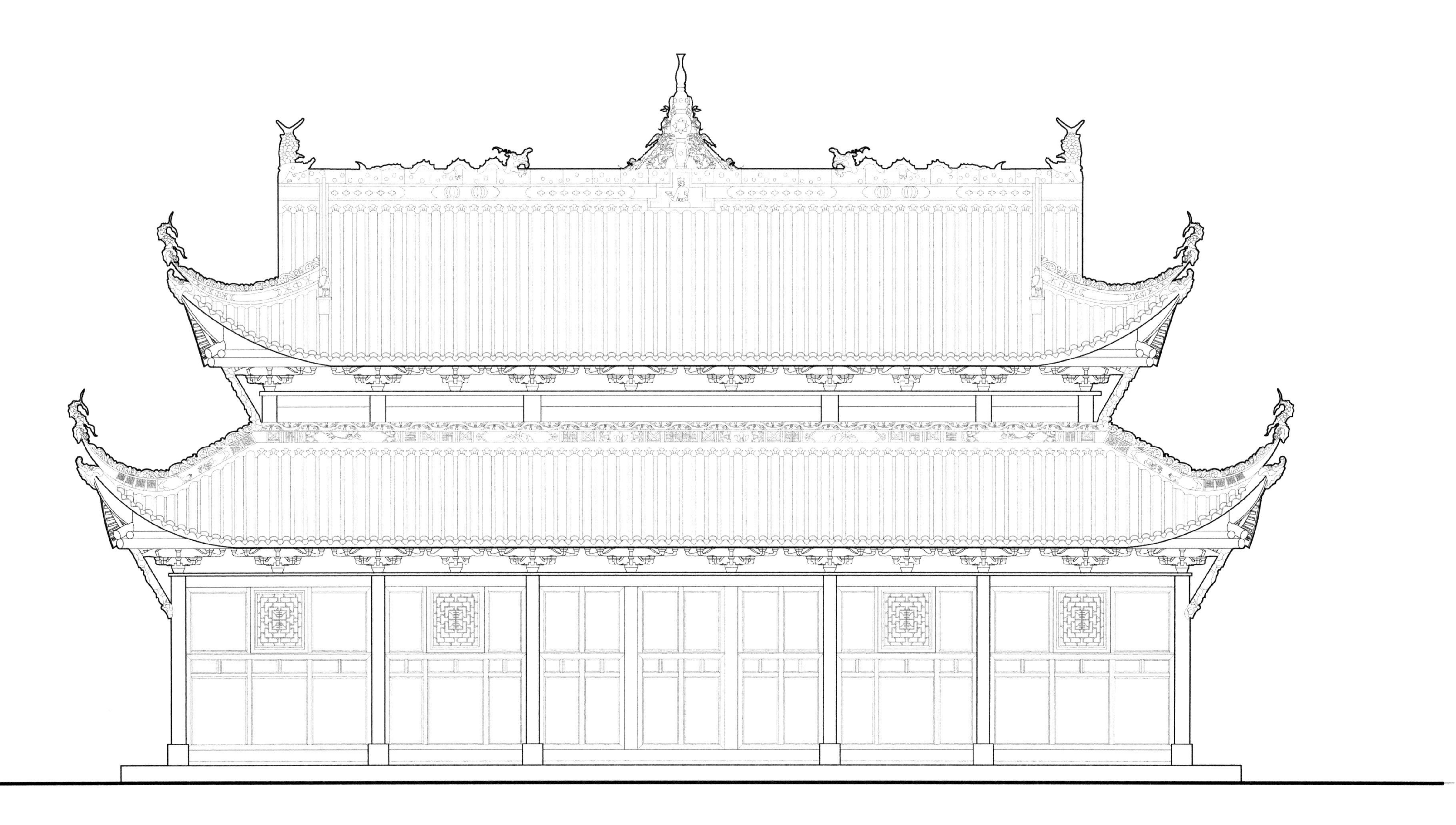
0 1 2 3m

大足圣寿寺维摩殿背立面图
Back Elevation of Vimalakirti Hall, Shengshou Temple, Dazu

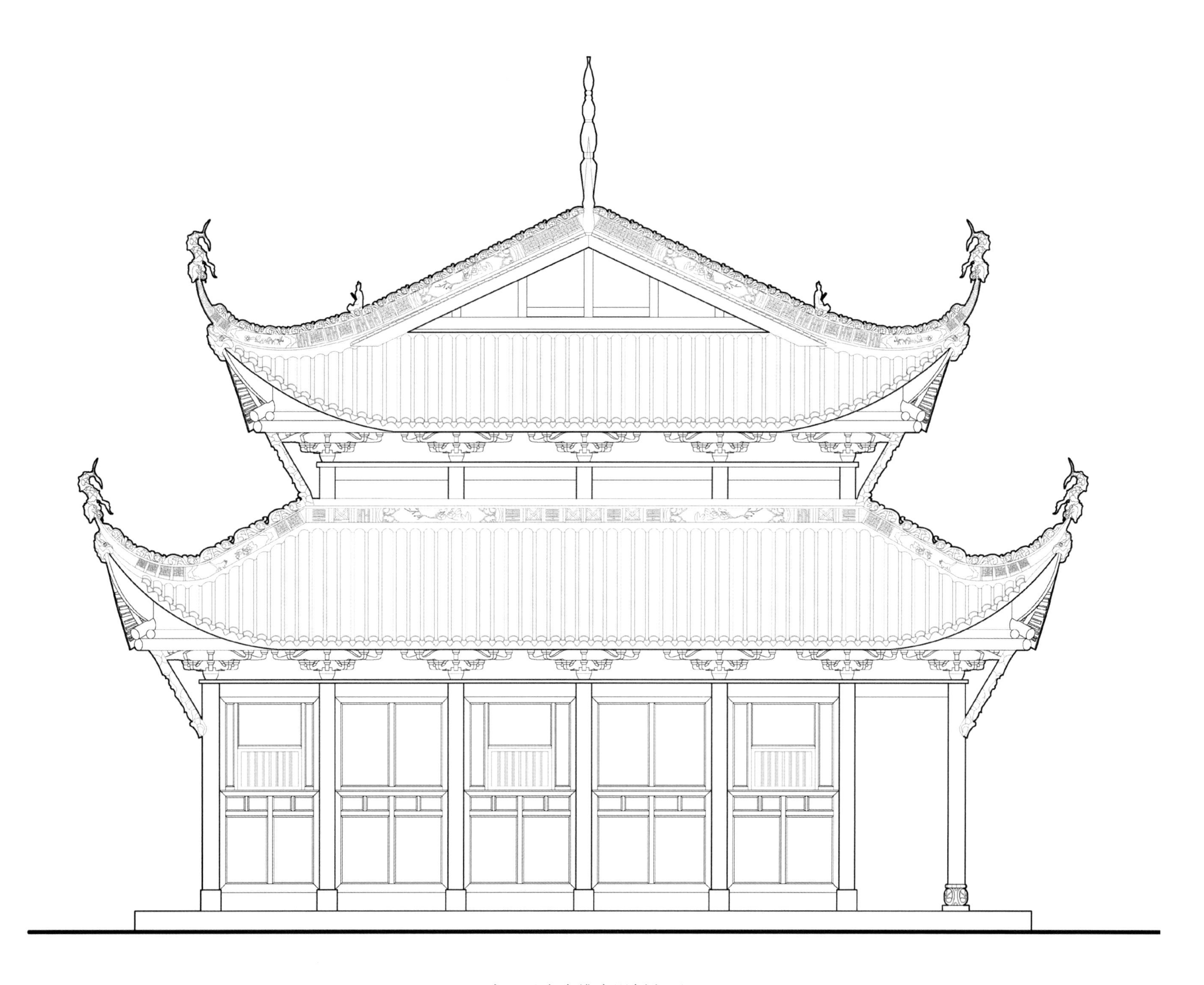
0
1
2
3m

大足圣寿寺维摩殿侧立面图
Side Elevation of Vimalakirti Hall, Shengshou Temple, Dazu

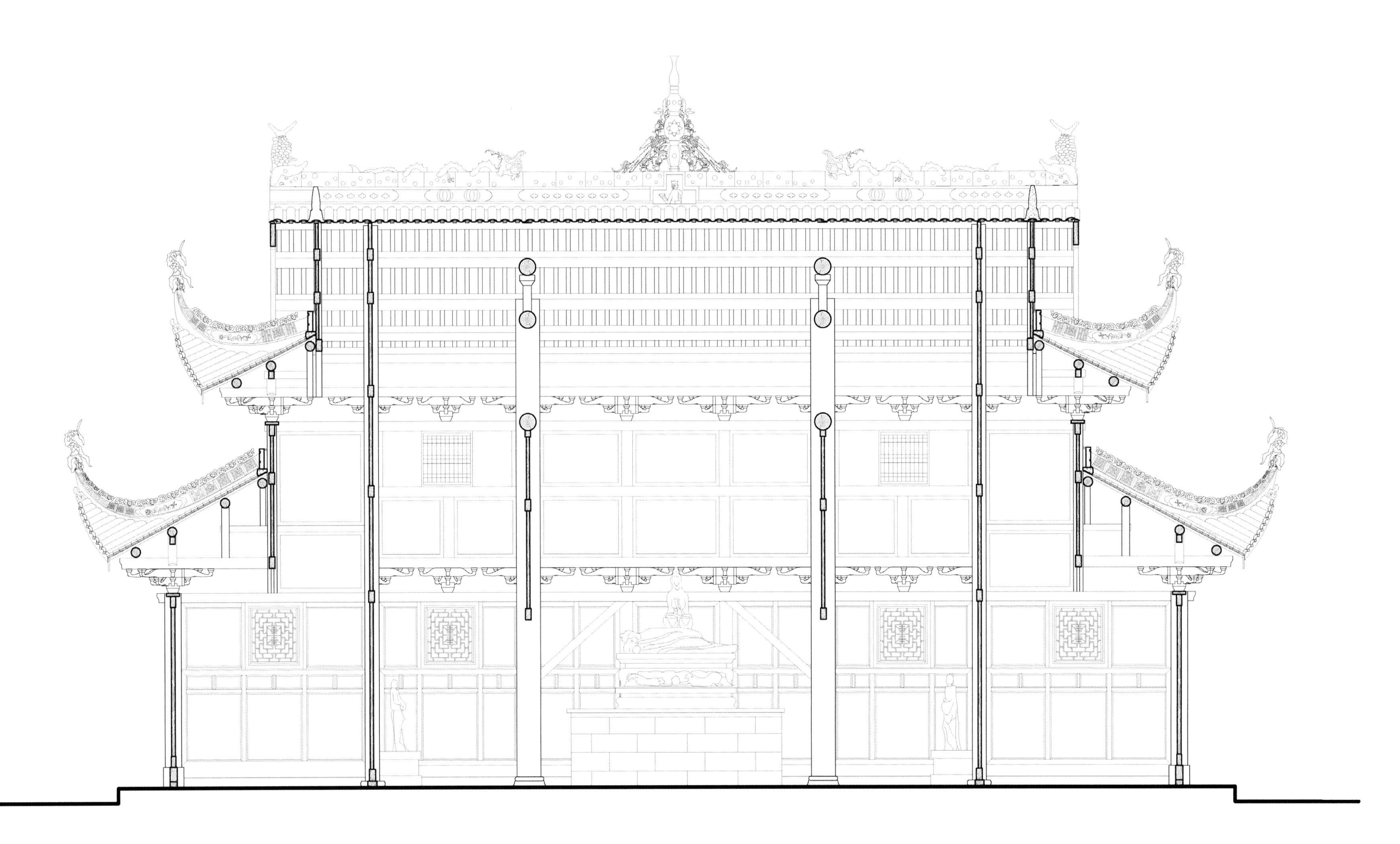
0 1 2 3m

大足圣寿寺维摩殿纵剖面图
Longitudinal Section of Vimalakirti Hall, Shengshou Temple, Dazu

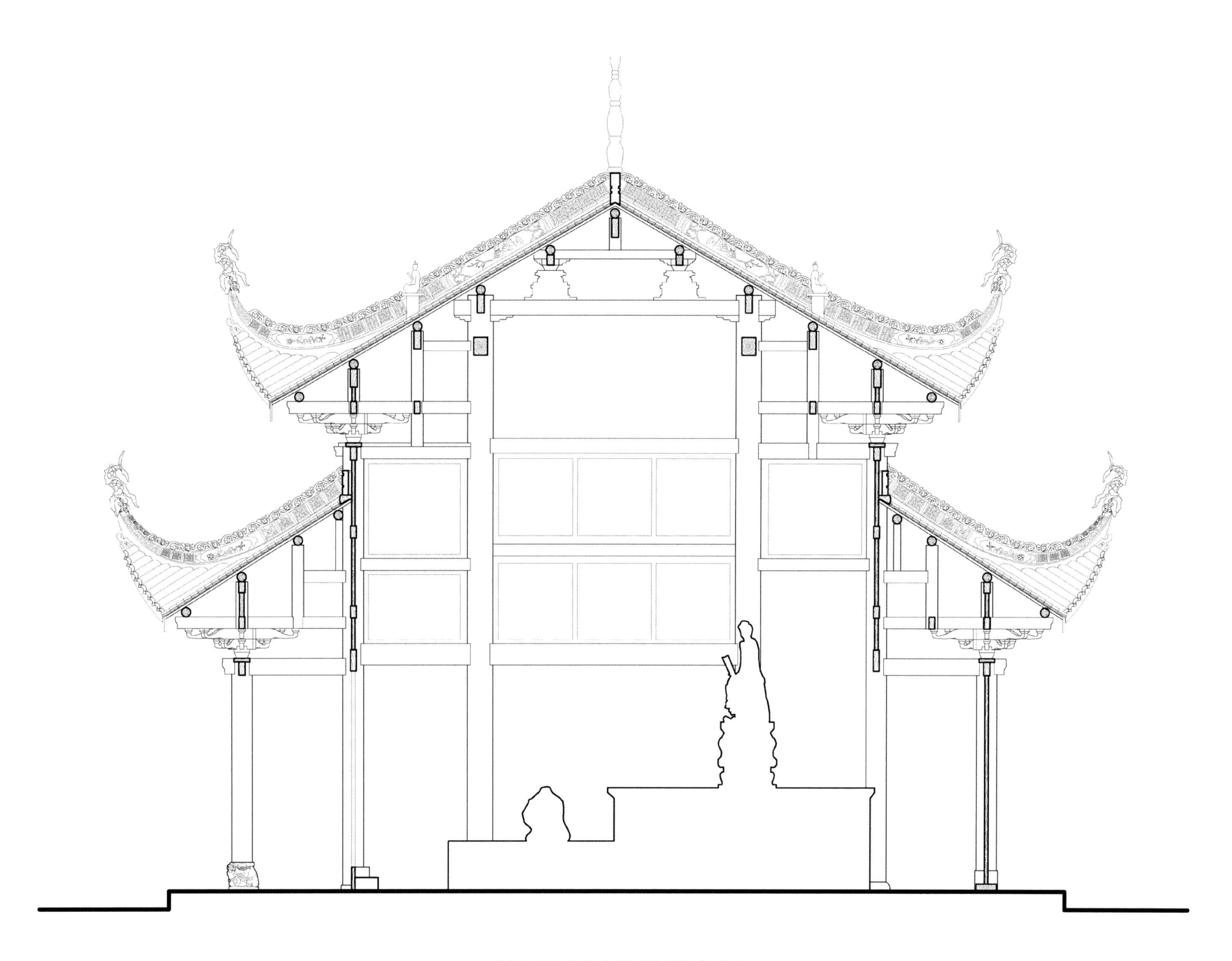

大足圣寿寺维摩殿明间横剖面图
Cross Section of Central Bay (Mingjian), Vimalakirti Hall, Shengshou Temple, Dazu

0 1 2 3m

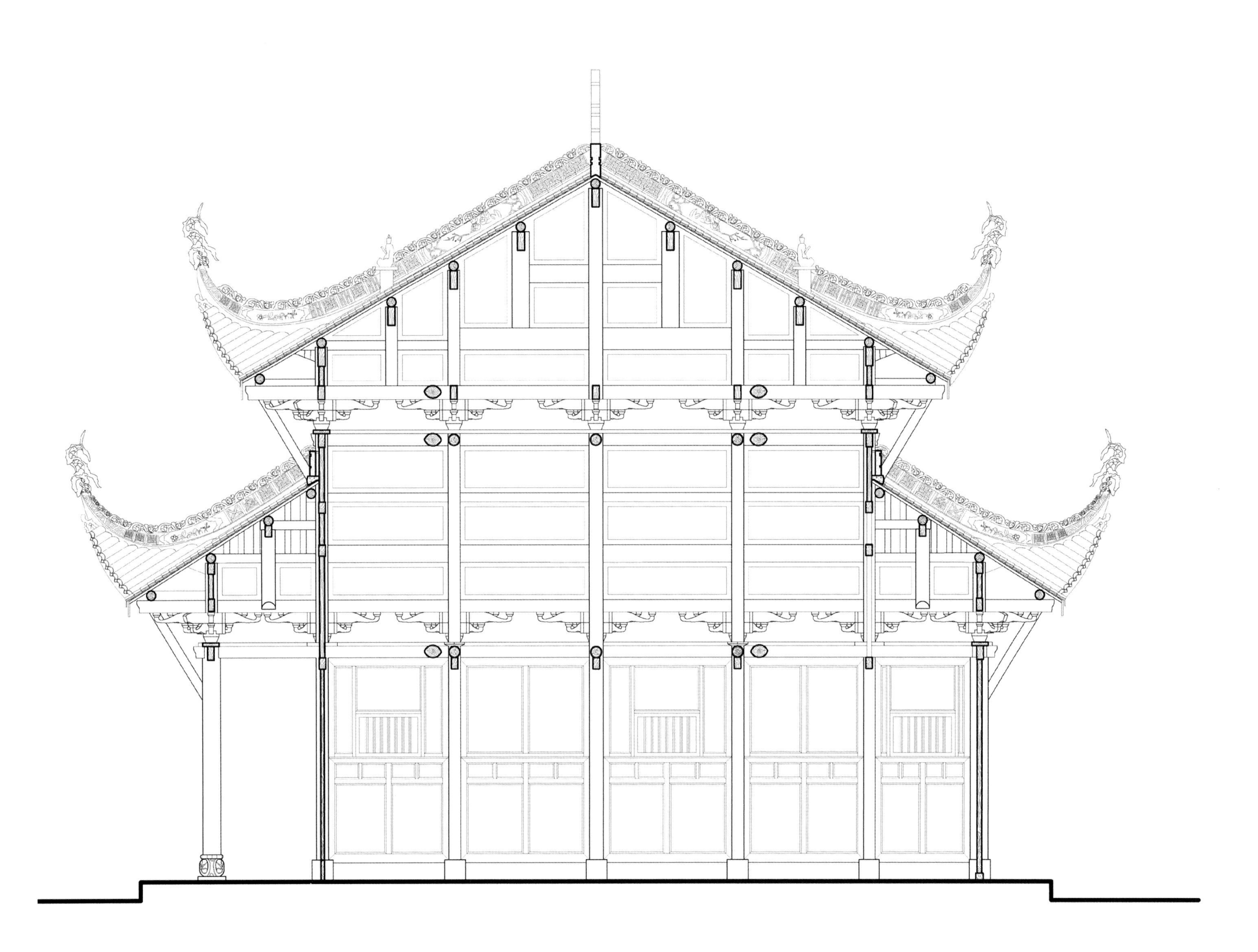
0 1 2 3m

大足圣寿寺维摩殿尽间横剖面图
Cross Section of Ending Bay (Jingjian), Vimalakirti Hall, Shengshou Temple, Dazu

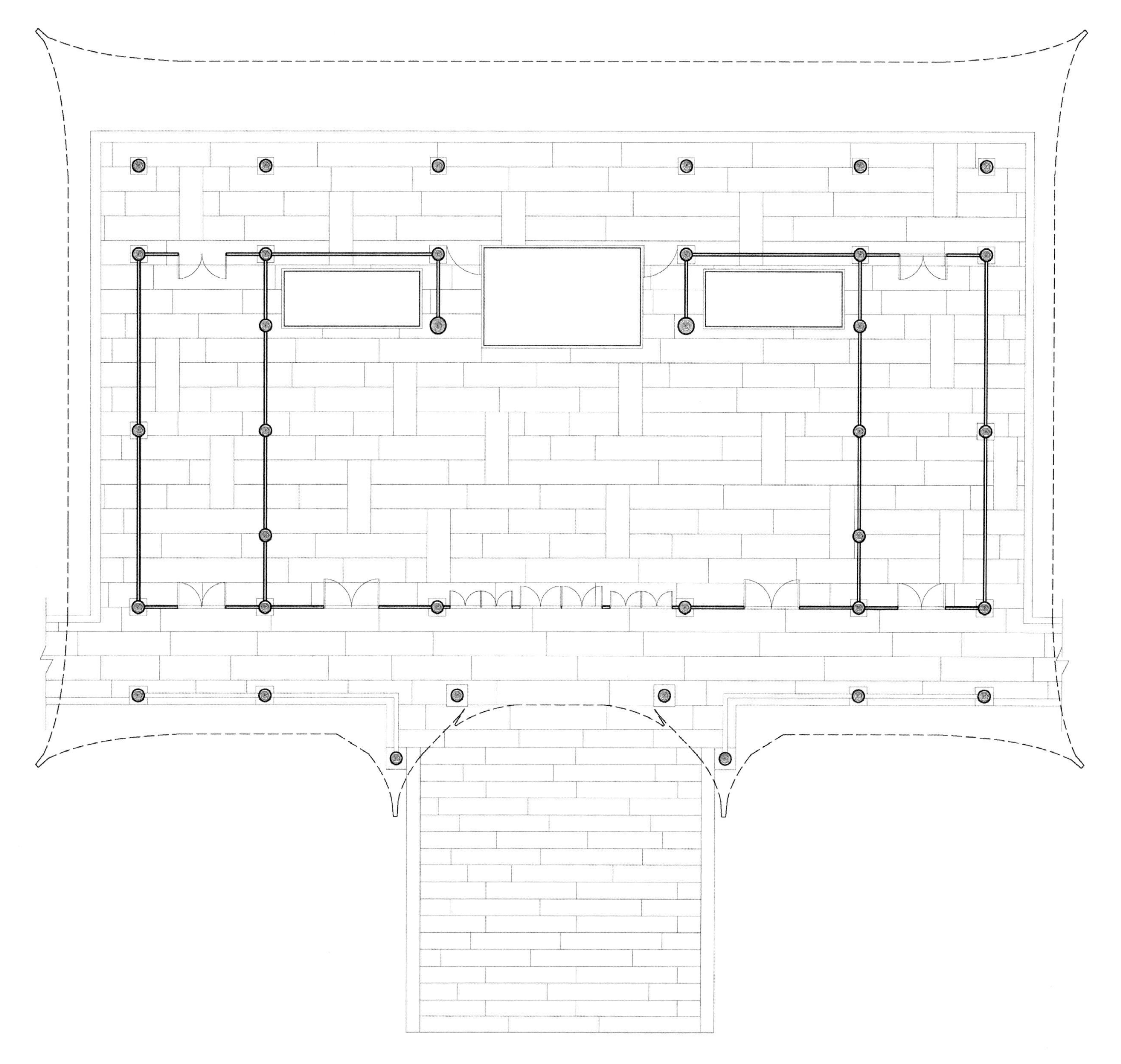
0 1 2 3 4 5m

大足圣寿寺帝释殿平面图
Ground Floor Plan of Śakra Hall, Shengshou Temple, Dazu

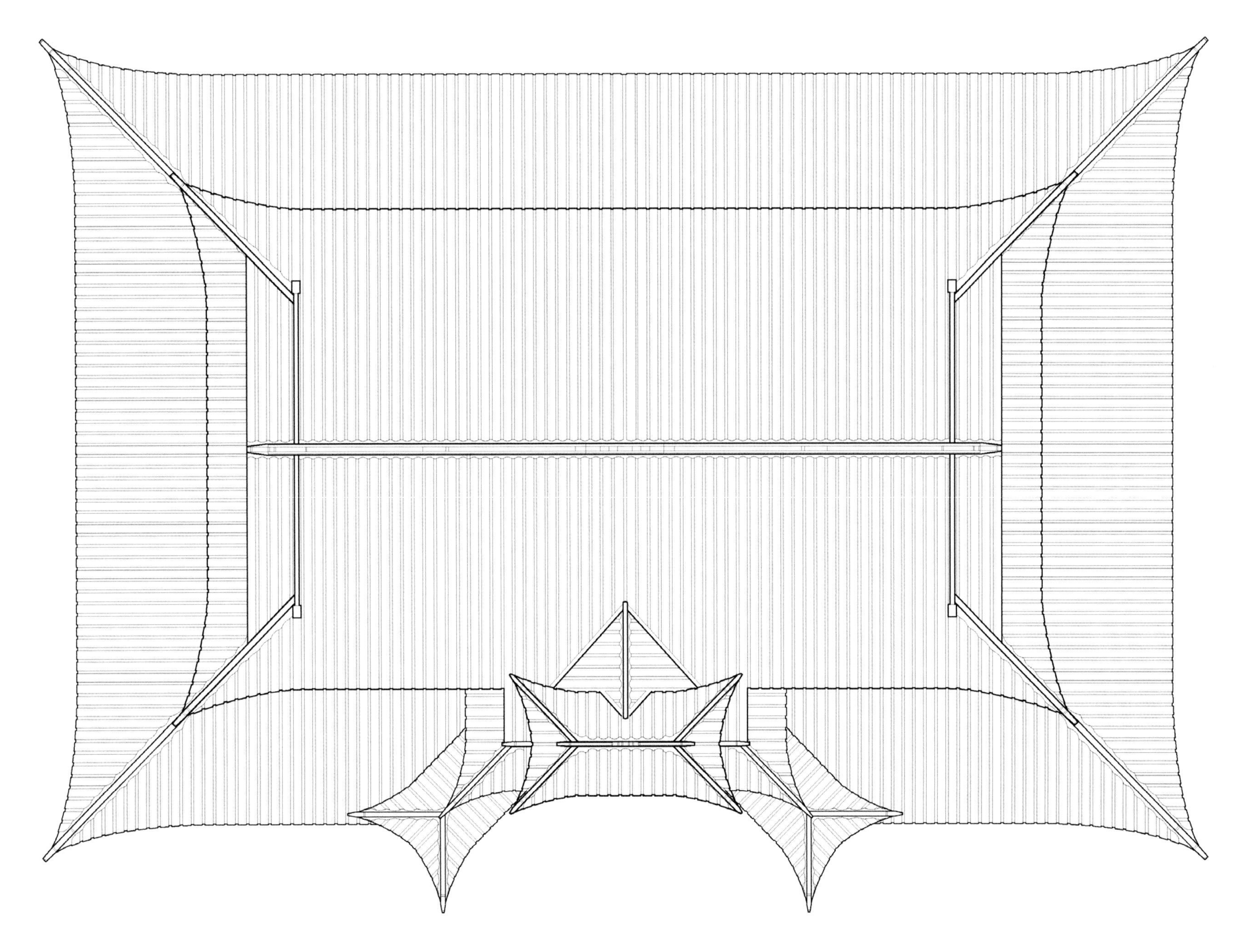

大足圣寿寺帝释殿屋顶平面图
Roof Plan of Śakra Hall, Shengshou Temple, Dazu

0 1 2 3m

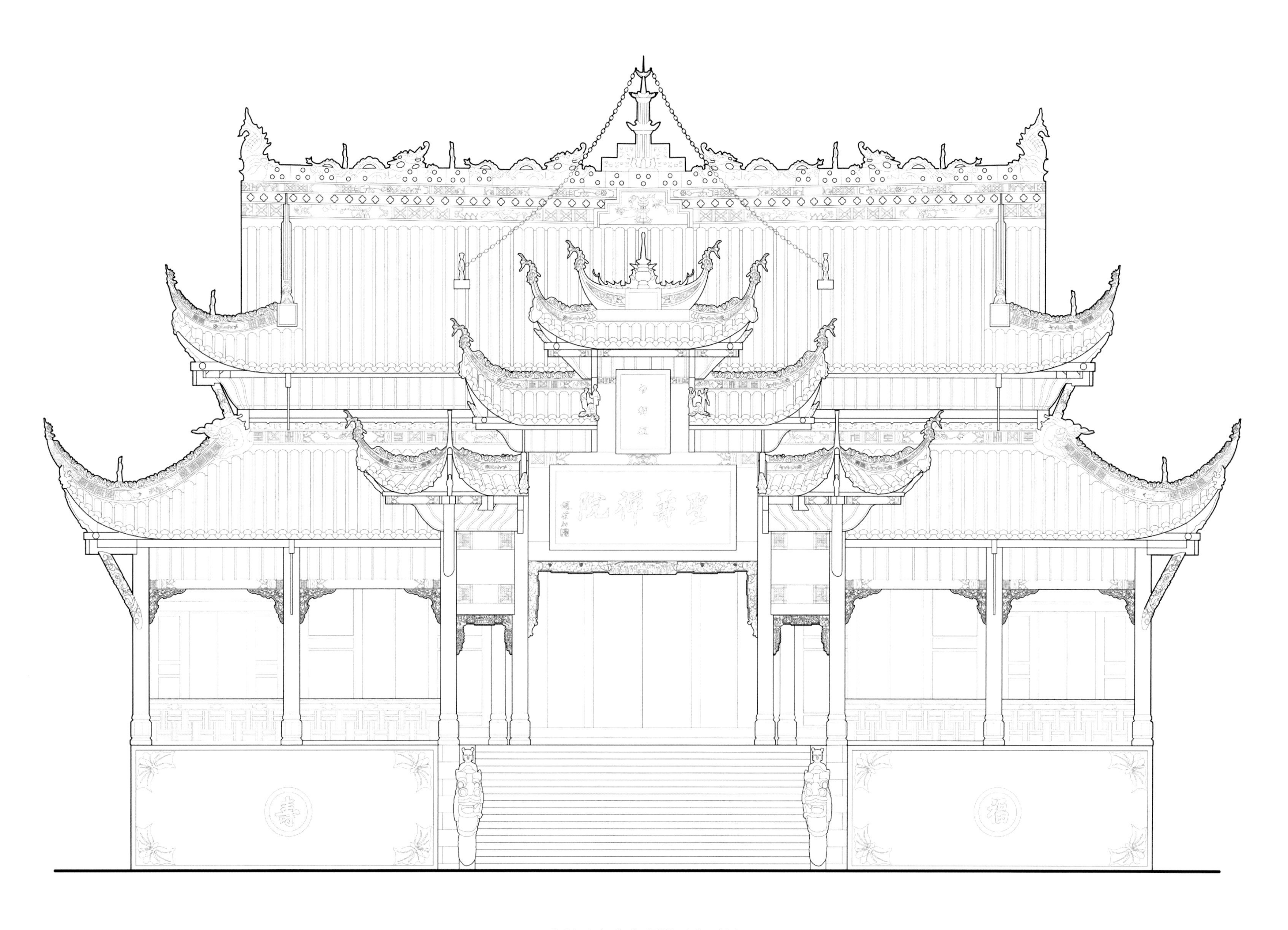
聖壽禪院
壽
福

大足圣寿寺帝释殿正立面图

Front Elevation of Śakra Hall, Shengshou Temple, Dazu

0 1 2 3 4 5m

0
1
2
3
4
5m

大足圣寿寺帝释殿背立面图
Back Elevation of Śakra Hall, Shengshou Temple, Dazu

0
1
2
3
4
5m

大足圣寿寺帝释殿纵剖面图
Longitudinal Section of Śakra Hall, Shengshou Temple, Dazu

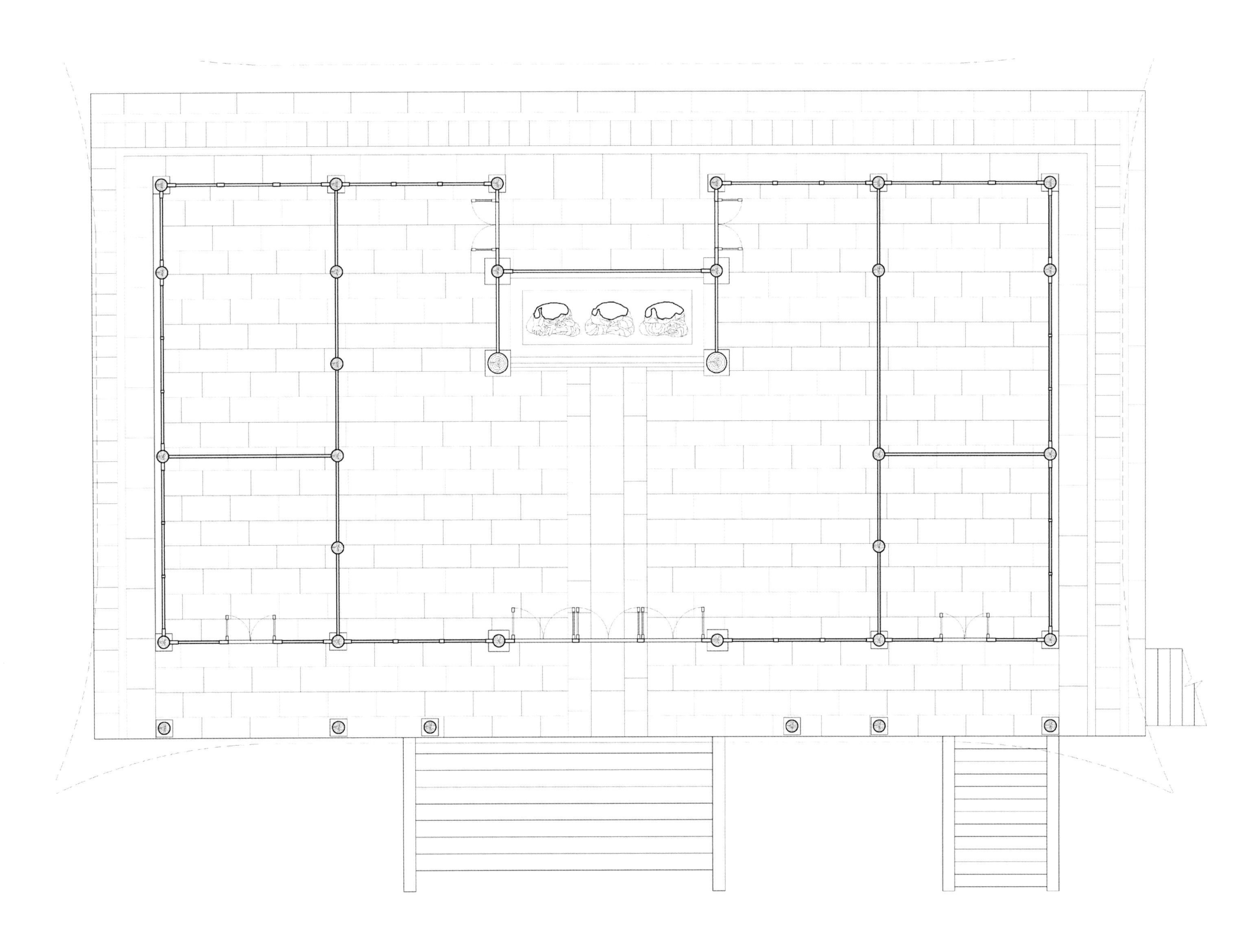

大足圣寿寺三世佛殿平面图
Ground Floor Plan of The Three Buddhas Hall, Shengshou Temple, Dazu

0 1 2 3m

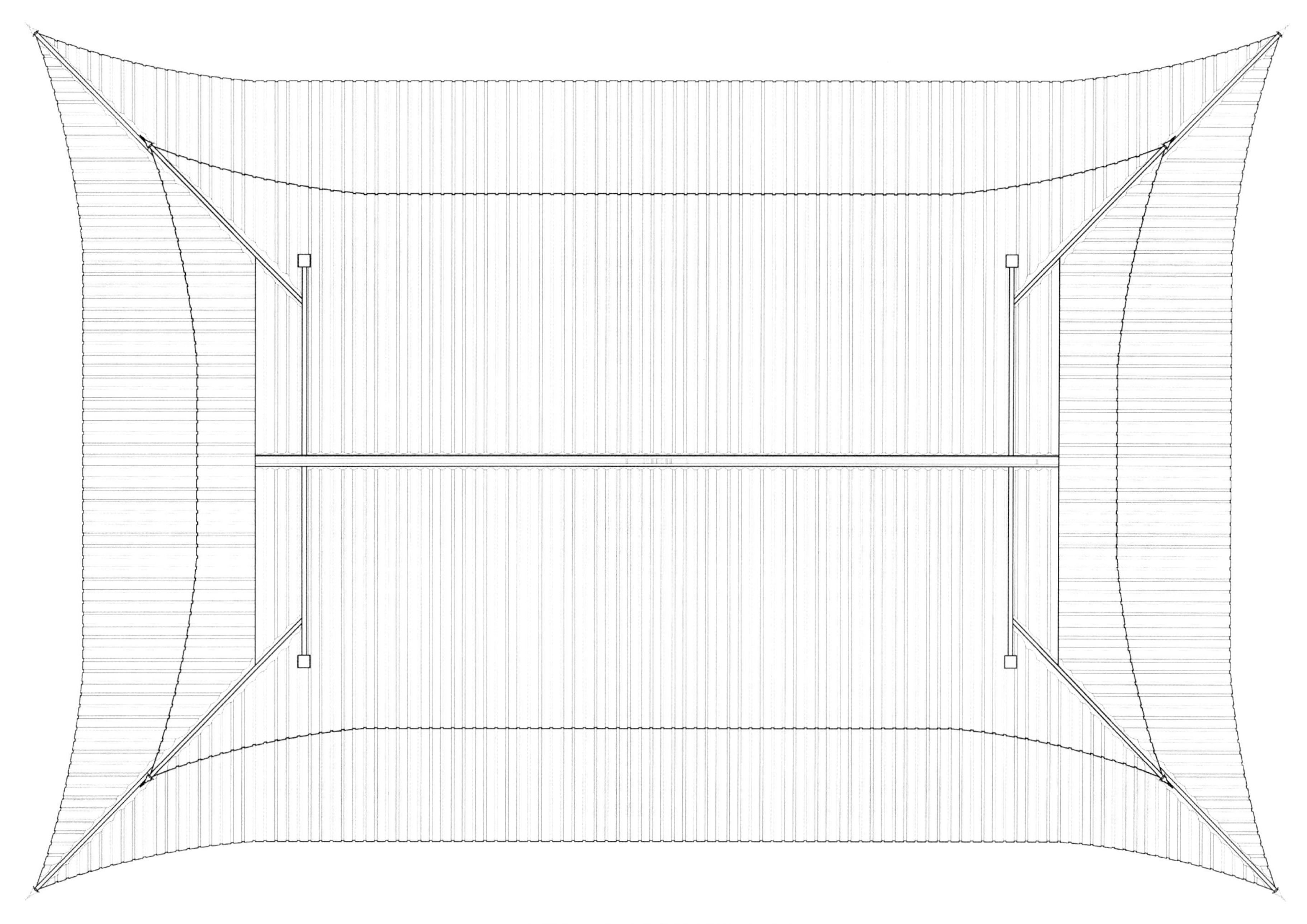
0 1 2 3m

大足圣寿寺三世佛殿屋顶平面图
Roof Plan of The Three Buddhas Hall, Shengshou Temple, Dazu

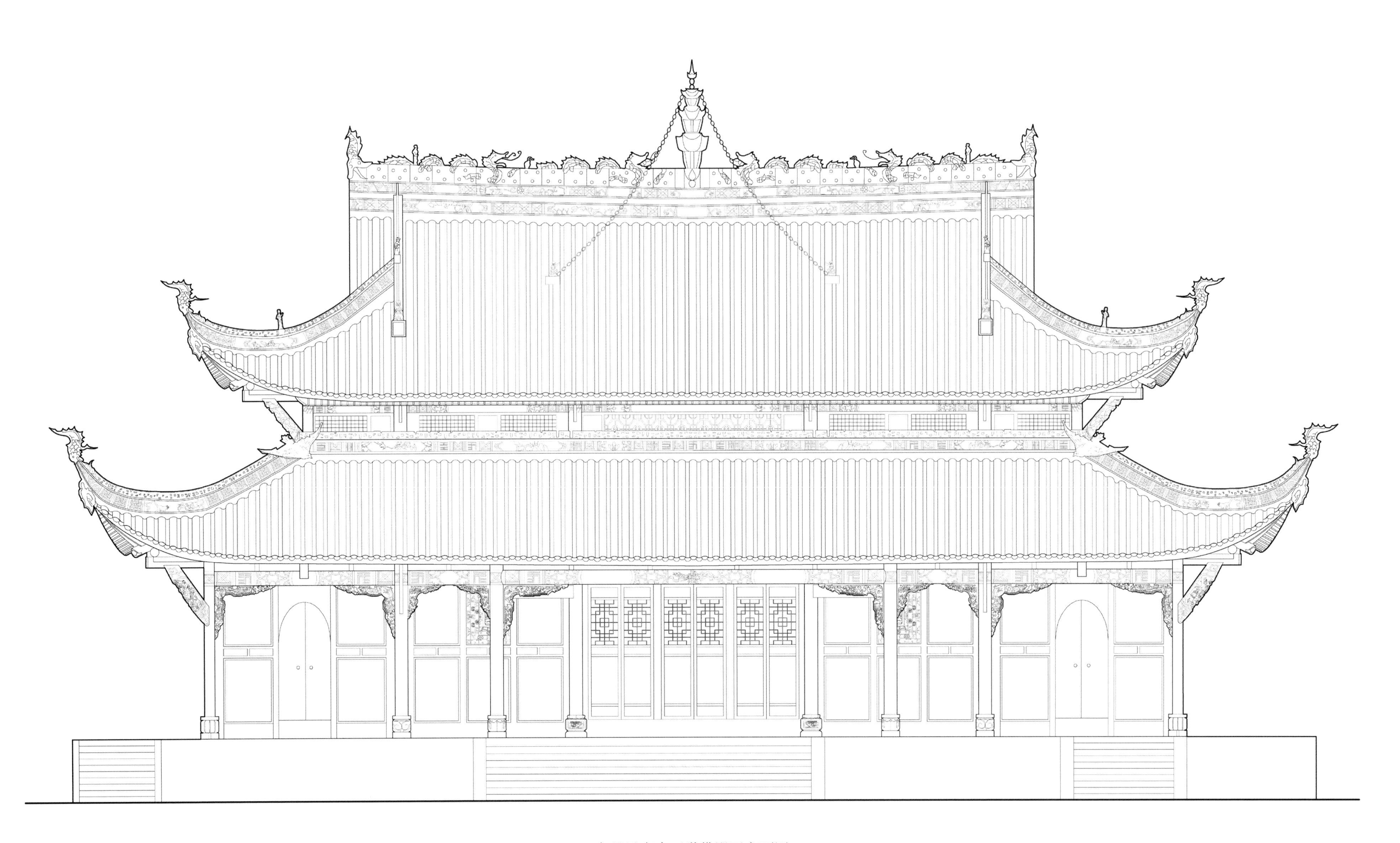
0
1
2
3m

大足圣寿寺三世佛殿正立面图
Front Elevation of The Three Buddhas Hall, Shengshou Temple, Dazu

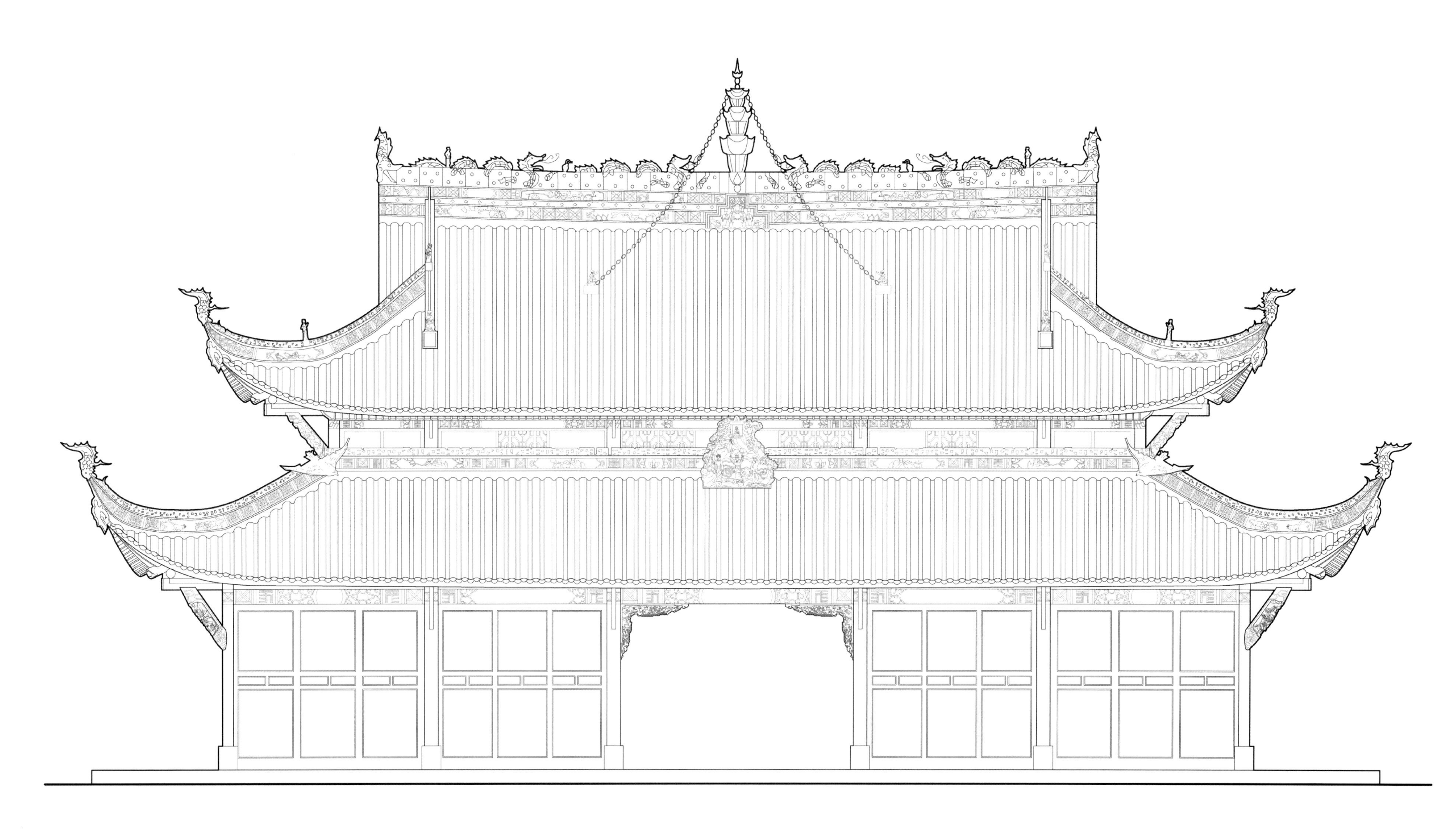
0
1
2
3m

大足圣寿寺三世佛殿背立面图
Back Elevation of The Three Buddhas Hall, Shengshou Temple, Dazu

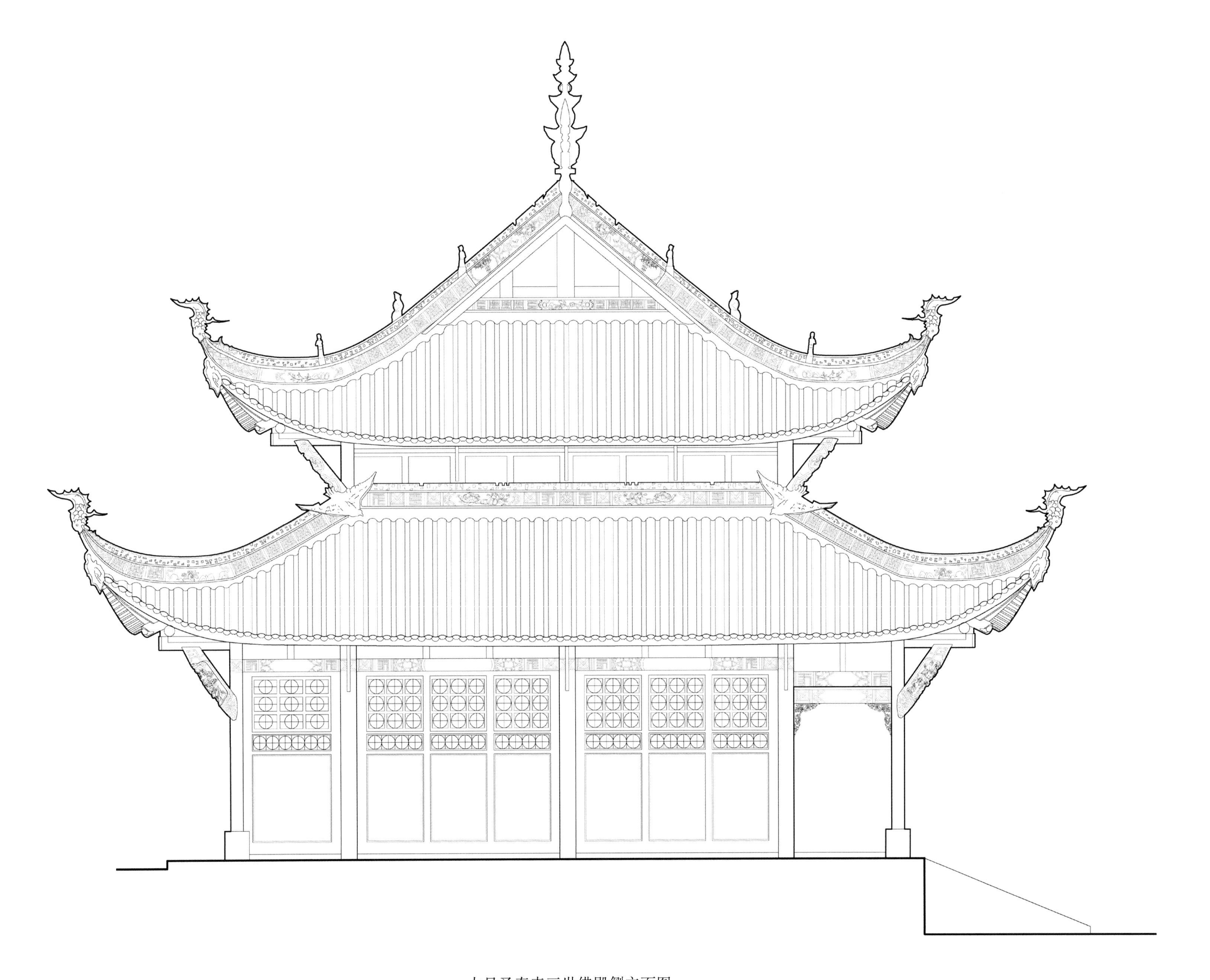
0
1
2
3m

大足圣寿寺三世佛殿侧立面图
Side Elevation of The Three Buddhas Hall, Shengshou Temple, Dazu

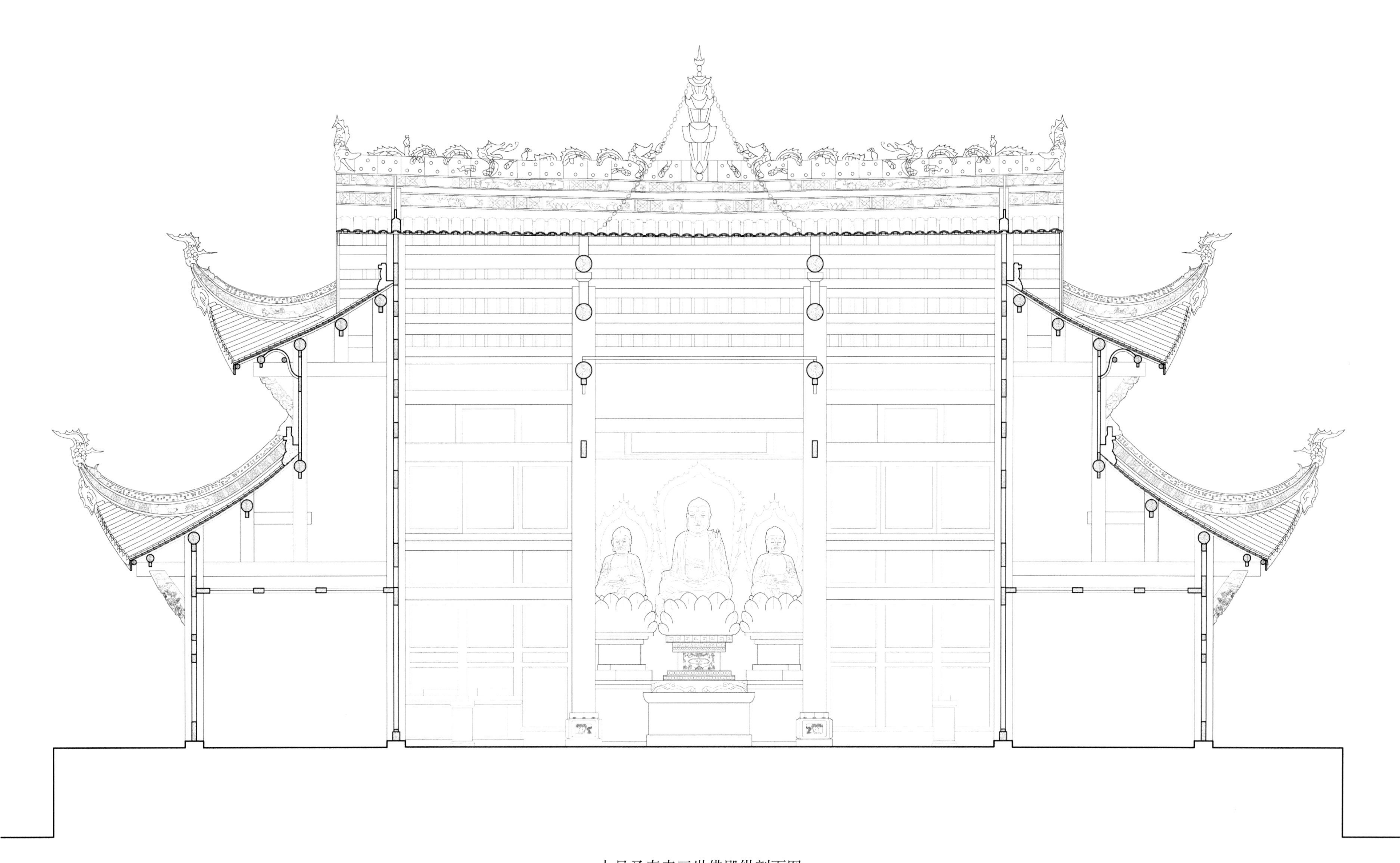
0 1 2 3m

大足圣寿寺三世佛殿纵剖面图

Longitudinal Section of The Three Buddhas Hall, Shengshou Temple, Dazu

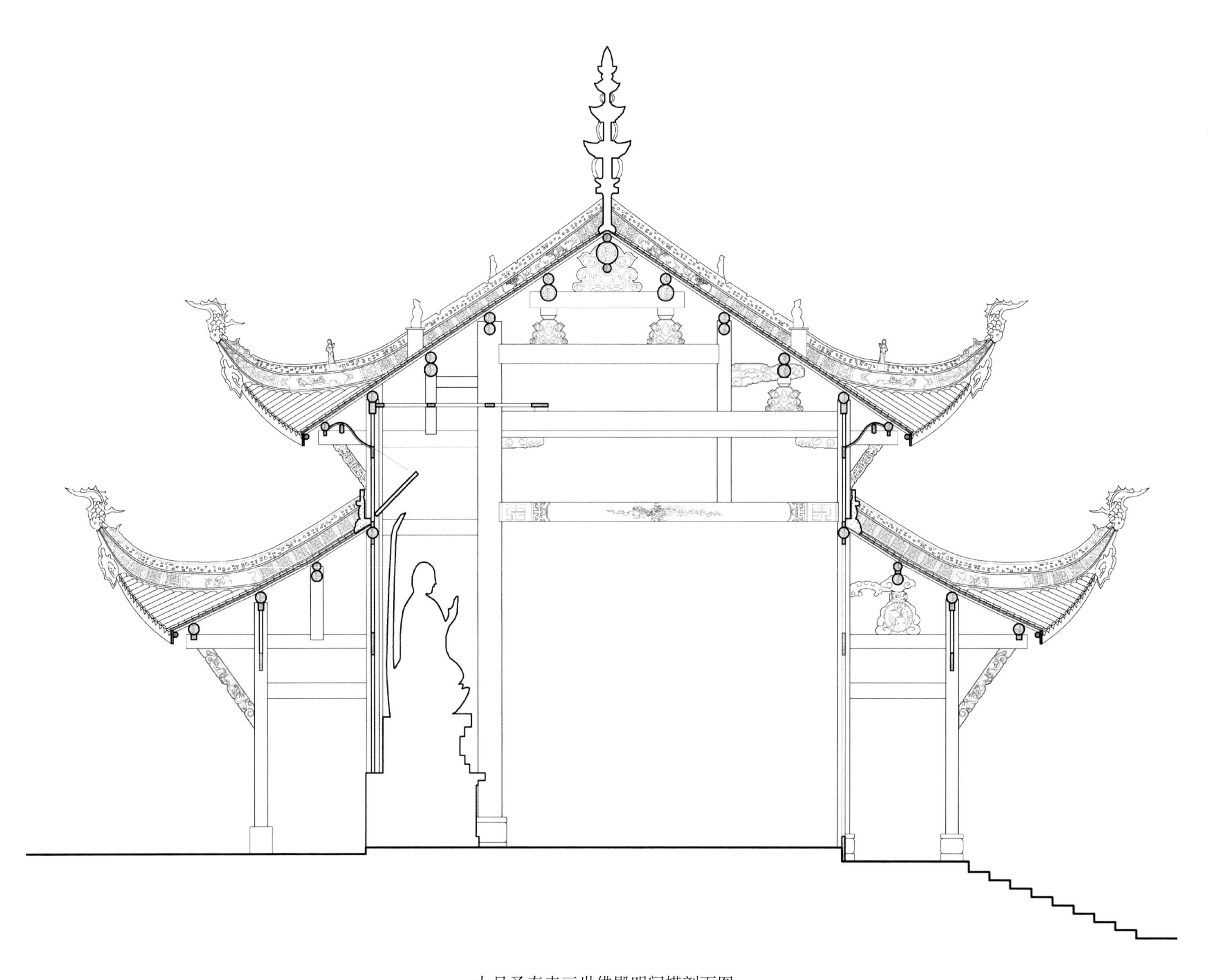

大足圣寿寺三世佛殿明间横剖面图

Cross Section of Central Bay (Mingjian), The Three Buddhas Hall, Shengshou Temple, Dazu

0 1 2 3m

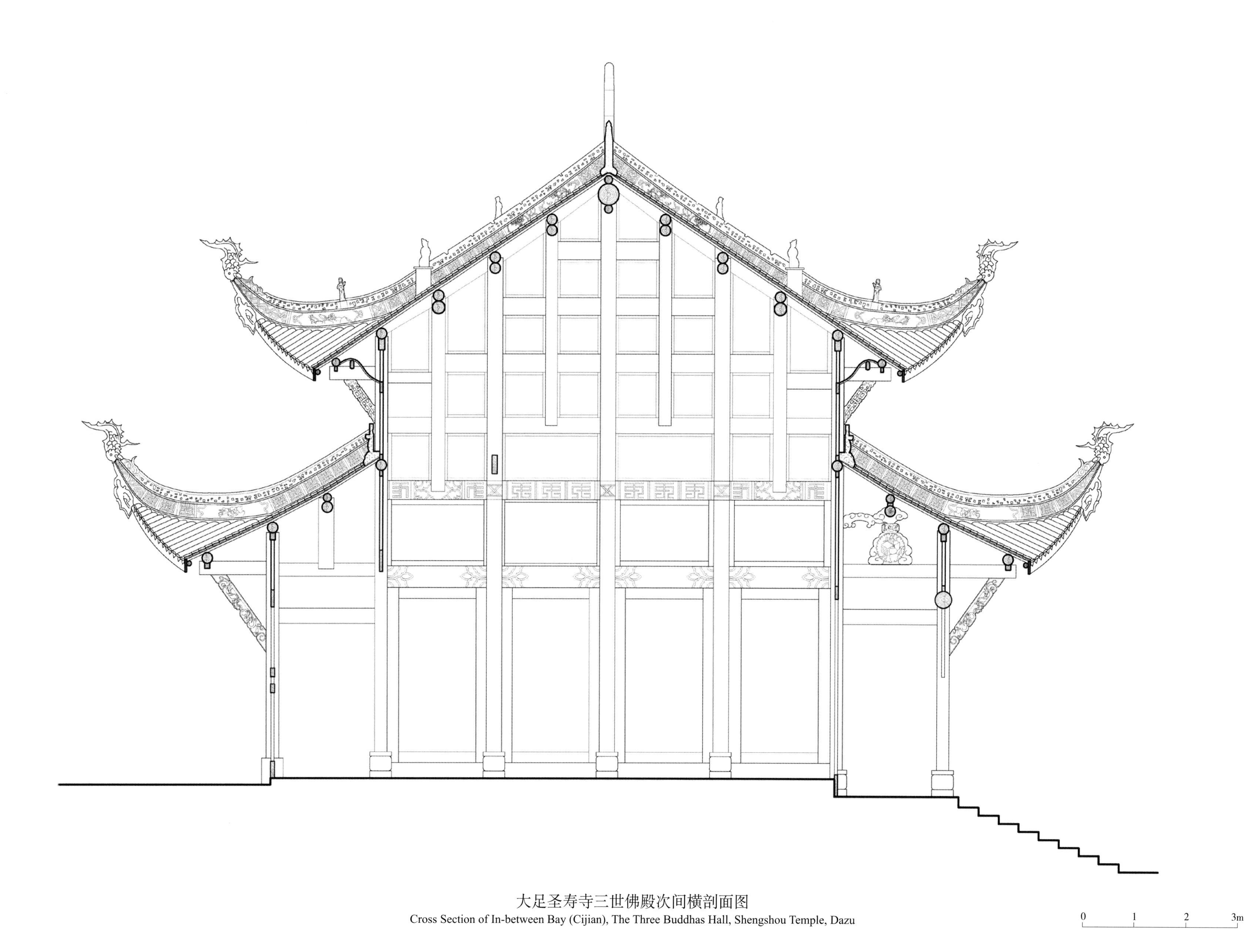
0
1
2
3m

大足圣寿寺三世佛殿次间横剖面图
Cross Section of In-between Bay (Cijian), The Three Buddhas Hall, Shengshou Temple, Dazu

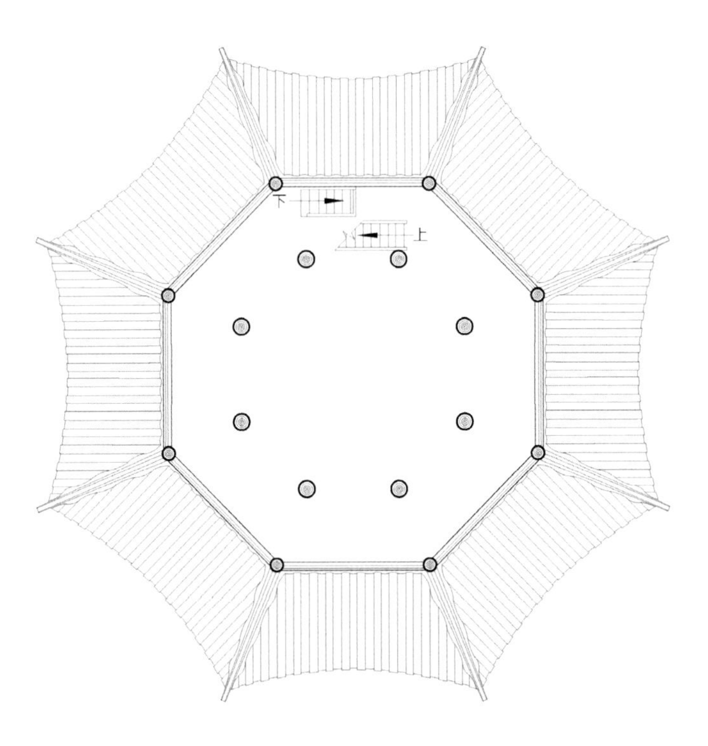
下
上

万岁楼二层平面图

万岁楼屋顶平面图

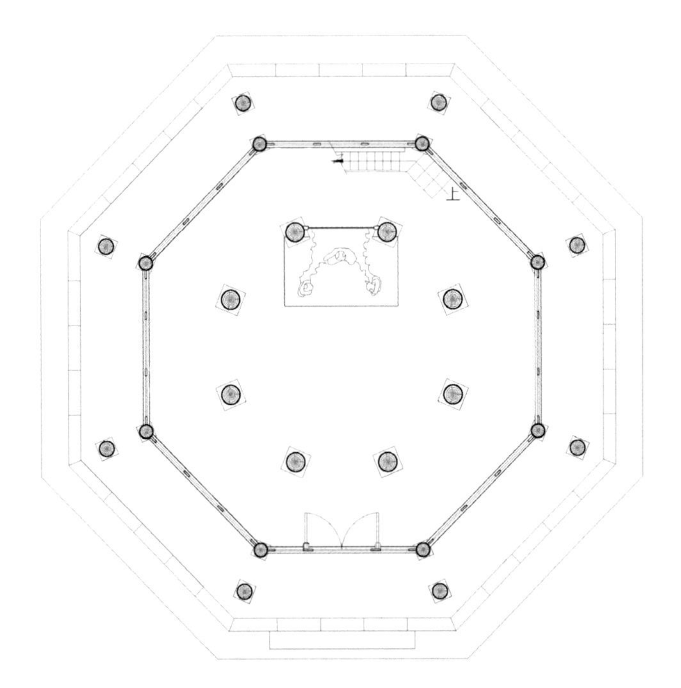
上

万岁楼一层平面图

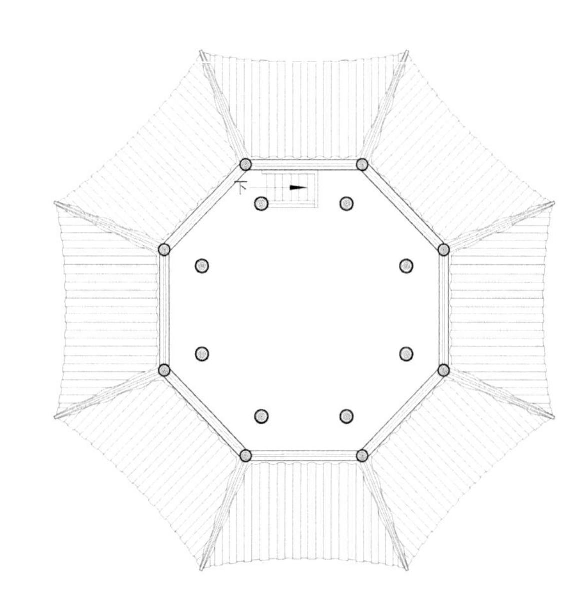
下

万岁楼三层平面图

大足圣寿寺万岁楼平面图

Ground Floor Plan of Wansui Building, Shengshou Temple, Dazu

0 1 2 3m

大足圣寿寺万岁楼立面图
Front Elevation of Wansui Building, Shengshou Temple, Dazu

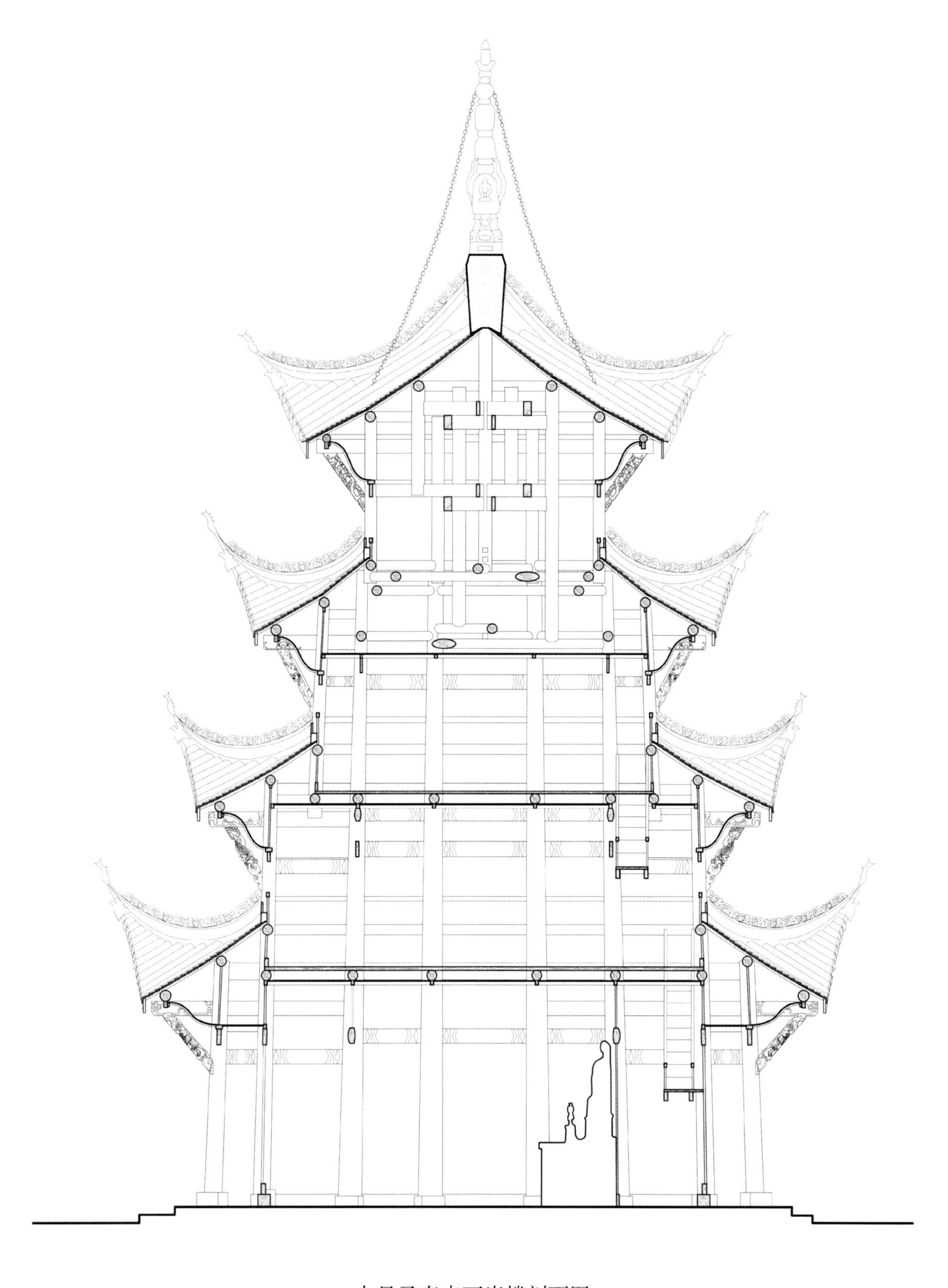
0 1 2 3m

大足圣寿寺万岁楼剖面图
Cross Section of Wansui Building, Shengshou Temple, Dazu

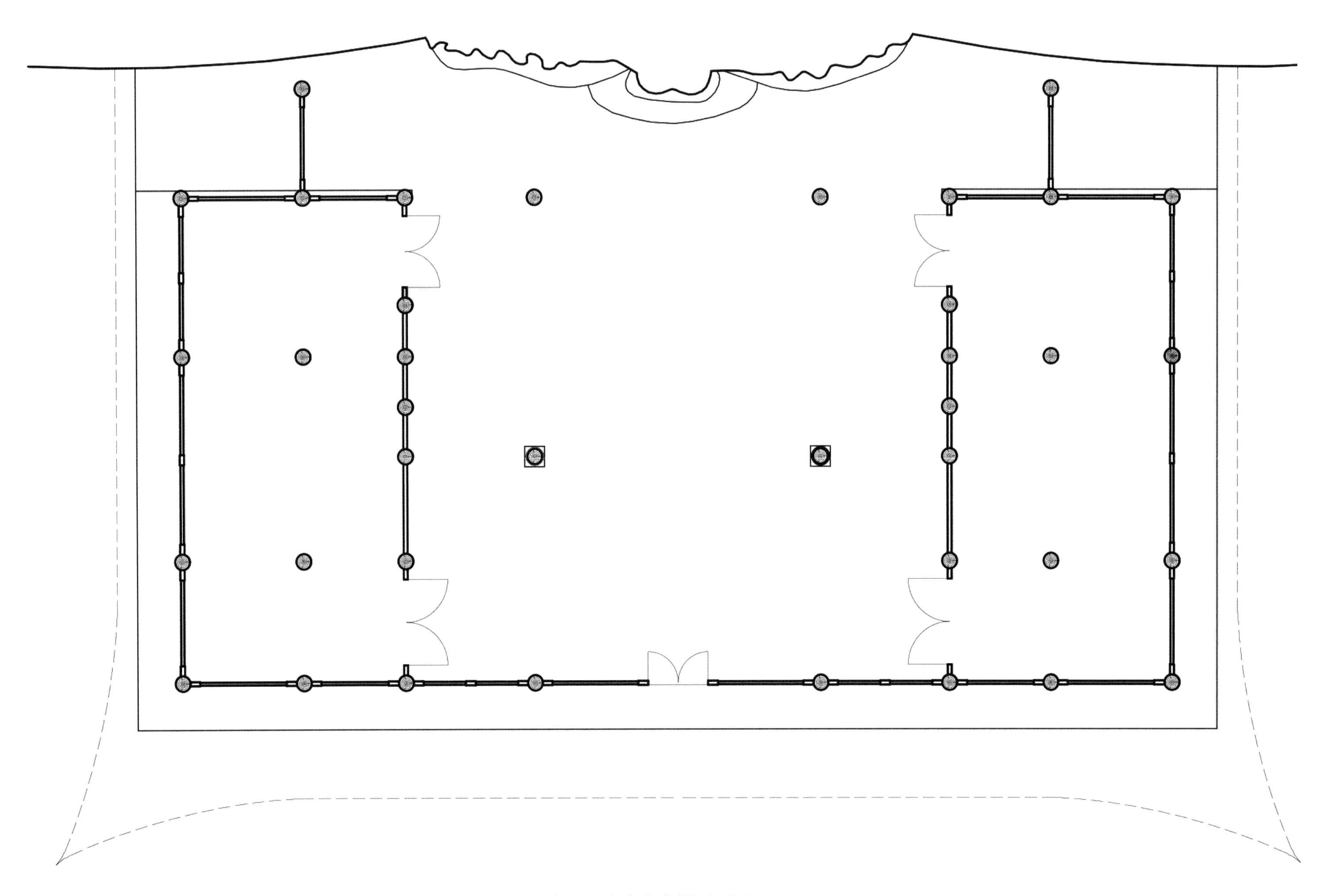

大足圣寿寺大悲殿平面图
Ground Floor Plan of Dabei Hall, Shengshou Temple, Dazu

0
1
2
3m

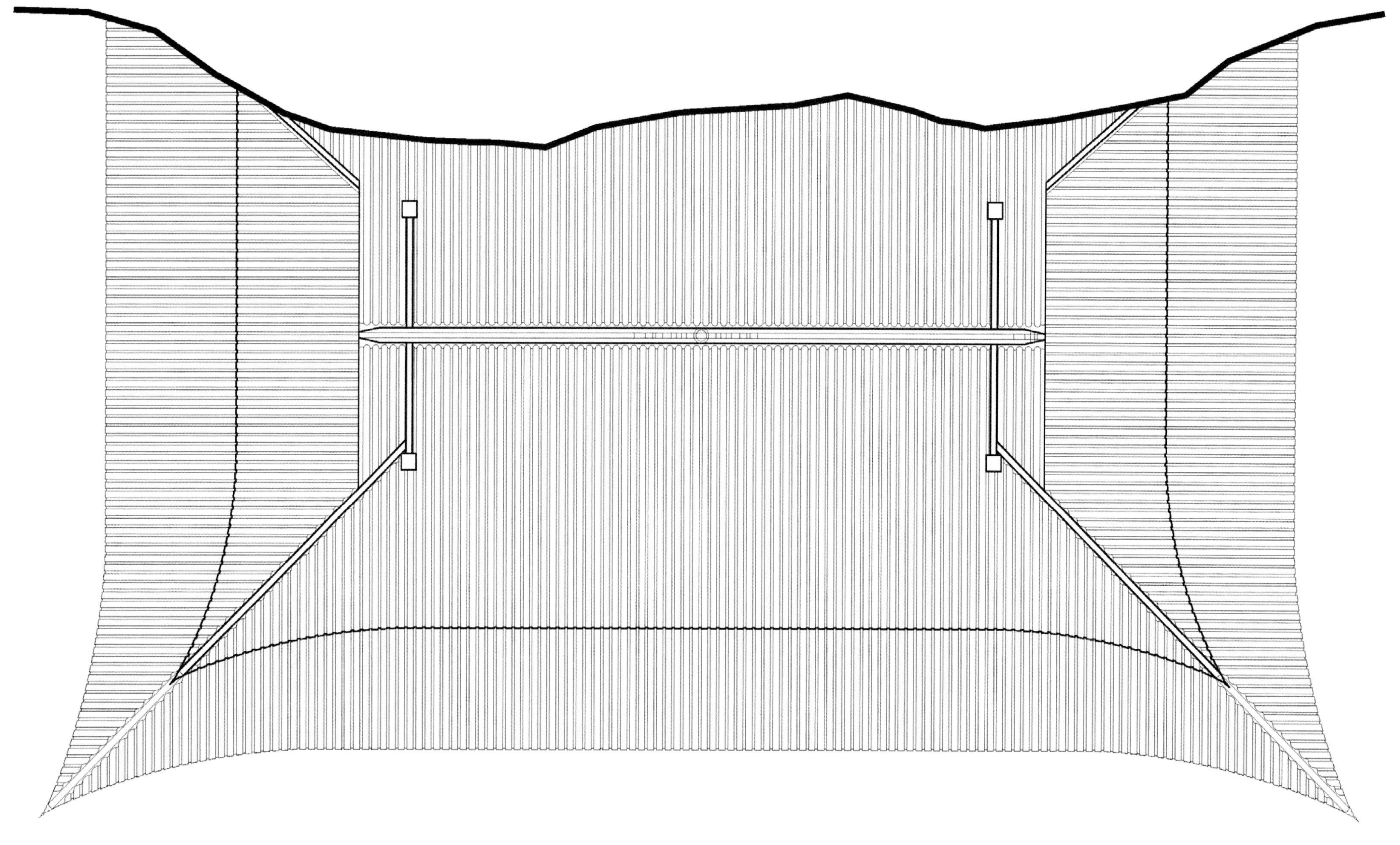
0 1 2 3m

大足圣寿寺大悲殿屋顶平面图
Roof Plan of Dabei Hall, Shengshou Temple, Dazu

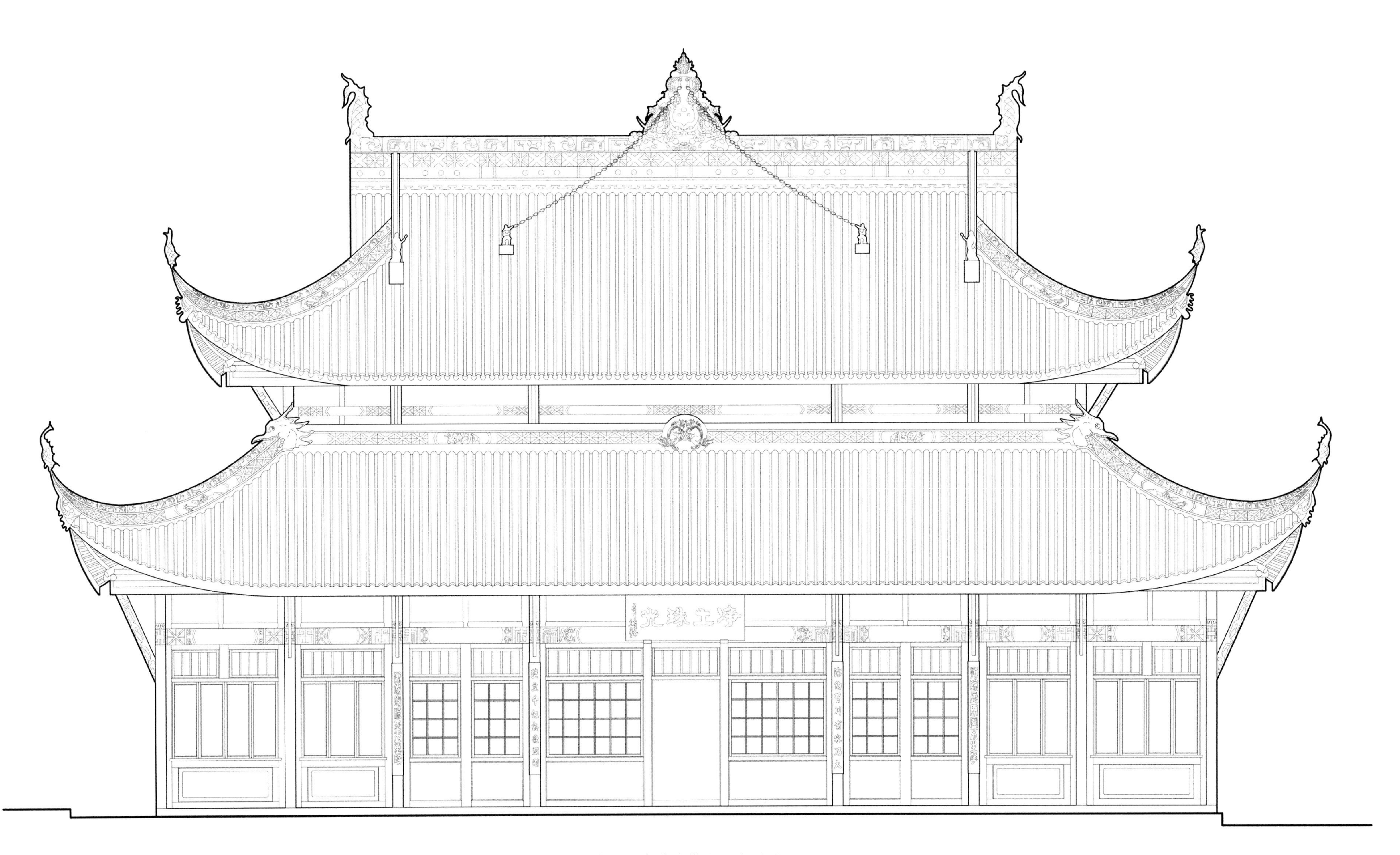
净土珠光

大足圣寿寺大悲殿正立面图
Front Elevation of Dabei Hall, Shengshou Temple, Dazu

0 1 2 3m

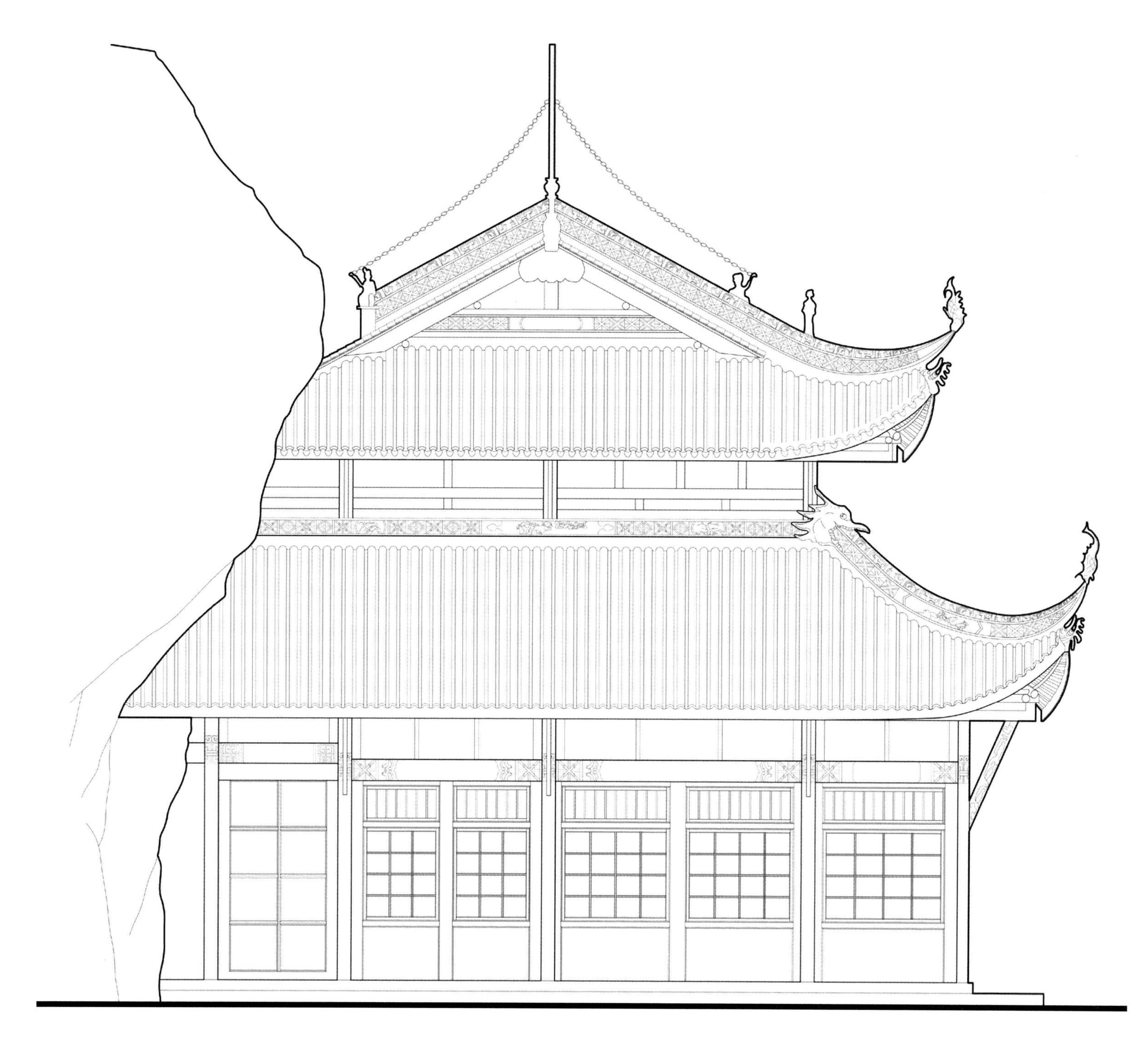
0
1
2
3m

大足圣寿寺大悲殿侧立面图
Side Elevation of Dabei Hall, Shengshou Temple, Dazu

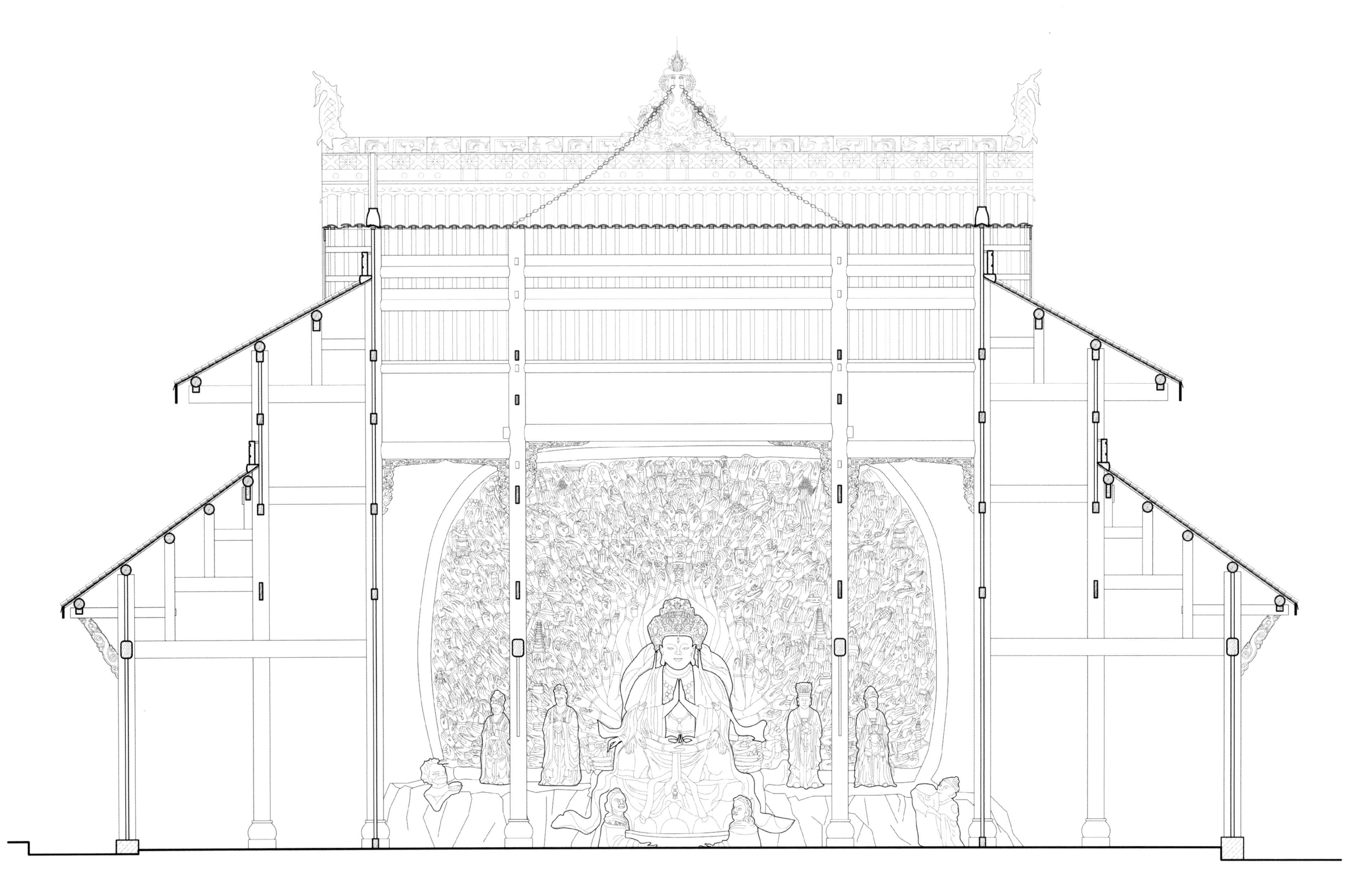
0
1
2
3m

大足圣寿寺大悲殿纵剖面图
Longitudinal Section of Dabei Hall, Shengshou Temple, Dazu

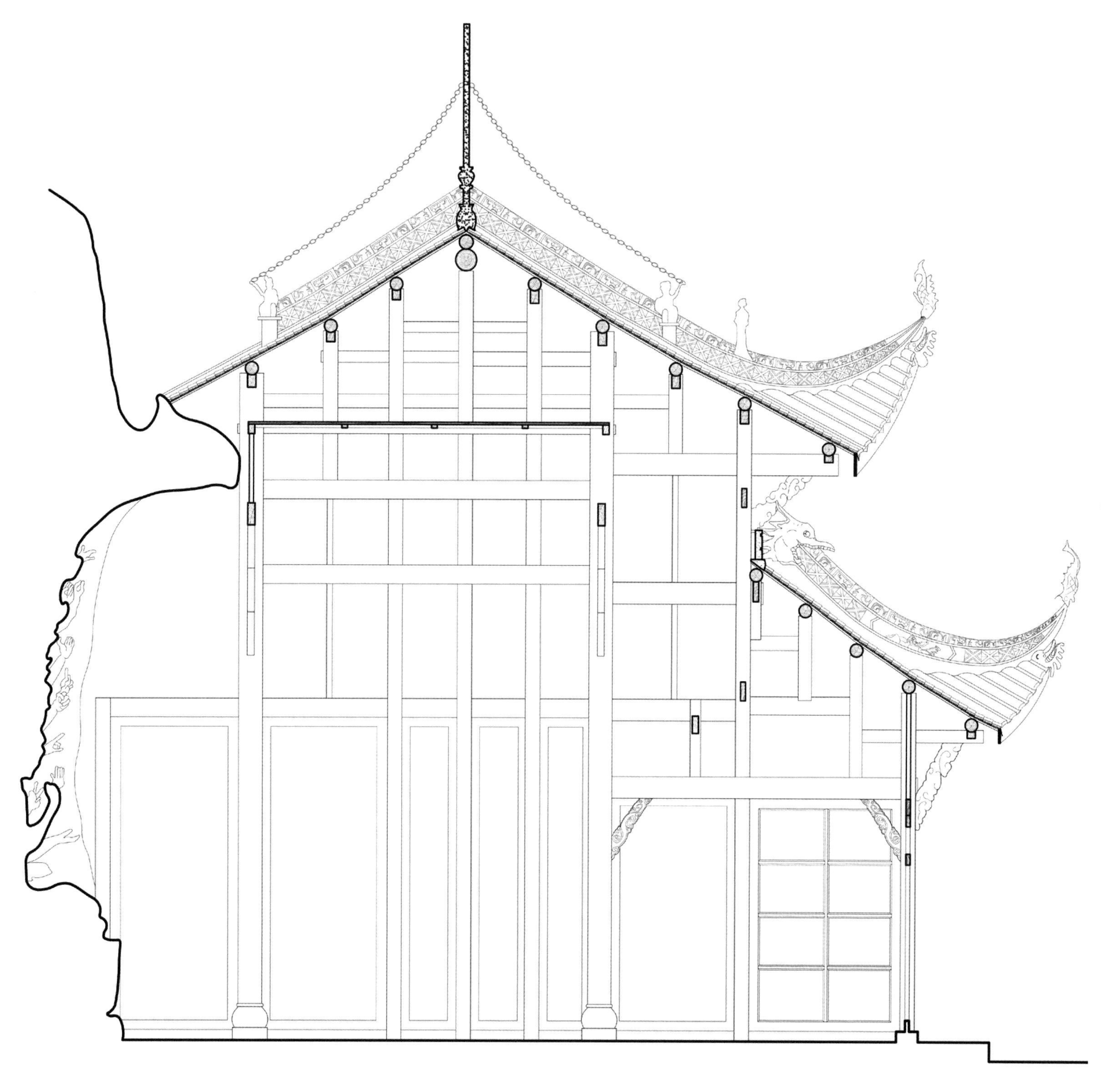
0
1
2
3m

大足圣寿寺大悲殿横剖面图
Cross Section of Dabei Hall, Shengshou Temple, Dazu

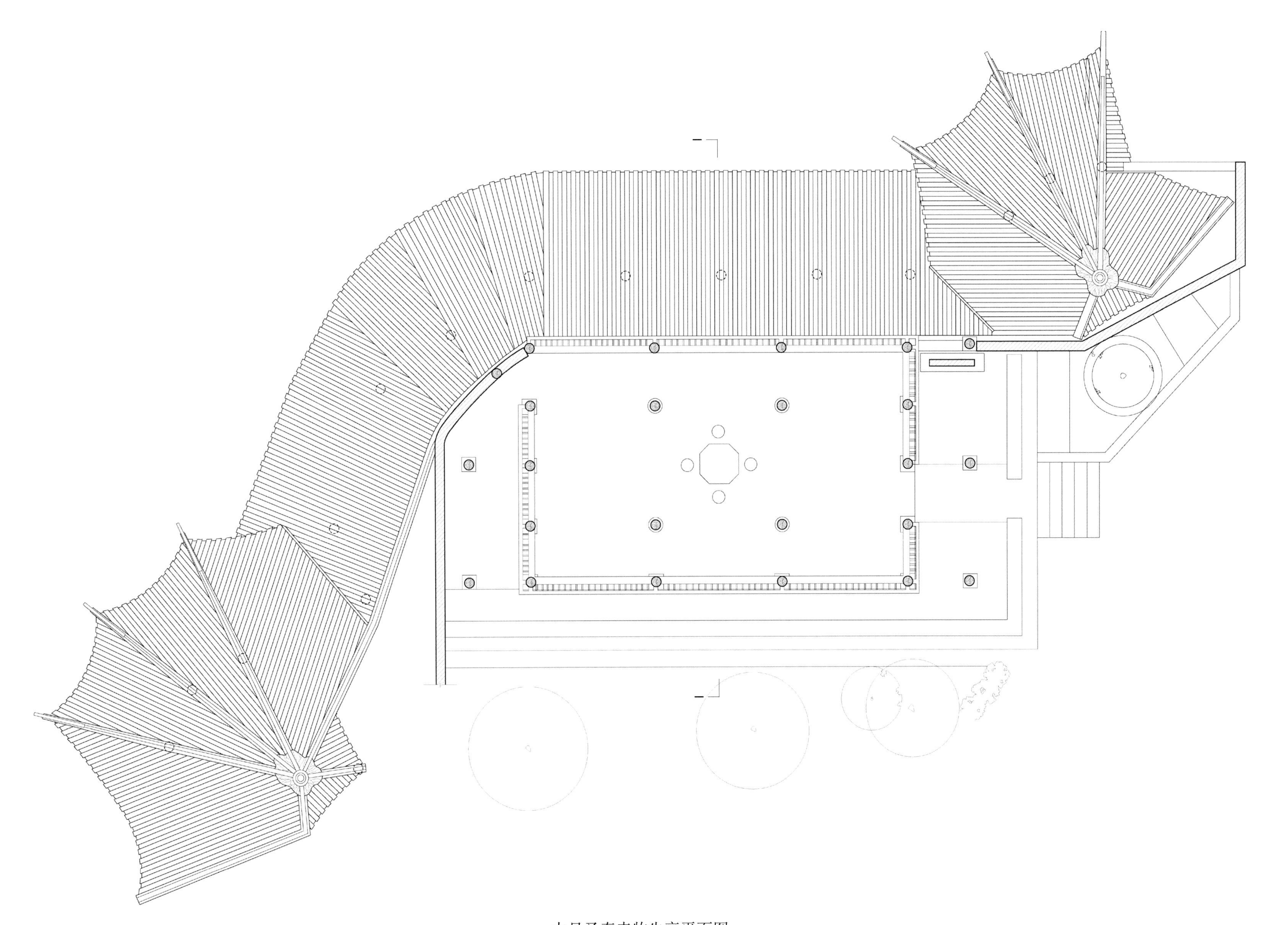
0 1 2 3m

大足圣寿寺牧牛亭平面图
Ground Floor Plan of Muniu Pavilion, Shengshou Temple, Dazu

0 1 2 3m

大足圣寿寺牧牛亭正立面图
Front Elevation of Muniu Pavilion, Shengshou Temple, Dazu

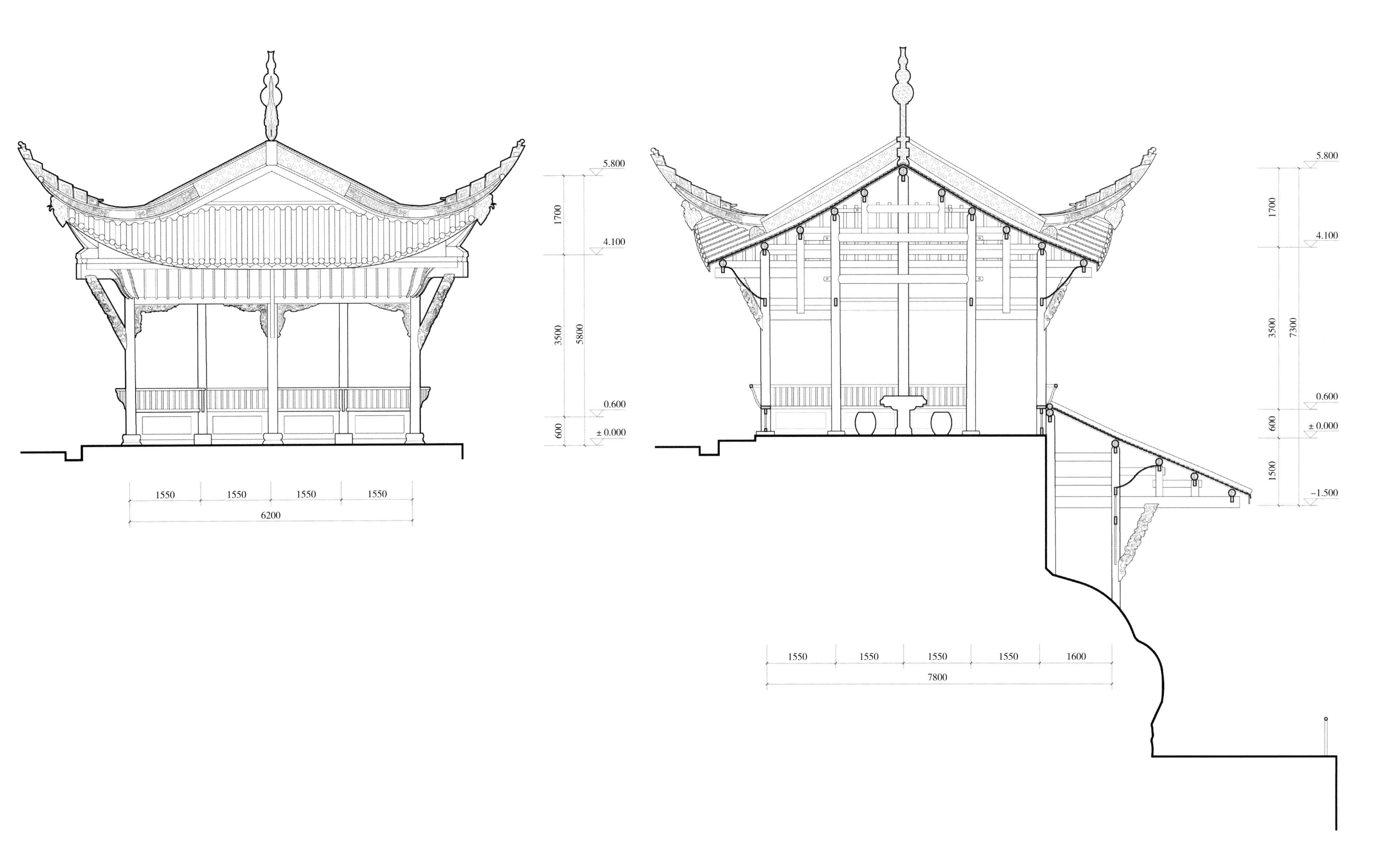
5.800
1700
4.100
3500
5800
0.600
600
±0.000
1550
1550
1550
1550
6200
5.800
1700
4.100
3500
7300
0.600
600
±0.000
1500
-1.500
1550
1550
1550
1550
1600
7800
0
1
2
3m

大足圣寿寺牧牛亭侧立面图
Side Elevation of Muniu Pavilion, Shengshou Temple, Dazu

大足圣寿寺牧牛亭剖面图
Cross Section of Muniu Pavilion, Shengshou Temple, Dazu

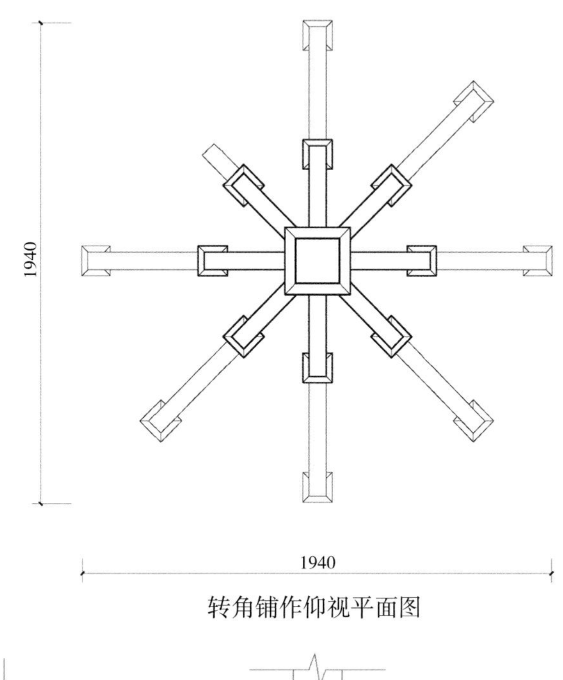
1940
1940

转角铺作仰视平面图

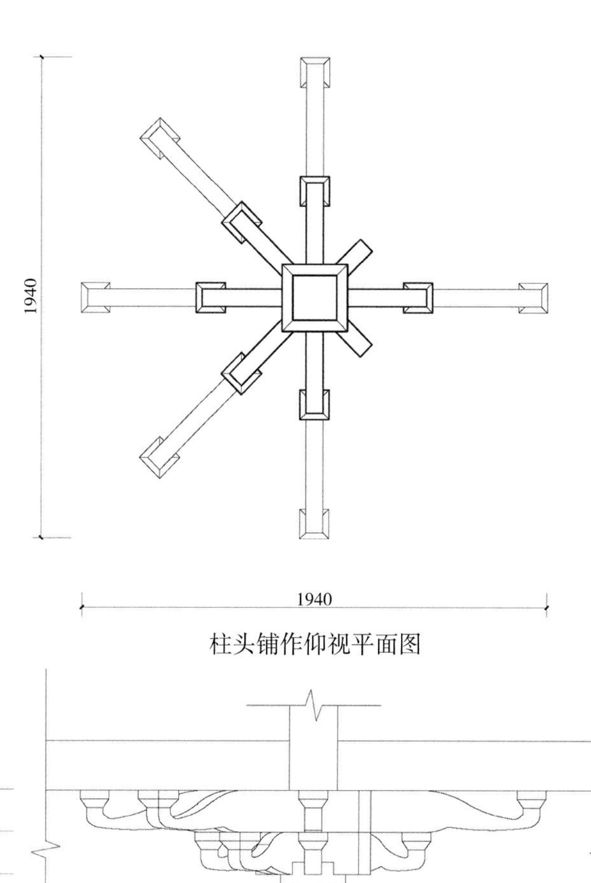
1940
1940

柱头铺作仰视平面图

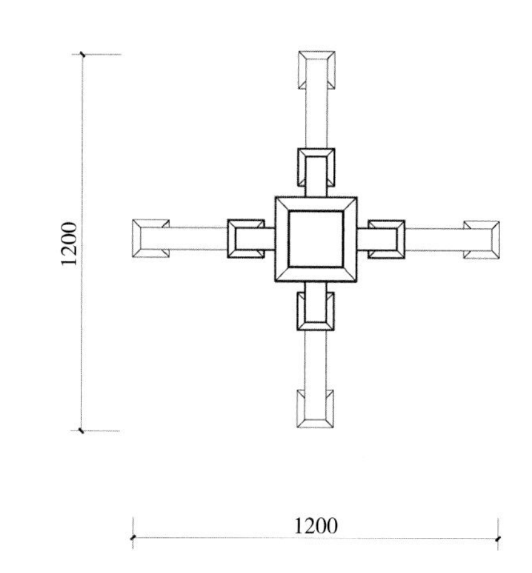
1200
1200

次间补间铺作仰视平面图

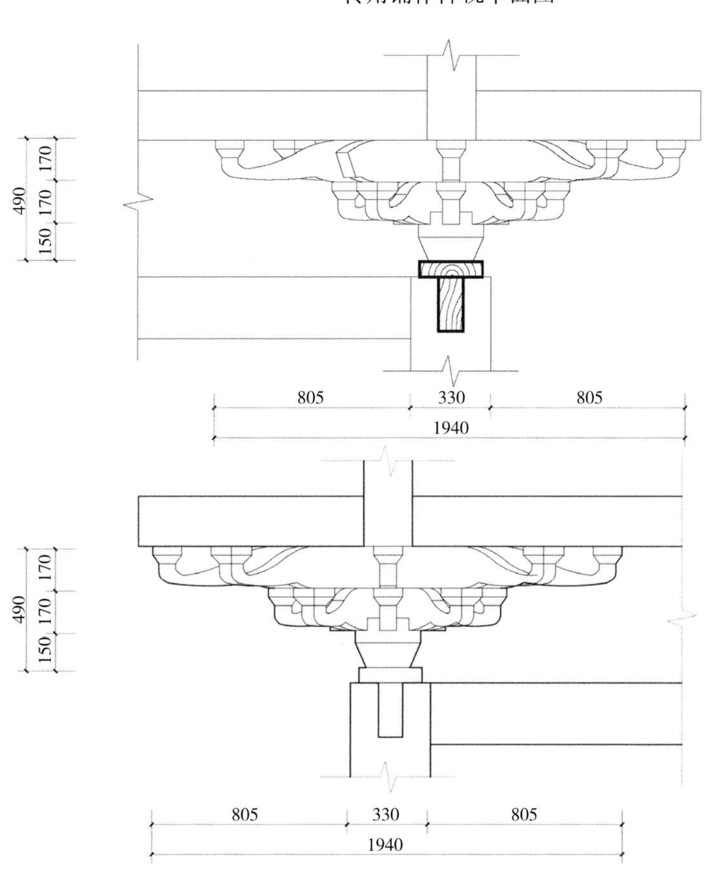
490
170
170
150
805
330
805
1940

转角铺作仰视立面图

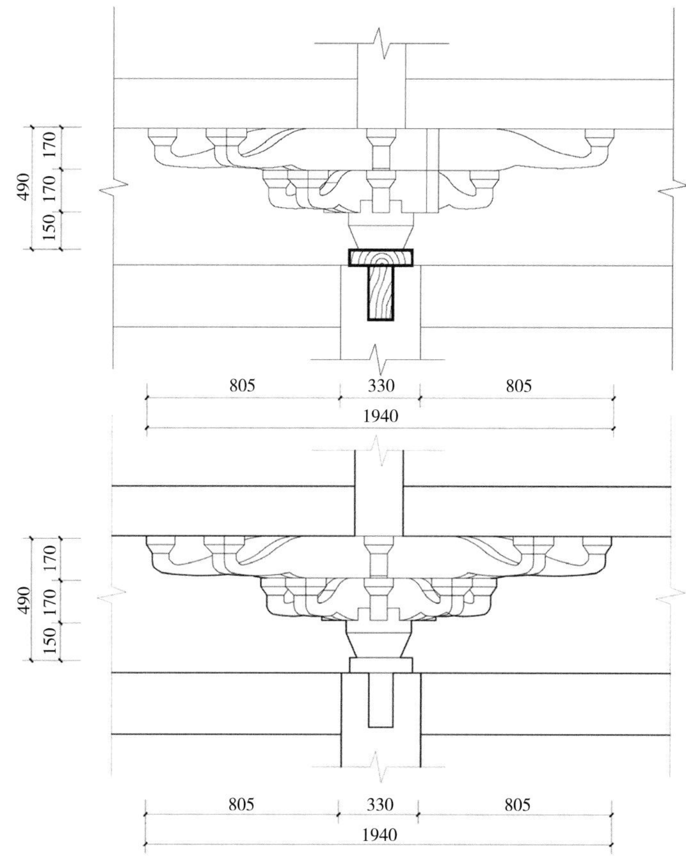
490
170
170
150
805
330
805
1940

柱头铺作仰视立面图

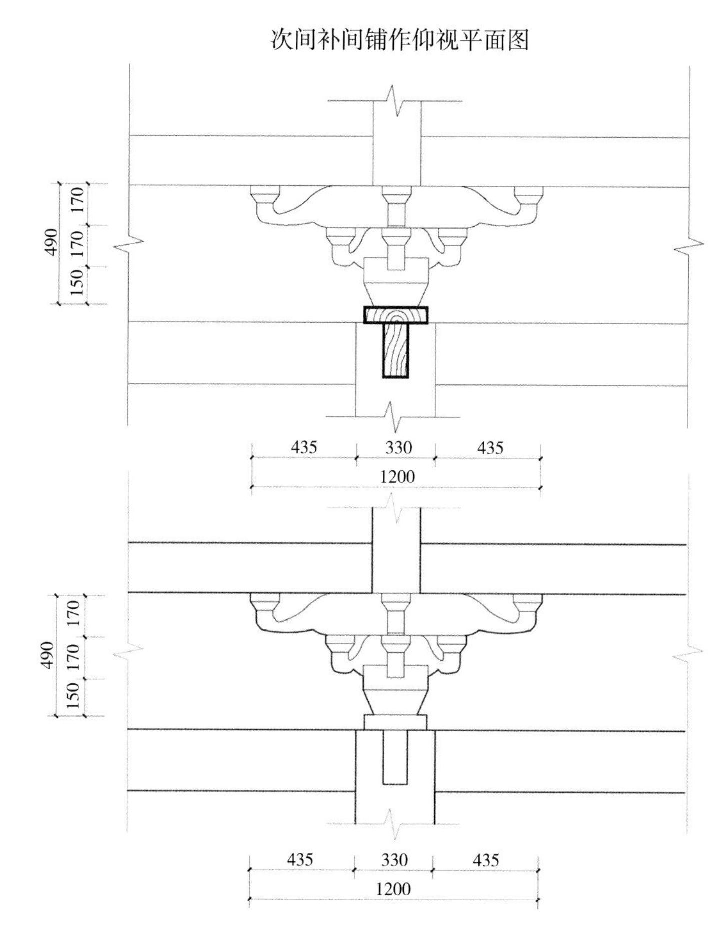
490
170
170
150
435
330
435
1200

次间补间铺作仰视立面图

大足圣寿寺维摩殿斗栱大样图

Detail, Dougong Sets , Vimalakirti Hall, Shengshou Temple, Dazu

0
10
20
30cm

大足圣寿寺大雄宝殿花牙子式样图
Detail, Huayazi, Mahavira Hall, Shengshou Temple, Dazu

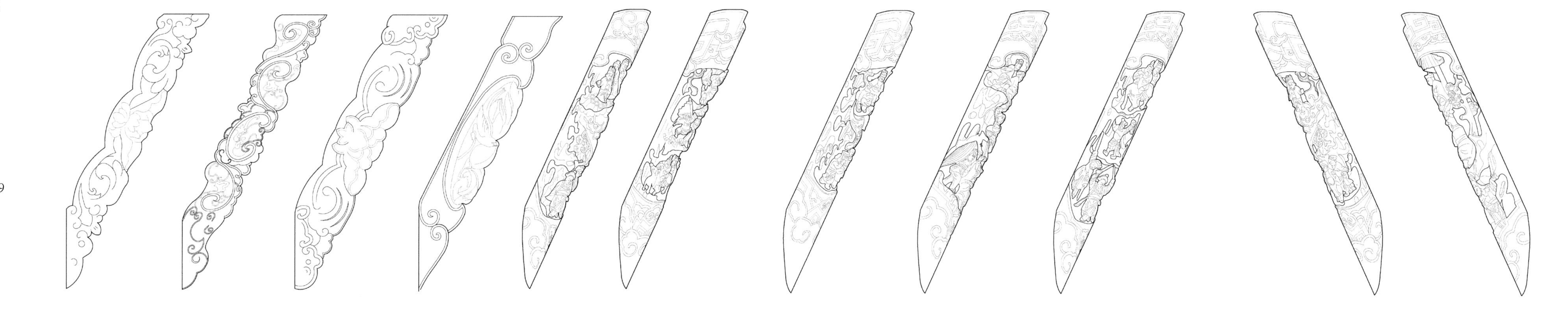

大足圣寿寺牧牛亭撑栱式样图
Detail, Supporting Gong, Muniu Pavilion, Shengshou Temple, Dazu

大足圣寿寺大雄宝殿撑栱式样图
Detail, Supporting Gong, Mahavira Hall, Shengshou Temple, Dazu

0 10 20 30m

大足圣寿寺万岁楼花牙子与梁枋式样图
Detail, Huayazi and Beams, Wansui Building, Shengshou Temple, Dazu

大足圣寿寺牧牛亭脊饰式样图
Detail, Ridge Decoration, Muniu Pavilion, Shengshou Temple, Dazu

梁平双桂堂

双桂堂位于重庆市梁平（古名梁山）区金带镇，是著名高僧破山海明禅师创建于清顺治十年（1653年）的禅宗寺庙。双桂堂的布局延续了明代以来典型的禅宗寺院空间格局，中轴线上是以祭拜为主的佛寺殿堂，依次有关圣殿、弥勒殿、大雄宝殿、文殊殿、破山塔、大悲殿、藏经楼七重殿堂；轴线左右两侧是有禅宗特色的禅堂、祖师殿、法堂、五观堂、客堂、伽蓝堂等，延续了禅宗寺院东西翼的空间布局特色。轴线上的殿堂建筑具有强烈的空间秩序，又有独自的建筑个性特色。大雄宝殿为三重檐庑殿顶，文殊殿为重檐楼阁但翼角和重檐处理有北方官式建筑的影响因素；藏经楼由三合院构成，重檐歇山丰富多姿，翼角高翘，巴蜀地域风格浓厚。因破山海明创建的双桂禅系在西南地区乃至在全国都有广泛的影响，因此双桂堂有『西南禅宗祖庭』之称。寺院内匾额楹联题刻丰富，并有雍正和乾隆年间奉旨钦定入藏的一千余册经书，这是双桂堂内涵丰富的历史文化资源。

Shuanggui Temple at Liangping

Located in Jindai Town of Liangping County (historically known as Liangshan) in Chongqing, Shuanggui is a Zen temple founded by the famous monk master Poshan Haiming in the 10th year of Shunzhi's reign in Qing Dynasty (AD1653). The layout of Shuanggui Temple is a continuation of the typical Zen temple spatial pattern from Ming Dynasty, where on the central axis are generally Buddhist structures for worship. The seven buildings are, successively, the Guansheng Hall, the Maitreya Hall, the Great Buddha Hall, the Manjushri Hall, the Poshan Pagoda, the Dabei Hall and the Scripture Library. On either side of the axis are buildings unique to Zen Buddhism–the meditation hall, ancestral hall, dhammasala, refectory, guesthouse, Qielan Hall and so on–which have maintained the layout features of Zen temples with east and west wings. Buildings on the axis reflect a strong spatial order, yet without losing their own architectural distinctiveness. The three-story Great Buddha Hall has a four-ridge roof. The Manjushri Hall is double eaved, while the treatment of upturned roof-ridge and double eave show influences of northern official style buildings. The Scripture Library stands in a courtyard enclosed on three sides, with vivid looking double eave and gabled roof and high upturned roof-ridge, illustrating strong Ba-Shu regional characteristics. Thanks to the impact of the Shuanggui Zen system created by Poshan Haiming in southwestern China and even the whole country, Shuanggui Temple is celebrated as the "ancestral Zen temple in the southwest". There are abundant plaques, couplets, and inscriptions in the temple, as well as over a thousand books of scripture collected here by imperial order during Yongzheng's and Qianlong's reign, which make up a significant historical and cultural resource of Shuanggui Temple.

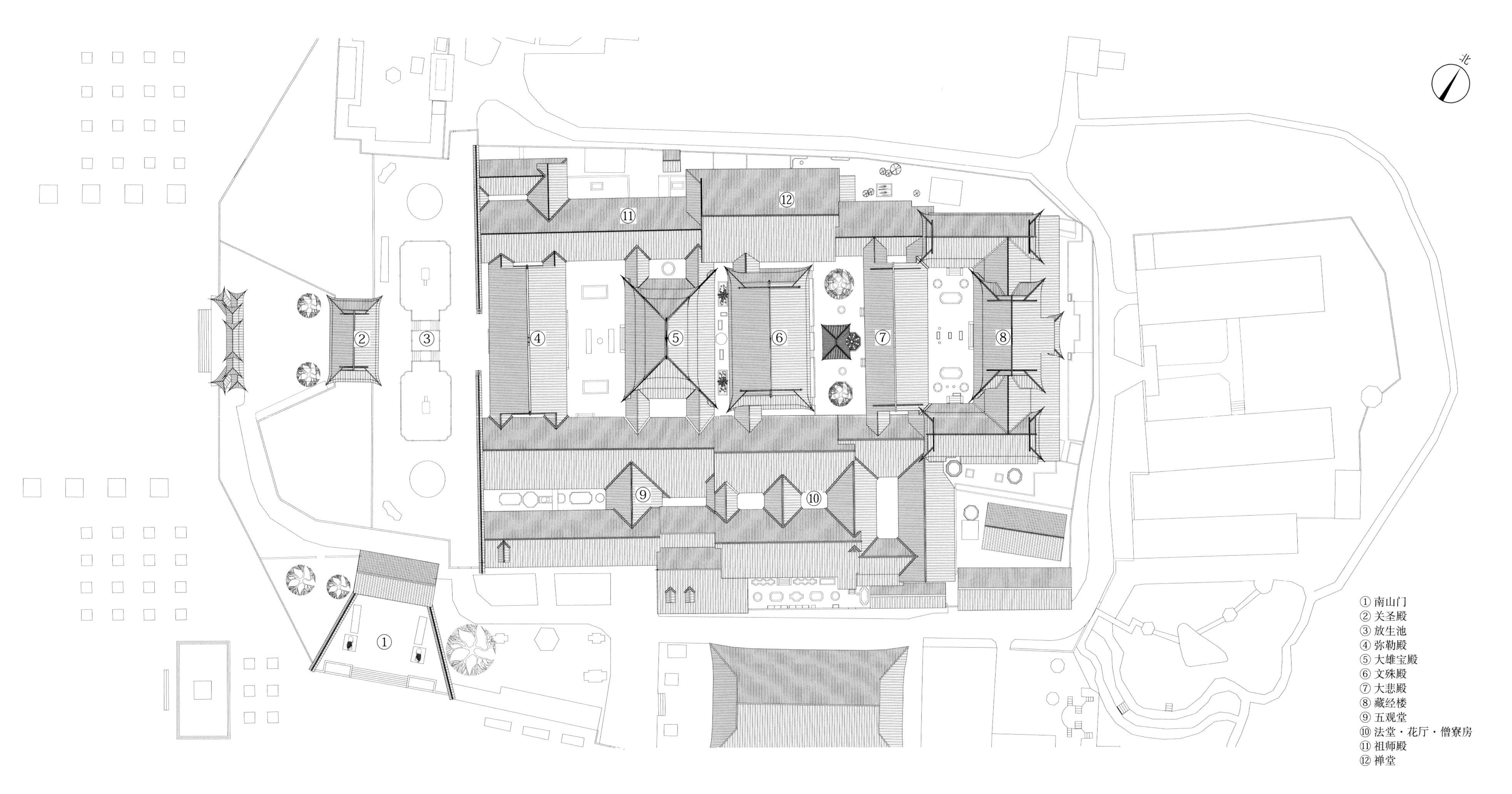
北
①
②
③
④
⑤
⑥
⑦
⑧
⑨
⑩
⑪
⑫

① 南山门
② 关圣殿
③ 放生池
④ 弥勒殿
⑤ 大雄宝殿
⑥ 文殊殿
⑦ 大悲殿
⑧ 藏经楼
⑨ 五观堂
⑩ 法堂·花厅·僧寮房
⑪ 祖师殿
⑫ 禅堂

梁平双桂堂总平面图
Site Plan of Shuanggui Temple, Liangping

0 5 10 15 20 25m

图一　梁平双桂堂建筑群鸟瞰图

图二　梁平双桂堂大雄宝殿

图三　梁平双桂堂大雄宝殿

Fig.1 Bird’s Eye View of Shuanggui Temple, Liangping
Fig.2 Mahavira Hall, Shuanggui Temple, Liangping
Fig.3 Mahavira Hall, Shuanggui Temple, Liangping

图四　梁平双桂堂南山门

图五　梁平双桂堂关圣殿正立面

图六　梁平双桂堂关圣殿背立面

Fig.4 South Main Gate (Shan Men), Shuanggui Temple, Liangping
Fig.5 Front Elevation of Guansheng Hall, Shuanggui Temple, Liangping
Fig.6 Back Elevation of Guansheng Hall, Shuanggui Temple, Liangping

图七 梁平双桂堂放生池庭院

图八 梁平双桂堂弥勒殿

图九 梁平双桂堂大悲殿

Fig.7 Courtyard of Free Life Pond, Shuanggui Temple, Liangping
Fig.8 Maitreya Hall, Shuanggui Temple, Liangping
Fig.9 Dabei Hall, Shuanggui Temple, Liangping

图十一　梁平双桂堂藏经楼

图十　梁平双桂堂文殊殿

图十二　梁平双桂堂藏经楼

Fig.10 Manjushri Hall, Shuanggui Temple, Liangping
Fig.11 Sutrua Library, Shuanggui Temple, Liangping
Fig.12 Sutrua Library, Shuanggui Temple, Liangping

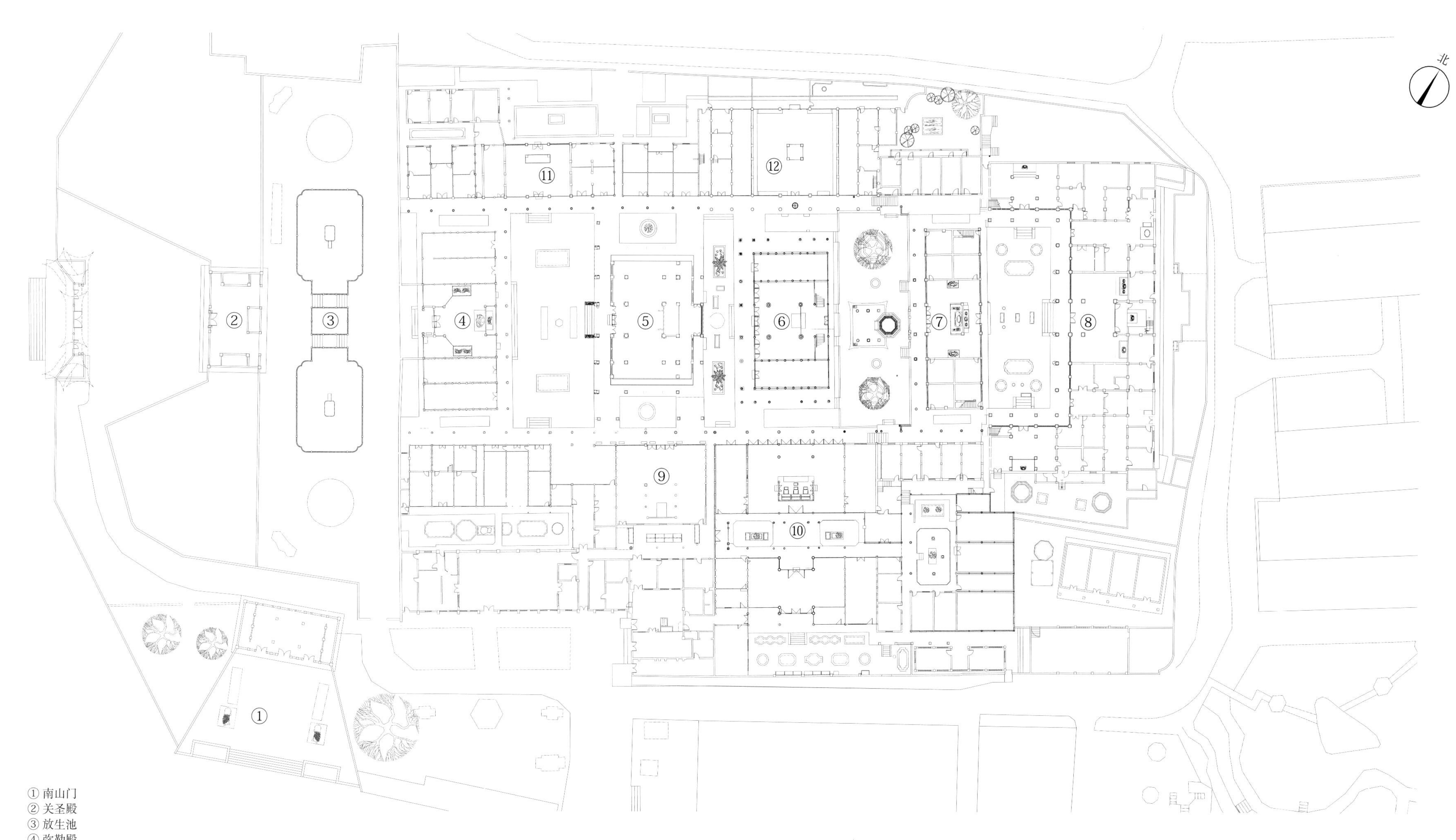
北
①
②
③
④
⑤
⑥
⑦
⑧
⑨
⑩
⑪
⑫
0 5 10 15 20 25m

① 南山门
② 关圣殿
③ 放生池
④ 弥勒殿
⑤ 大雄宝殿
⑥ 文殊殿
⑦ 大悲殿
⑧ 藏经楼
⑨ 五观堂
⑩ 法堂・花厅・僧寮房
⑪ 祖师殿
⑫ 禅堂

梁平双桂堂建筑群平面图

Ground Floor Plan of Entire Compound, Shuanggui Temple, Liangping

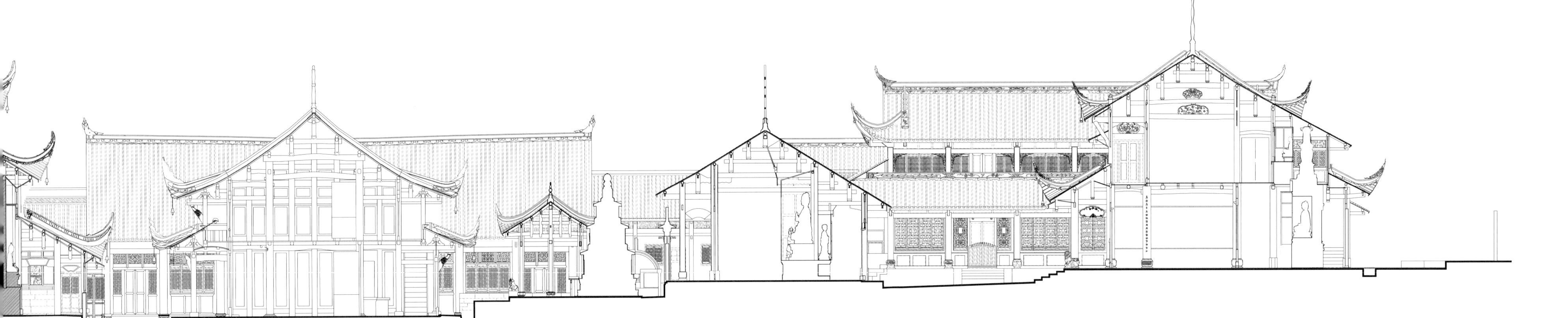
文殊殿
破山塔
大悲殿
藏经楼
0
5
10m

梁平双桂堂建筑群剖面图
Section of Entire Compound, Shuanggui Temple, Liangping

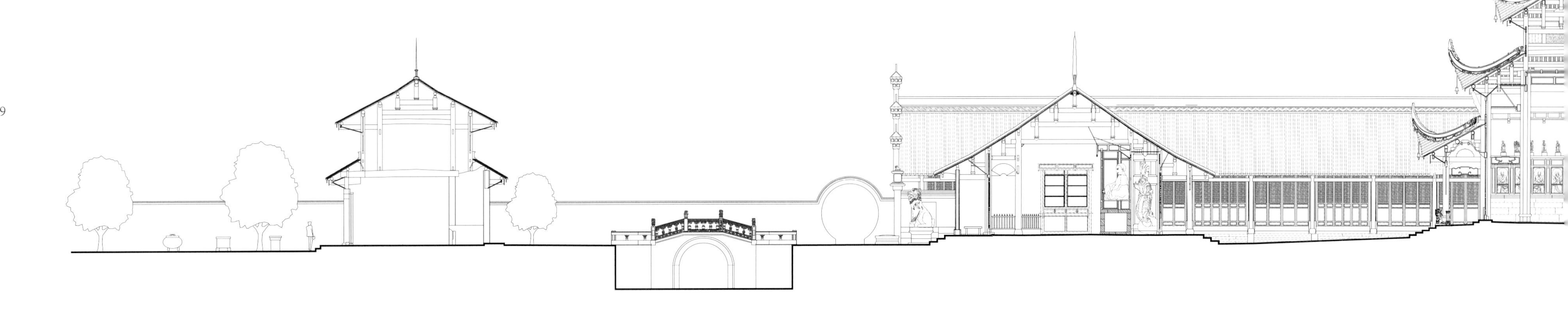
关圣殿
放生池
弥勒殿
大雄宝殿

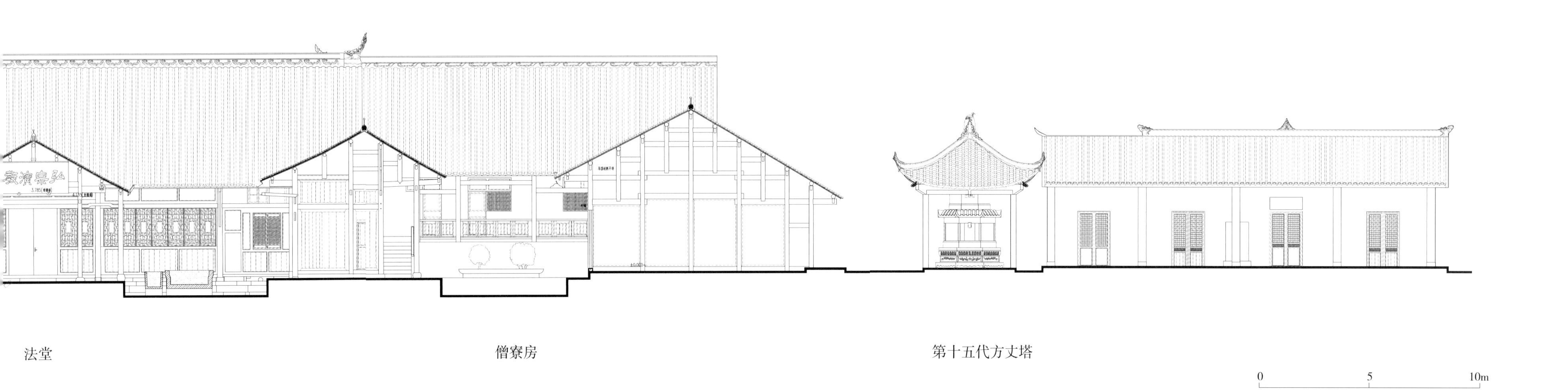
法堂
僧寮房
第十五代方丈塔
0
5
10m

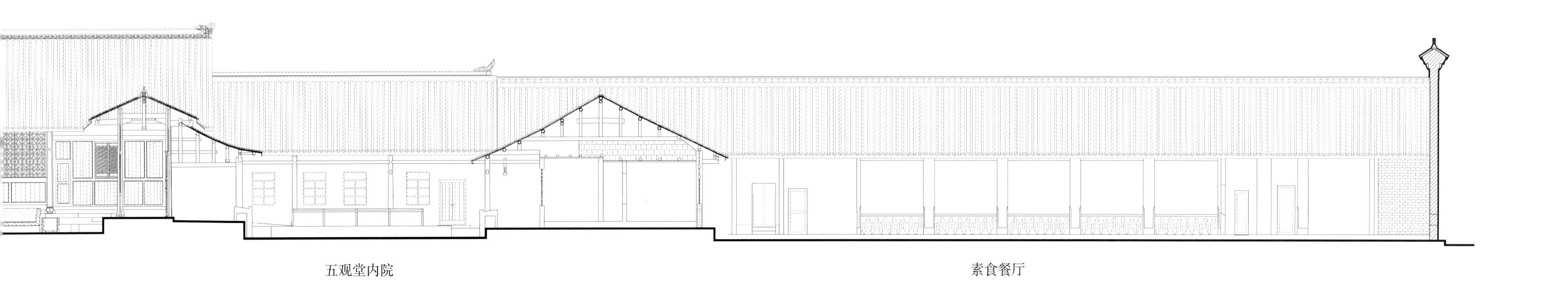
五观堂内院
素食餐厅
0
5
10m

梁平双桂堂建筑群厢房剖面图
Section of Wings, Shuanggui Temple, Liangping

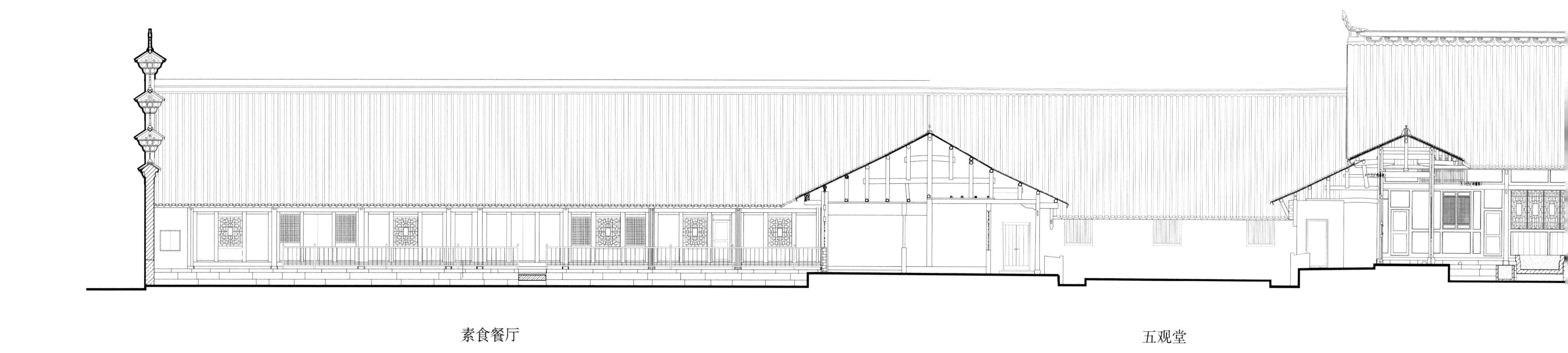
素食餐厅
五观堂

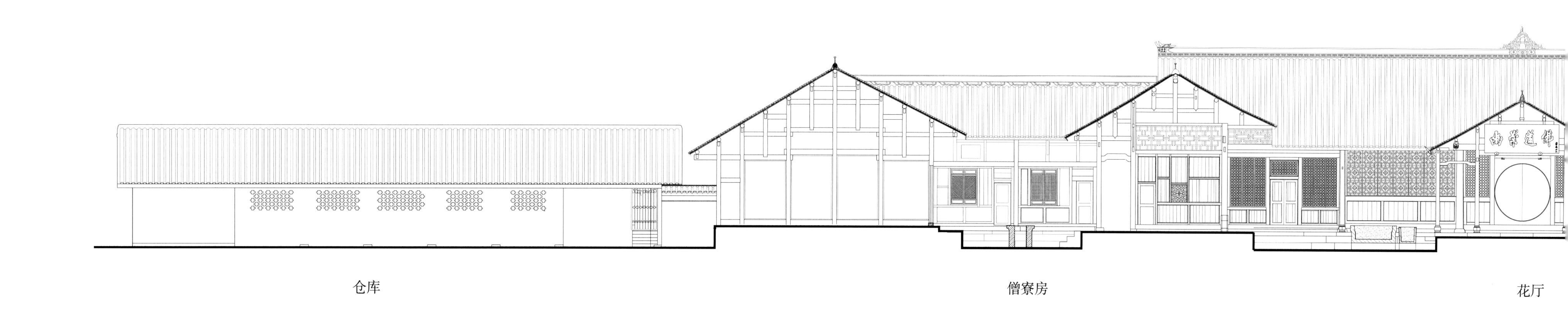
仓库
僧寮房
花厅

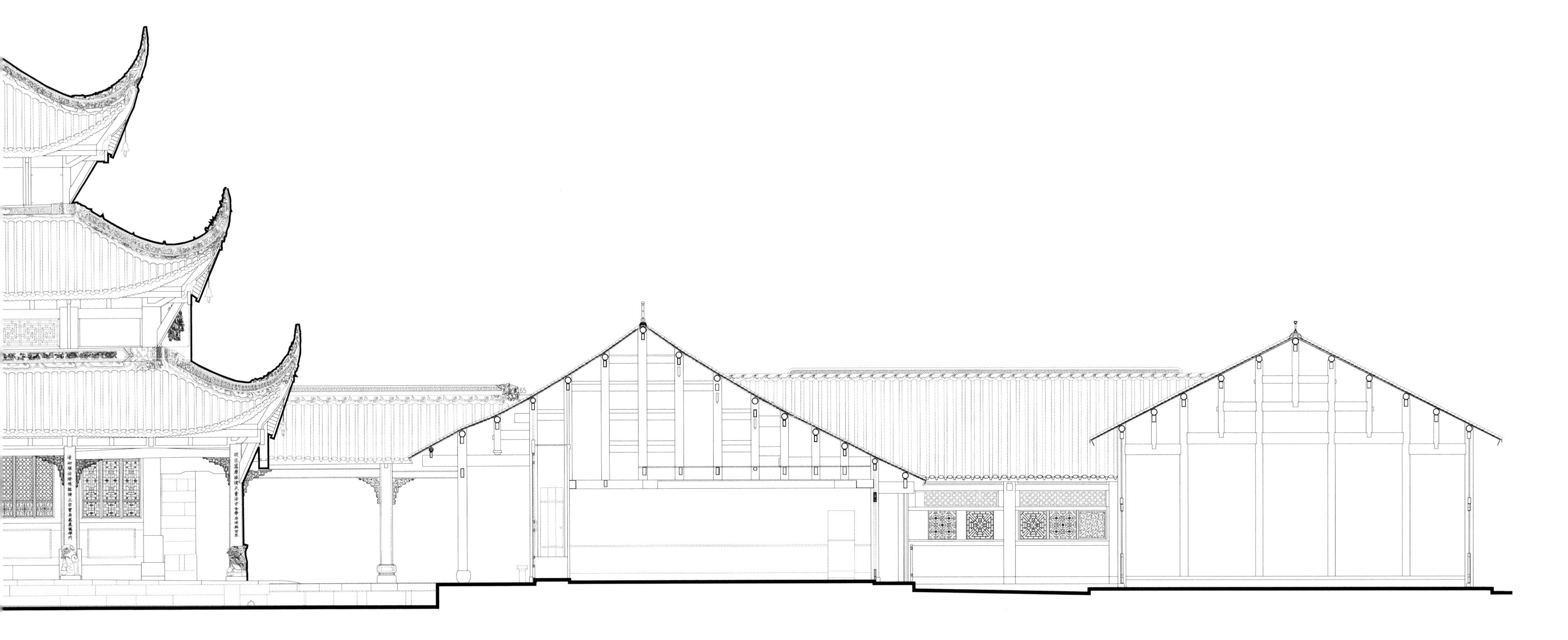

五观堂

北僧寮房

梁平双桂堂第一进院落横剖面图

Cross Section of the First Courtyard, Shuanggui Temple, Liangping

0 5 10m

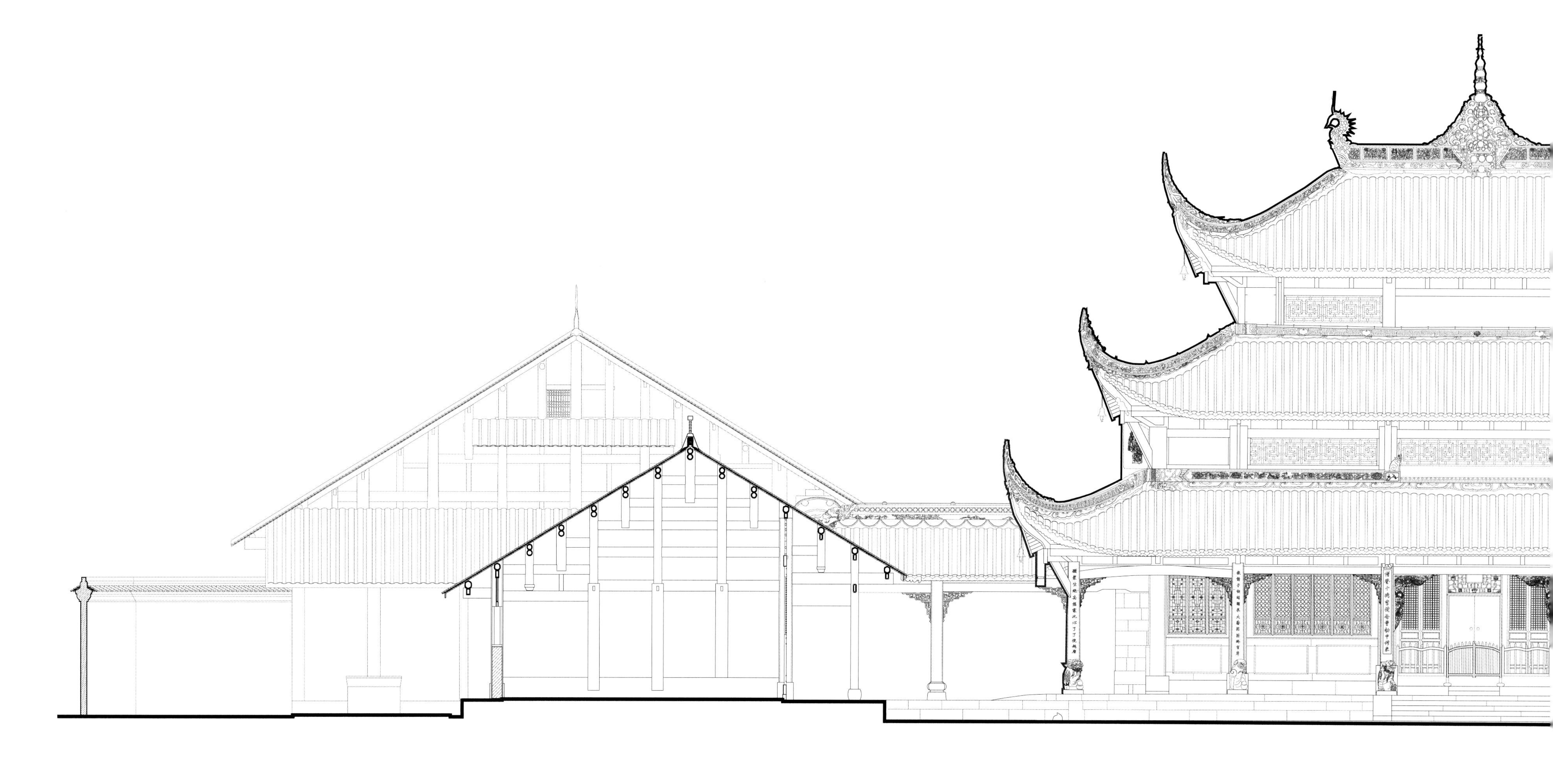

祖师殿　　　　大雄宝殿

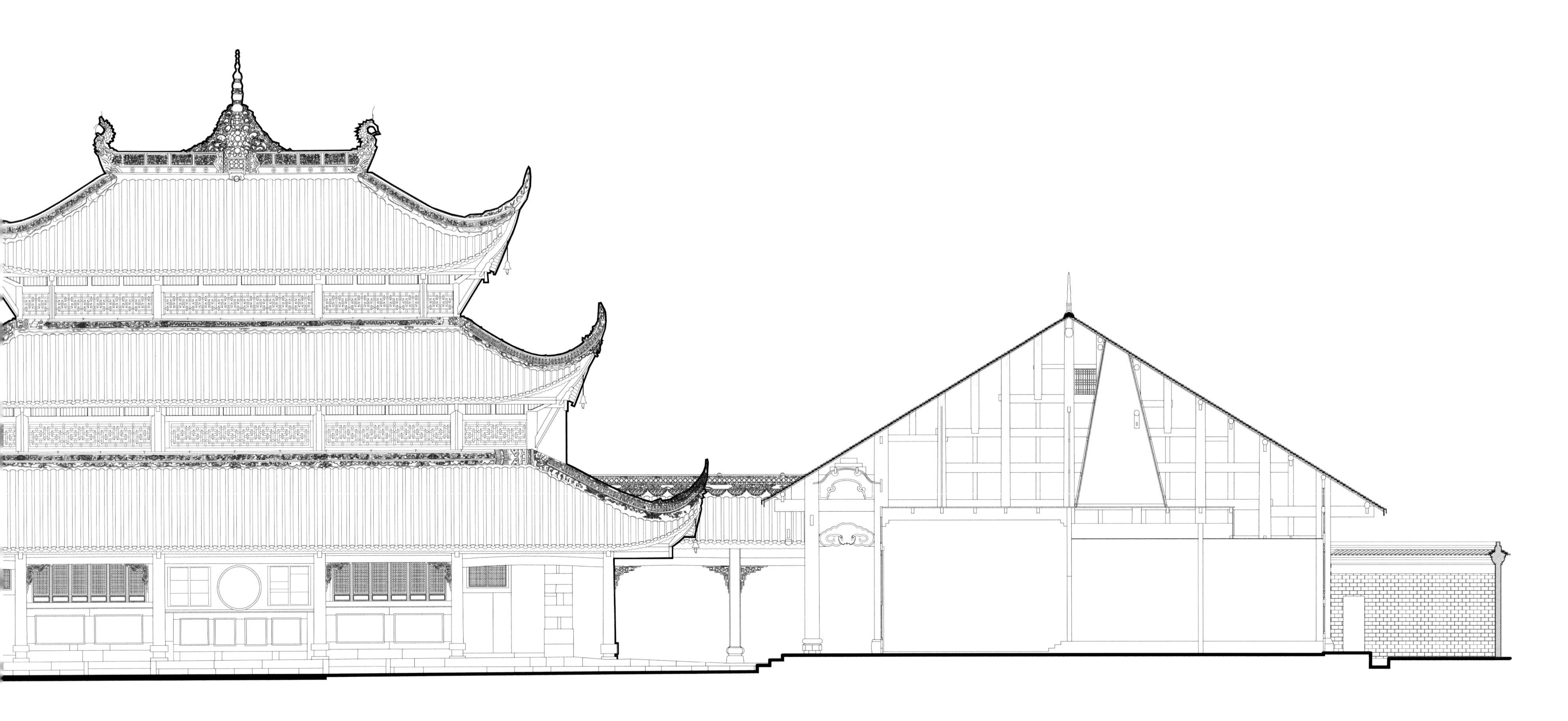
大雄宝殿
禅堂
0
5
10m

梁平双桂堂第二进院落横剖面图
Cross Section of the Second Courtyard, Shuanggui Temple, Liangping

200×160

花厅　　法堂

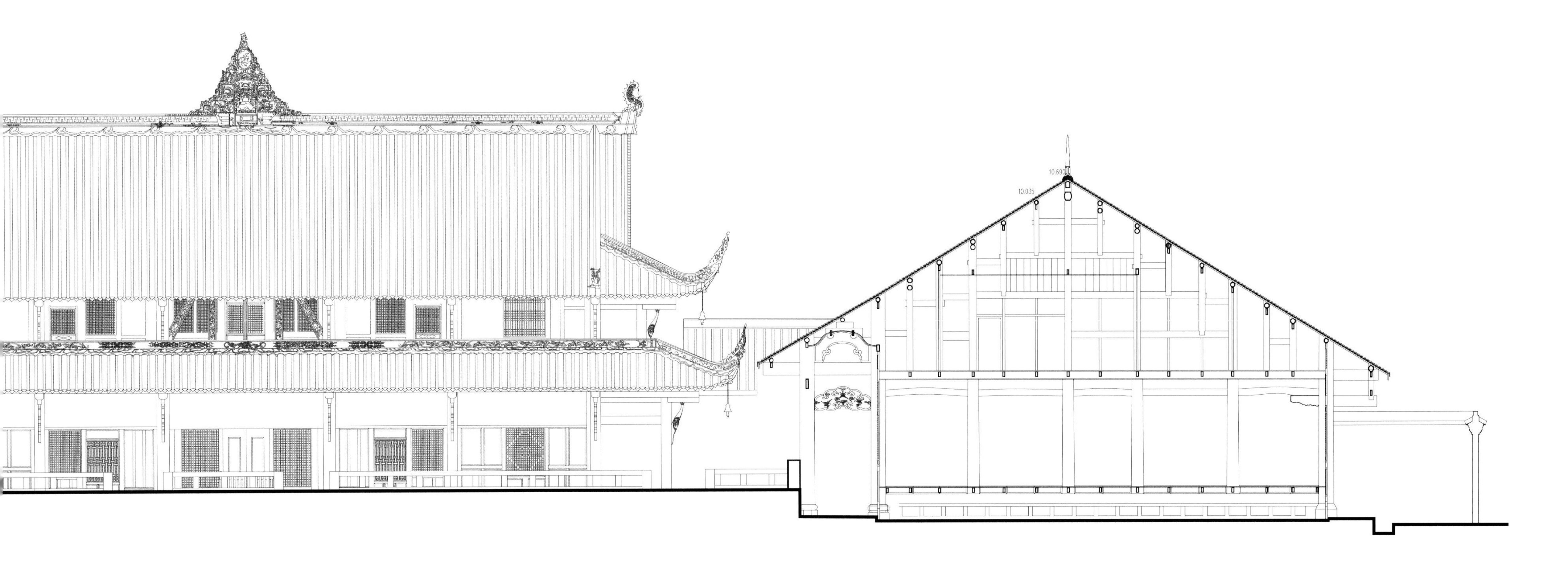
10.690
10.035
文殊殿
禅堂
0
5
10m

梁平双桂堂第三进院落横剖面图
Cross Section of the Third Courtyard, Shuanggui Temple, Liangping

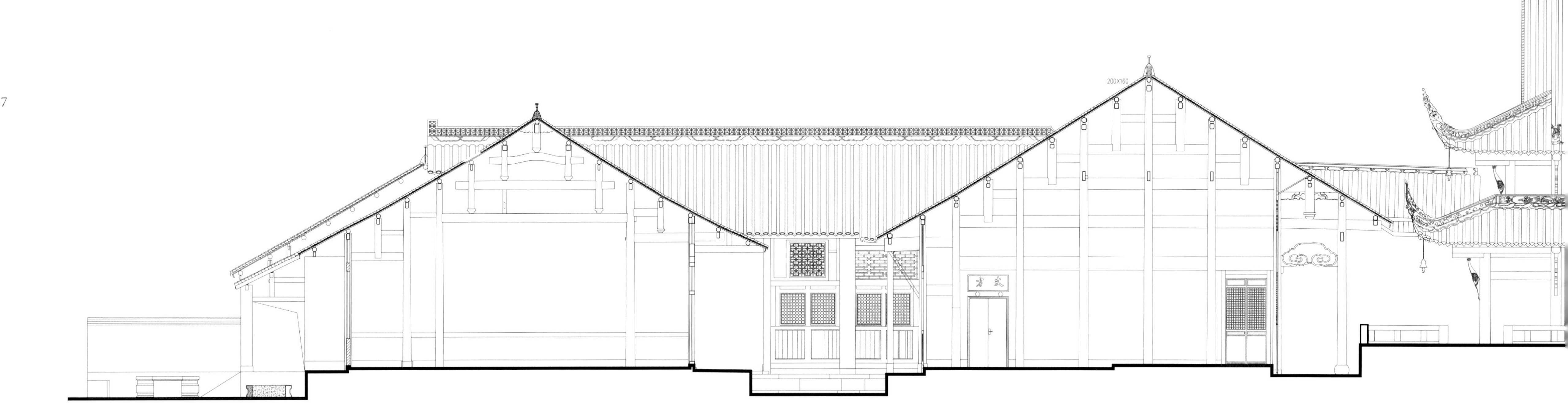
200×160

花厅

法堂

大悲殿

藏经楼北侧殿

0 5 10m

梁平双桂堂第四进院落横剖面图
Cross Section of the Fourth Courtyard, Shuanggui Temple, Liangping

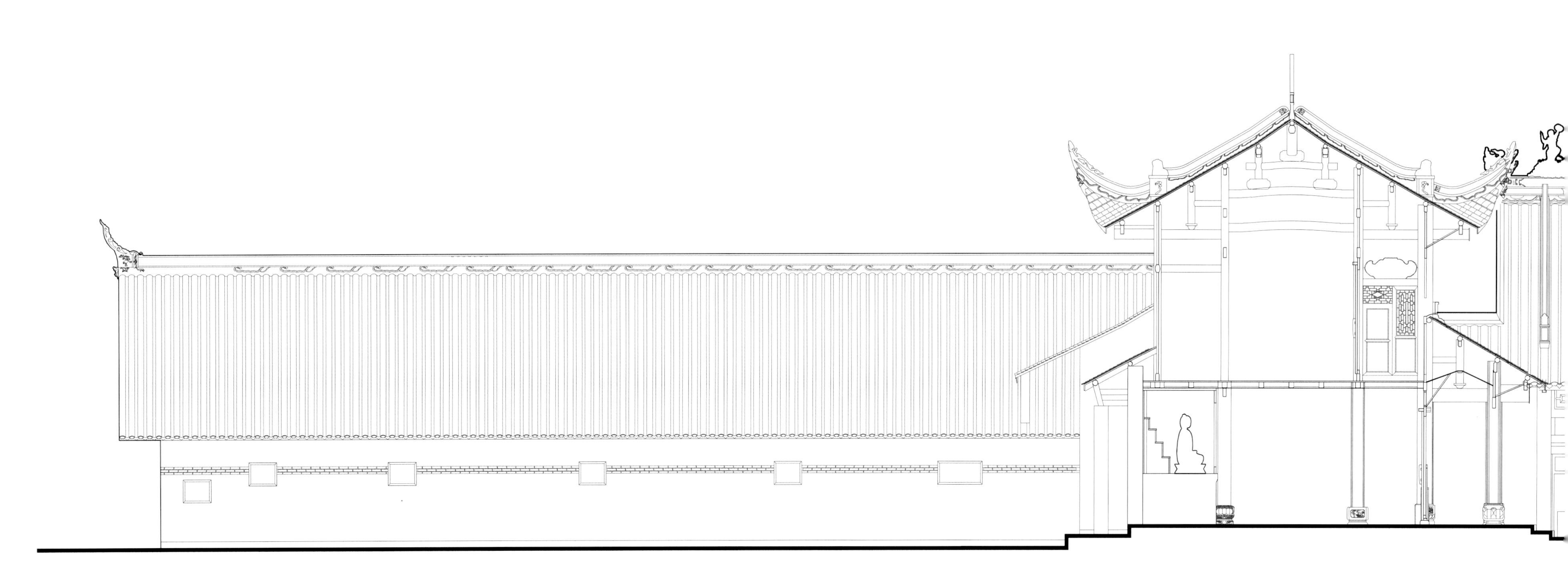

藏经楼南侧殿

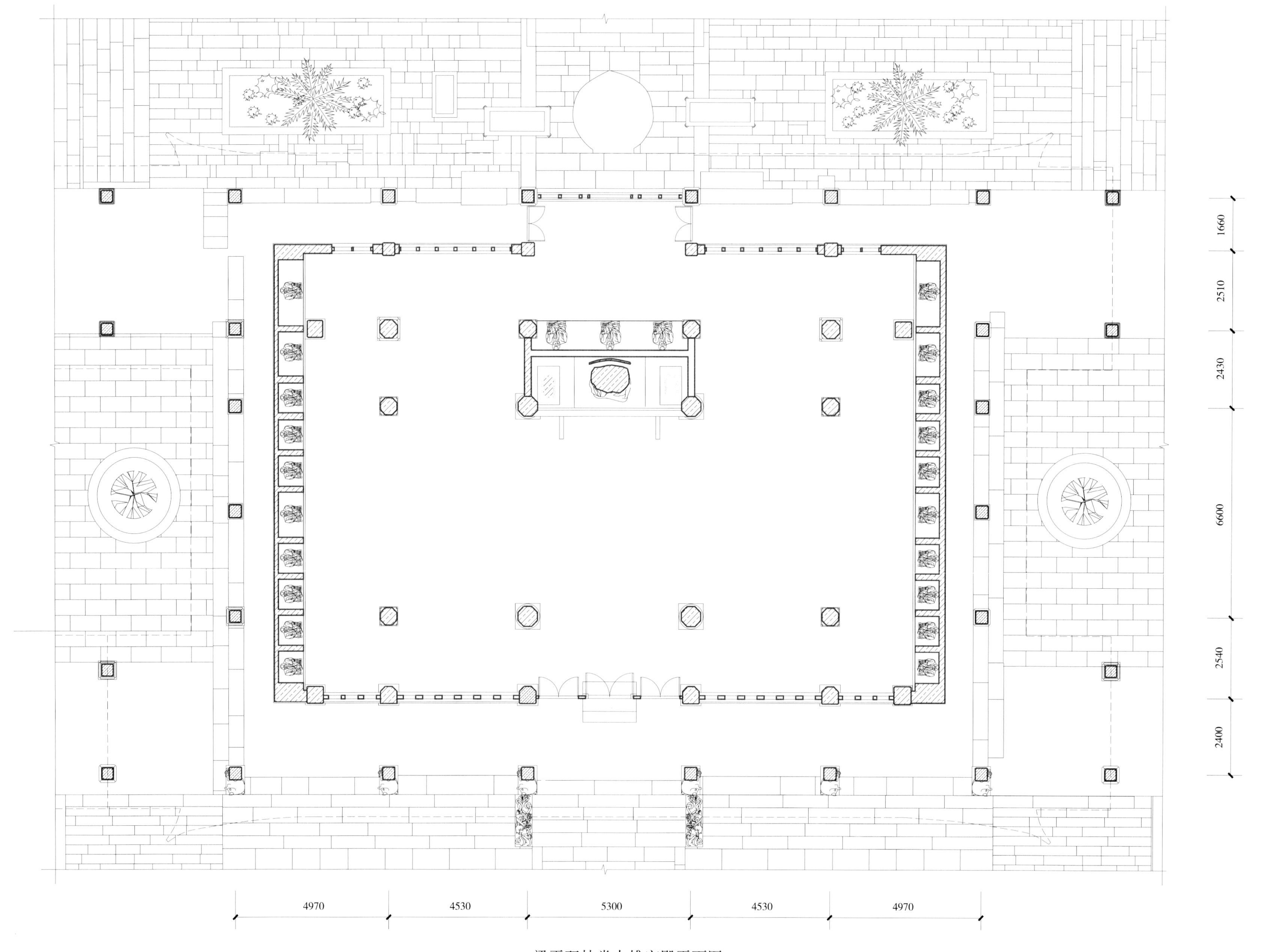
1660
2510
2430
6600
2540
2400
4970
4530
5300
4530
4970
0 1 2 3 4 5m

梁平双桂堂大雄宝殿平面图
Ground Floor Plan of Mahavira Hall, Shuanggui Temple, Liangping

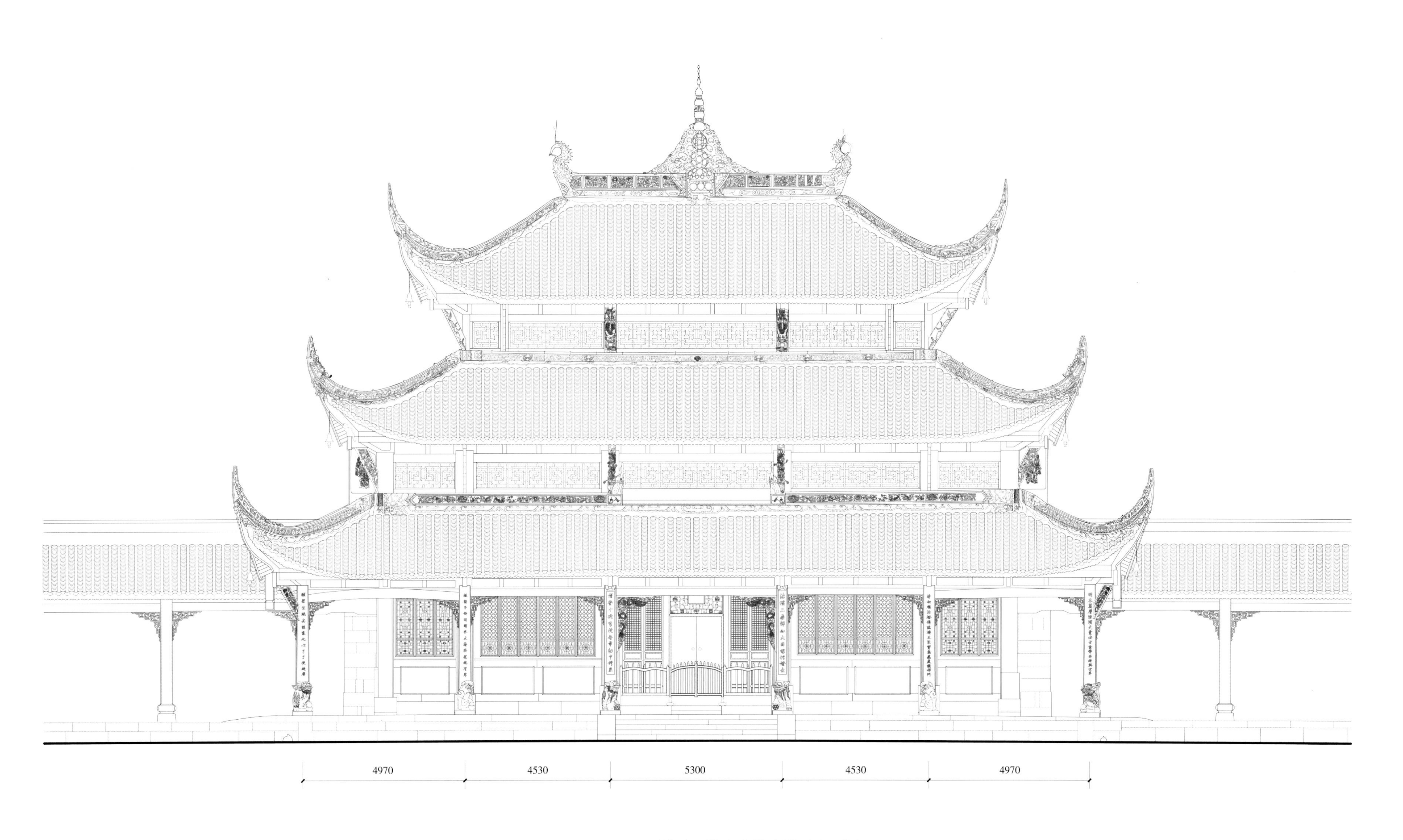
2920
1960
2300
2000
1920
4310
750
4970
4530
5300
4530
4970
0 1 2 3 4 5m

梁平双桂堂大雄宝殿正立面图
Front Elevation of Mahavira Hall, Shuanggui Temple, Liangping

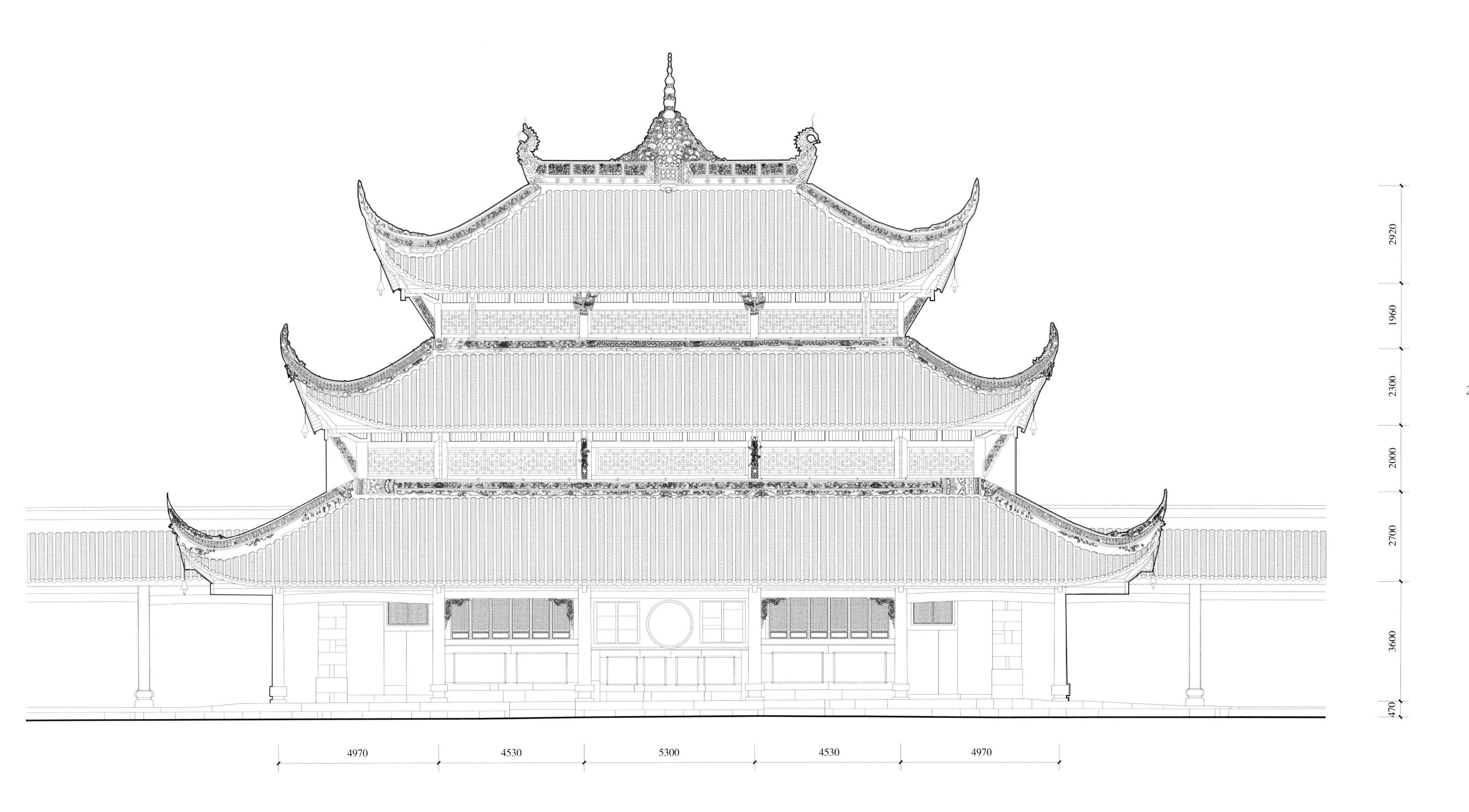
2920
1960
2300
2000
2700
3600
470
4970
4530
5300
4530
4970

梁平双桂堂大雄宝殿背立面图
Back Elevation of Mahavira Hall, Shuanggui Temple, Liangping

0 1 2 3 4 5m

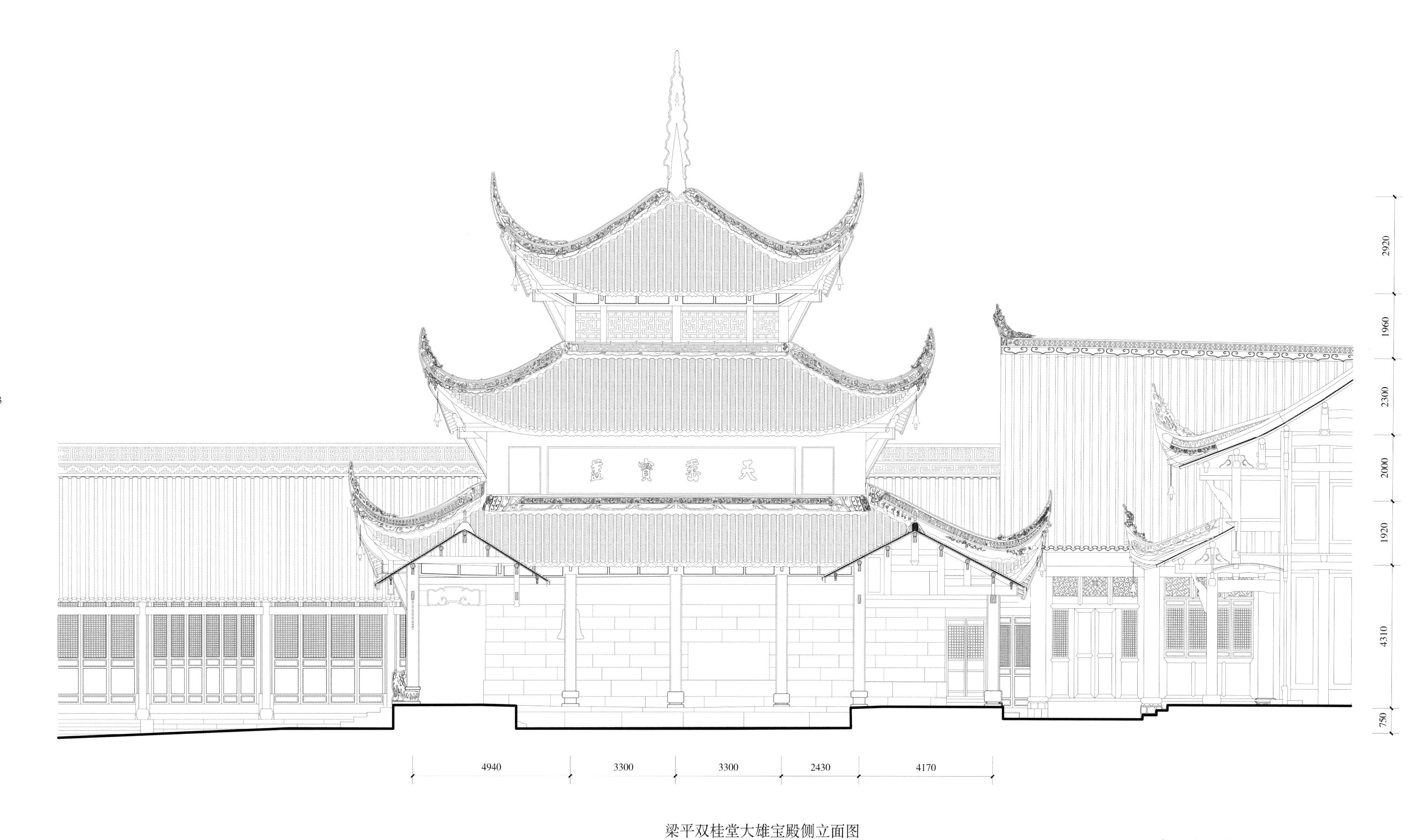
天雲寶蓋
2920
1960
2300
2000
1920
4310
750
4940
3300
3300
2430
4170
0
1
2
3
4
5m

梁平双桂堂大雄宝殿侧立面图
Side Elevation of Mahavira Hall, Shuanggui Temple, Liangping

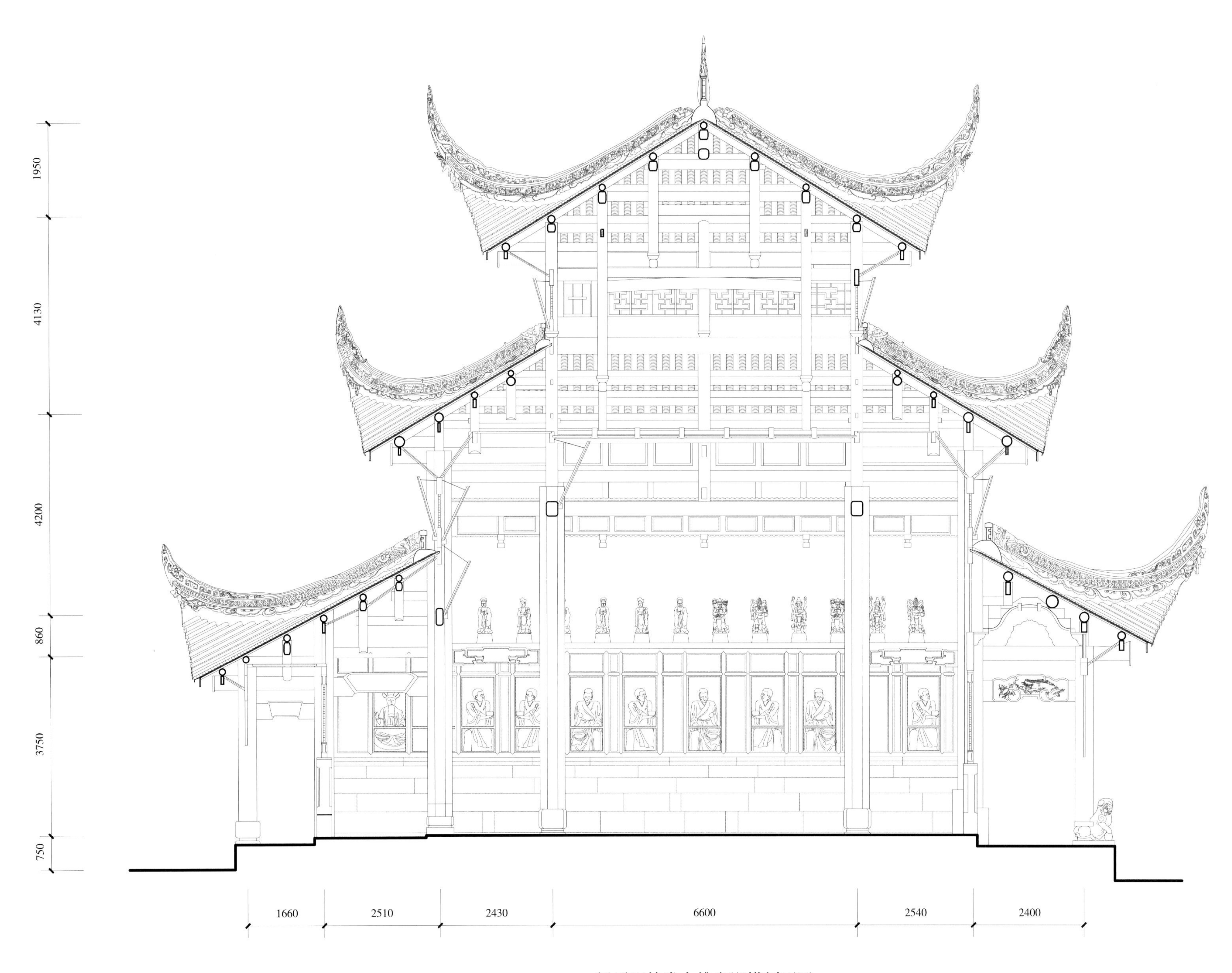
1950
4130
4200
860
3750
750
1660
2510
2430
6600
2540
2400
0
1
2
3
4
5m

梁平双桂堂大雄宝殿横剖面图

Cross Section of Mahavira Hall, Shuanggui Temple, Liangping

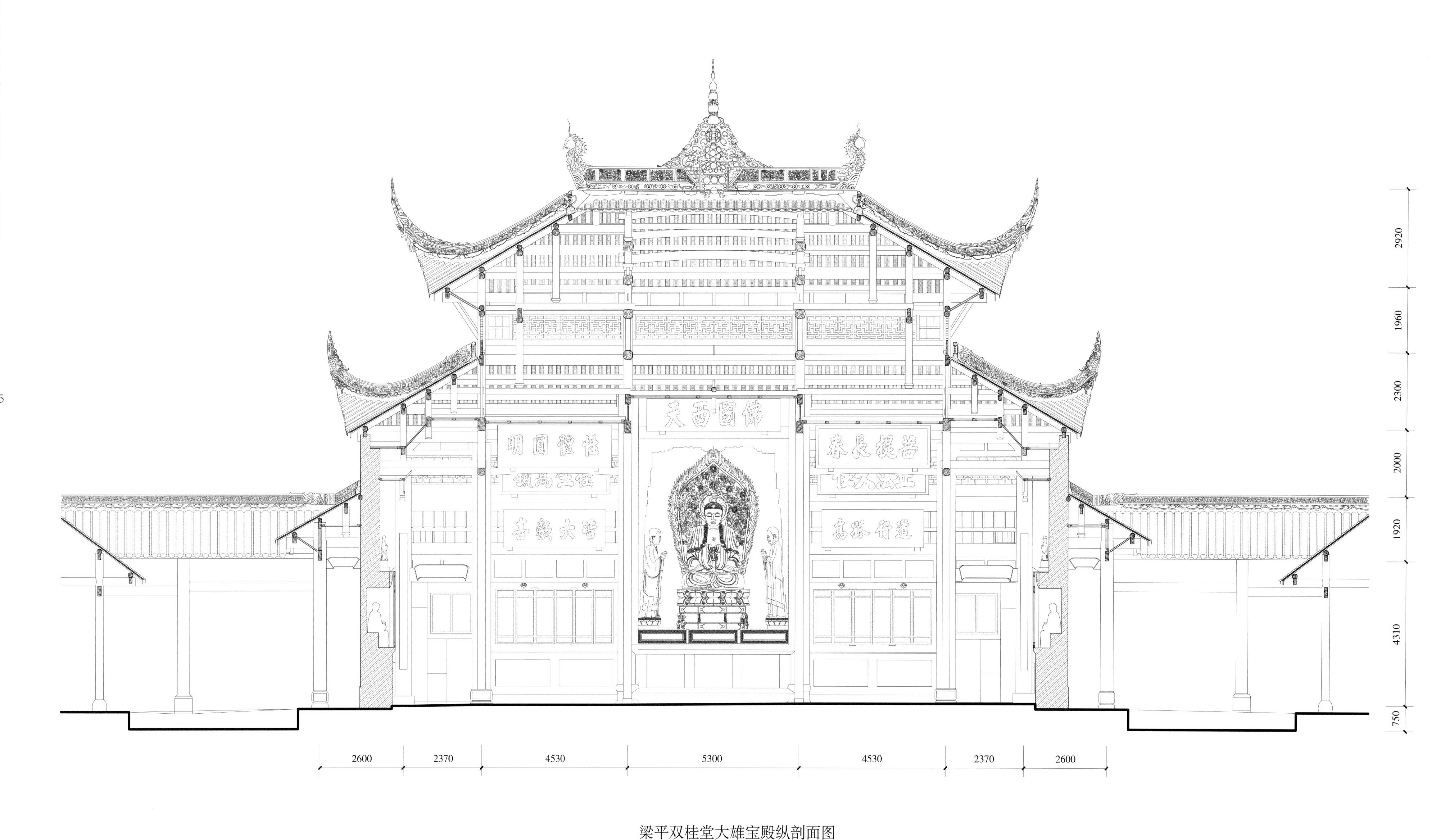
天西國佛
明圓體性
春長提菩
吉祥大吉
高深行道
2920
1960
2300
2000
1920
4310
750
2600
2370
4530
5300
4530
2370
2600
0
1
2
3
4
5m

梁平双桂堂大雄宝殿纵剖面图

Longitudinal Section of Mahavira Hall, Shuanggui Temple, Liangping

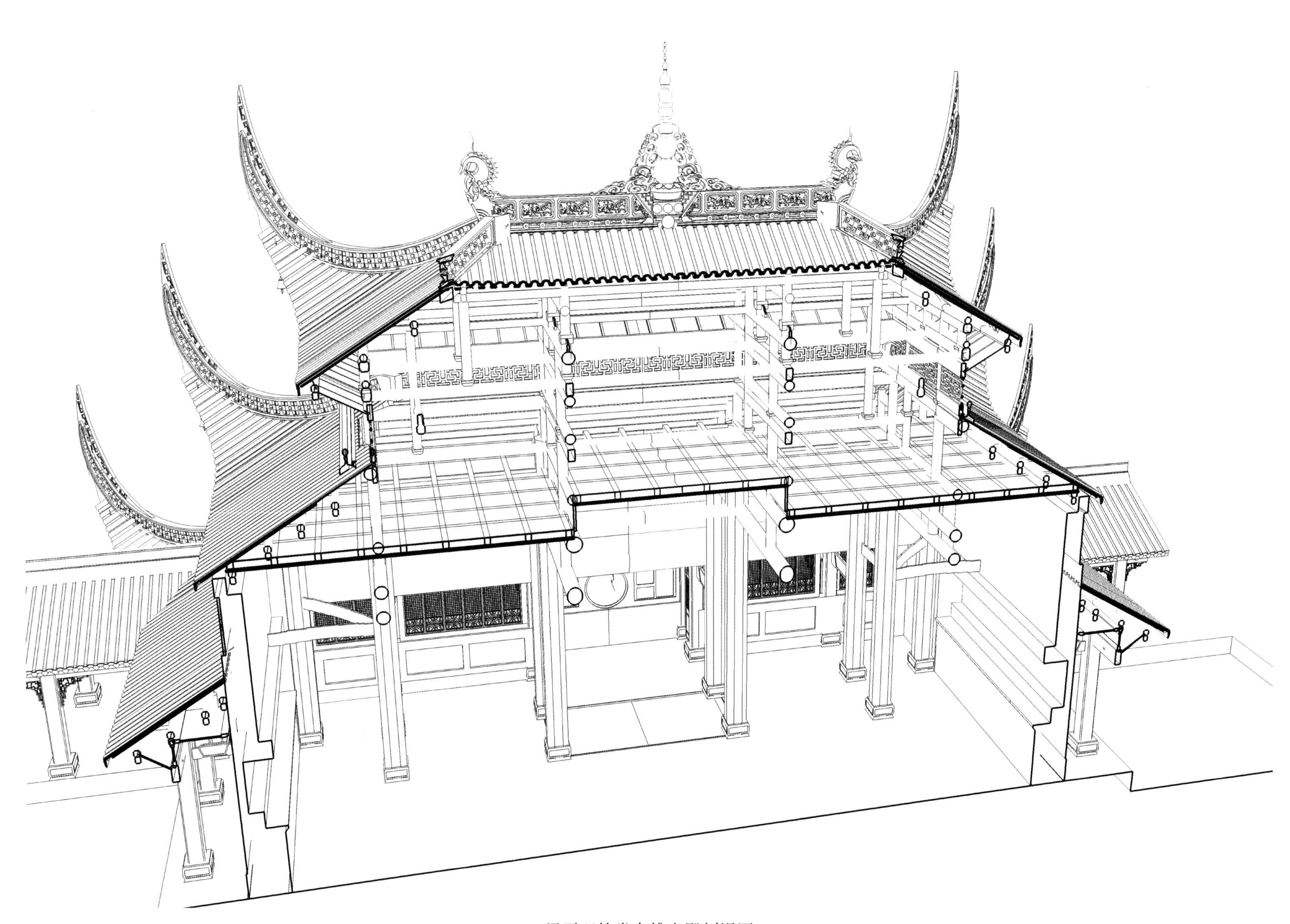

梁平双桂堂大雄宝殿剖视图
Cross Section of Mahavira Hall, Shuanggui Temple, Liangping

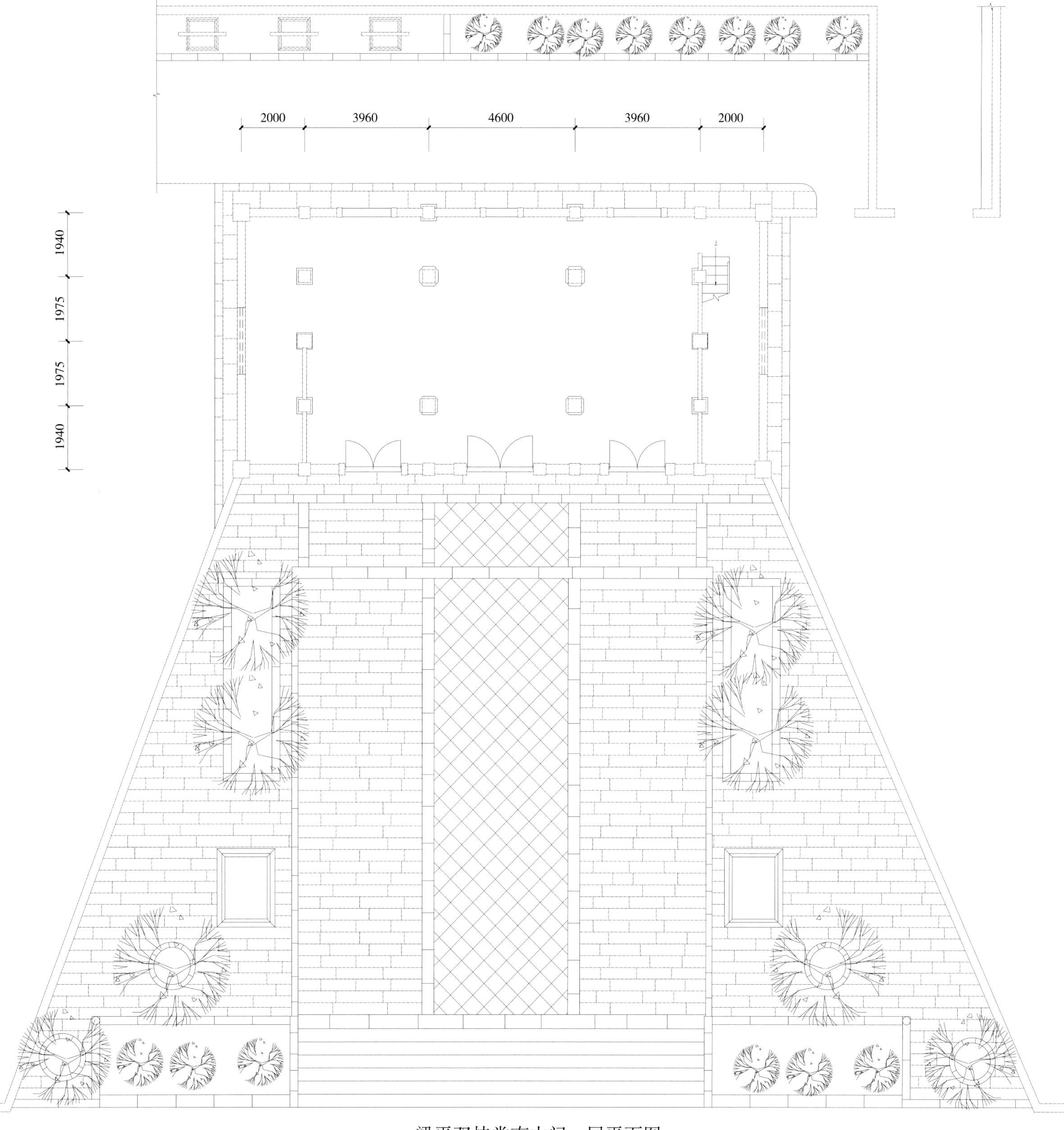
2000
3960
4600
3960
2000
1940
1975
1975
1940
0 1 2 3m

梁平双桂堂南山门一层平面图

Ground Floor Plan of South Main Gate (Shan Men), Shuanggui Temple, Liangping

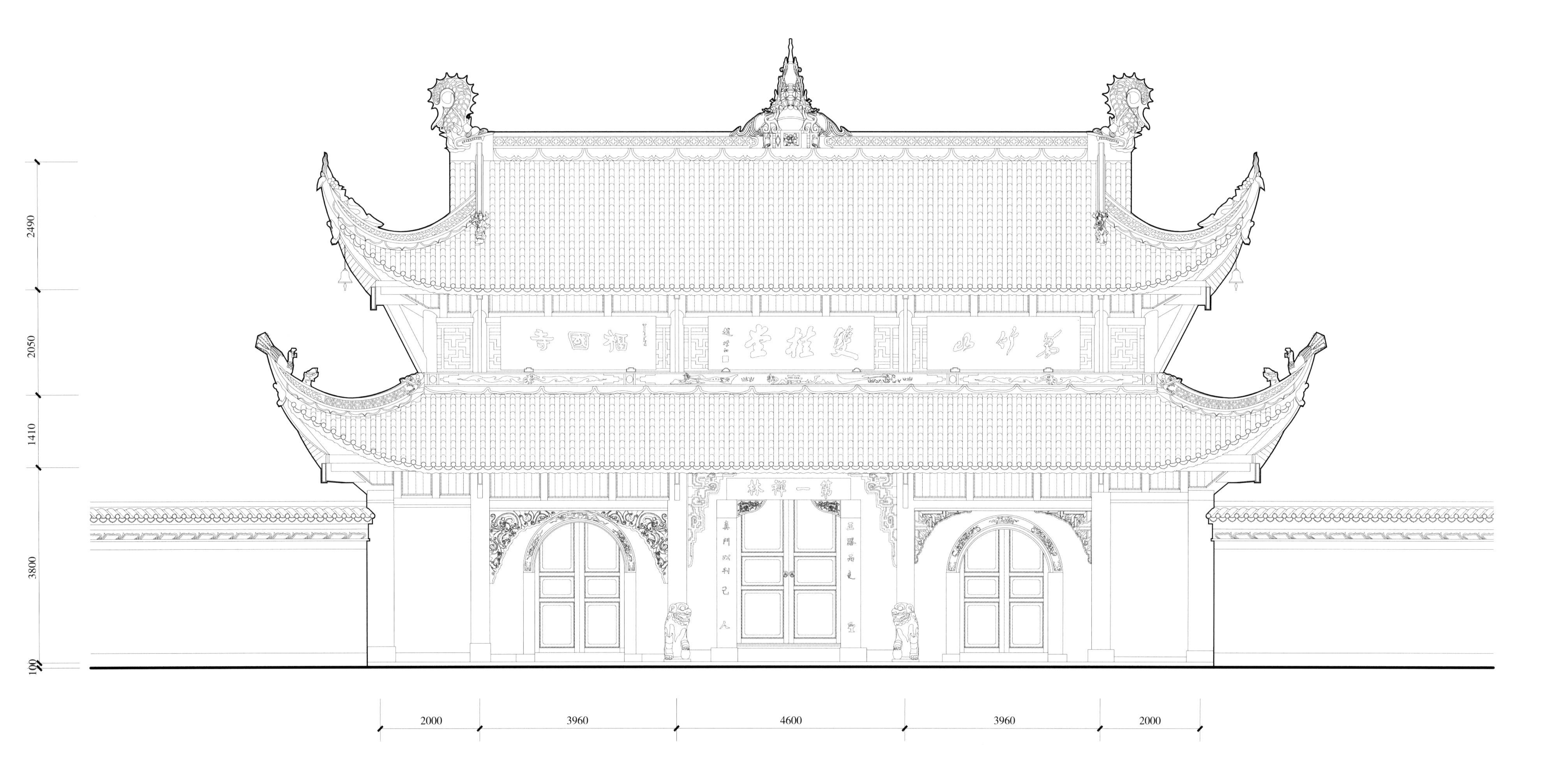
2490
2050
1410
3800
100
2000
3960
4600
3960
2000
0
1
2
3m

梁平双桂堂南山门正立面图

Front Elevation of South Main Gate (Shan Men), Shuanggui Temple, Liangping

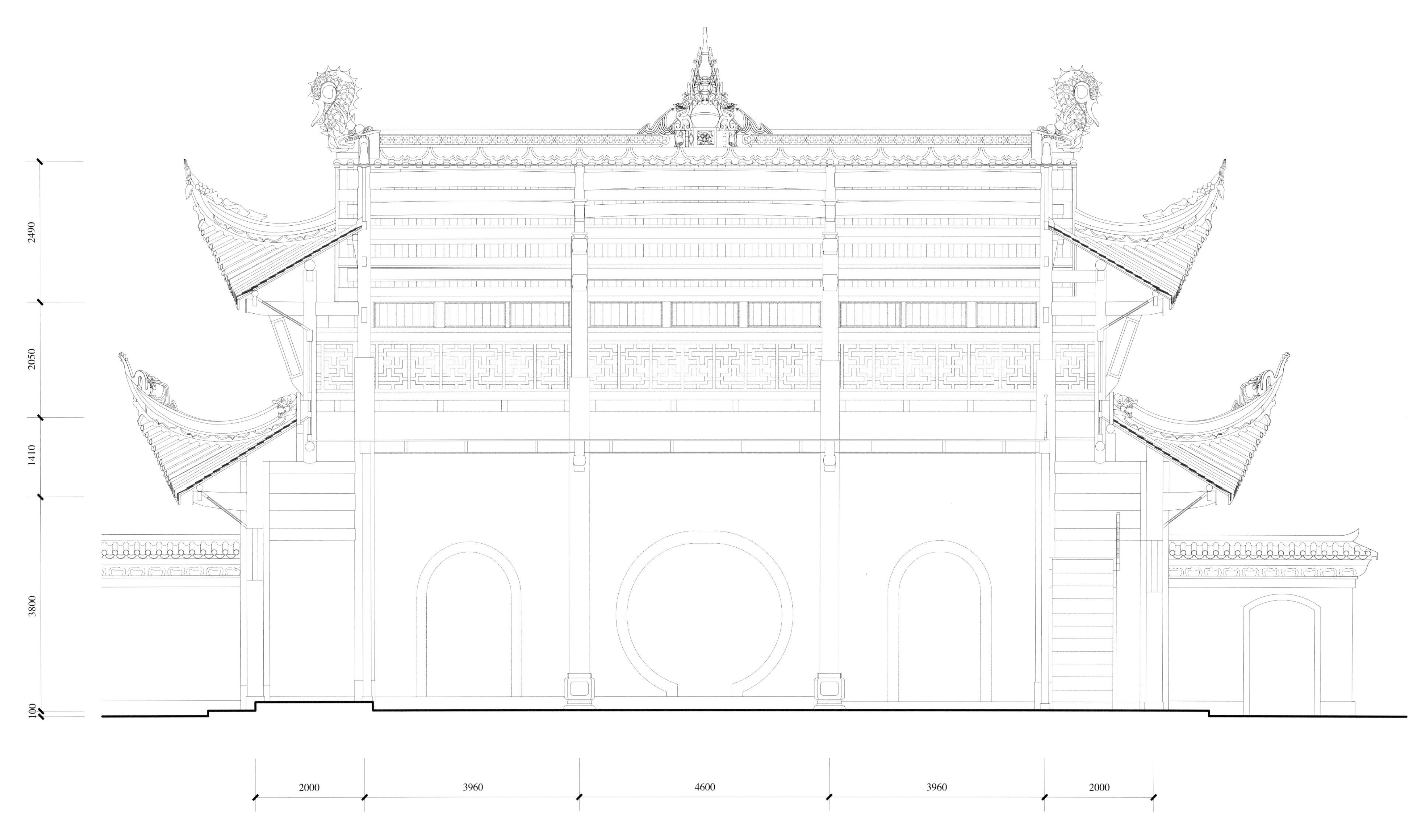
2490
2050
1410
3800
100
2000
3960
4600
3960
2000
0
1
2
3m

梁平双桂堂南山门纵剖面图

Longitudinal Section of South Main Gate (Shan Men), Shuanggui Temple, Liangping

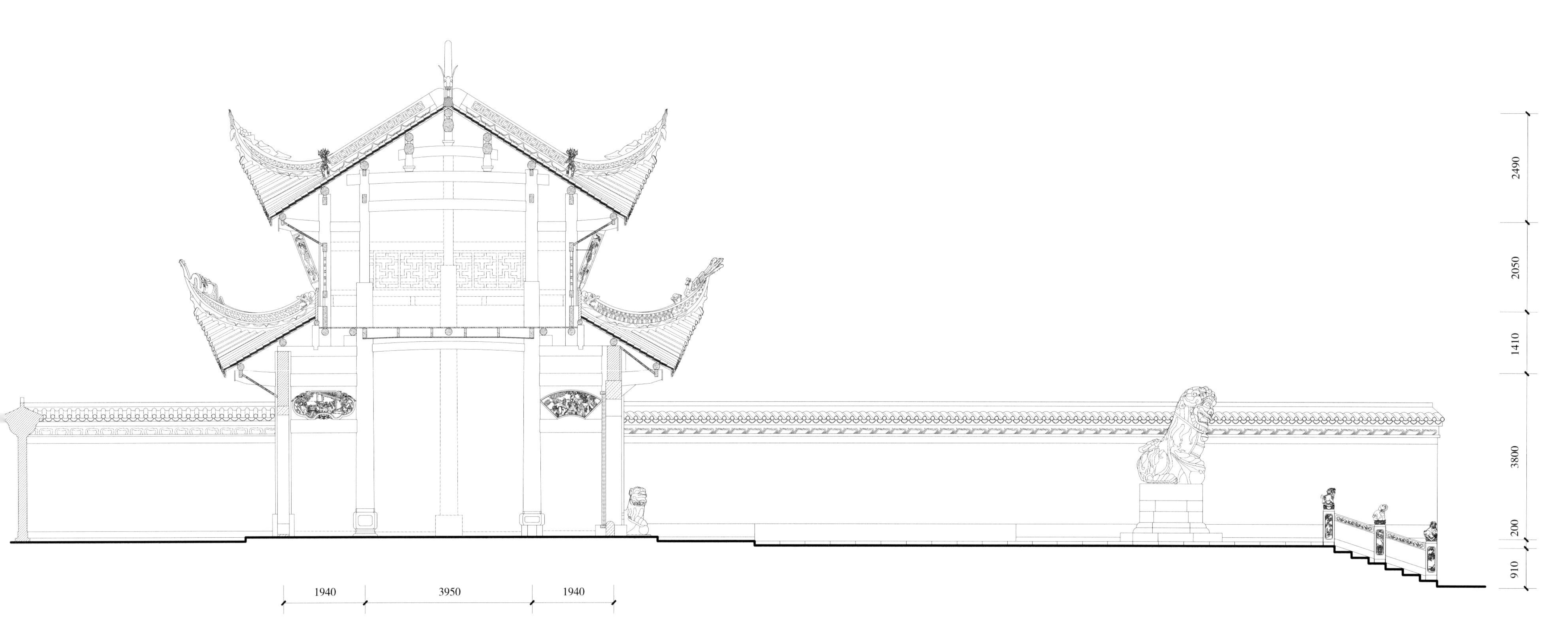
2490
2050
1410
3800
200
910
1940
3950
1940

梁平双桂堂南山门明间横剖面图

Cross Section of Central Bay (Mingjian), South Main Gate (Shan Men), Shuanggui Temple, Liangping

0 1 2 3 4 5m

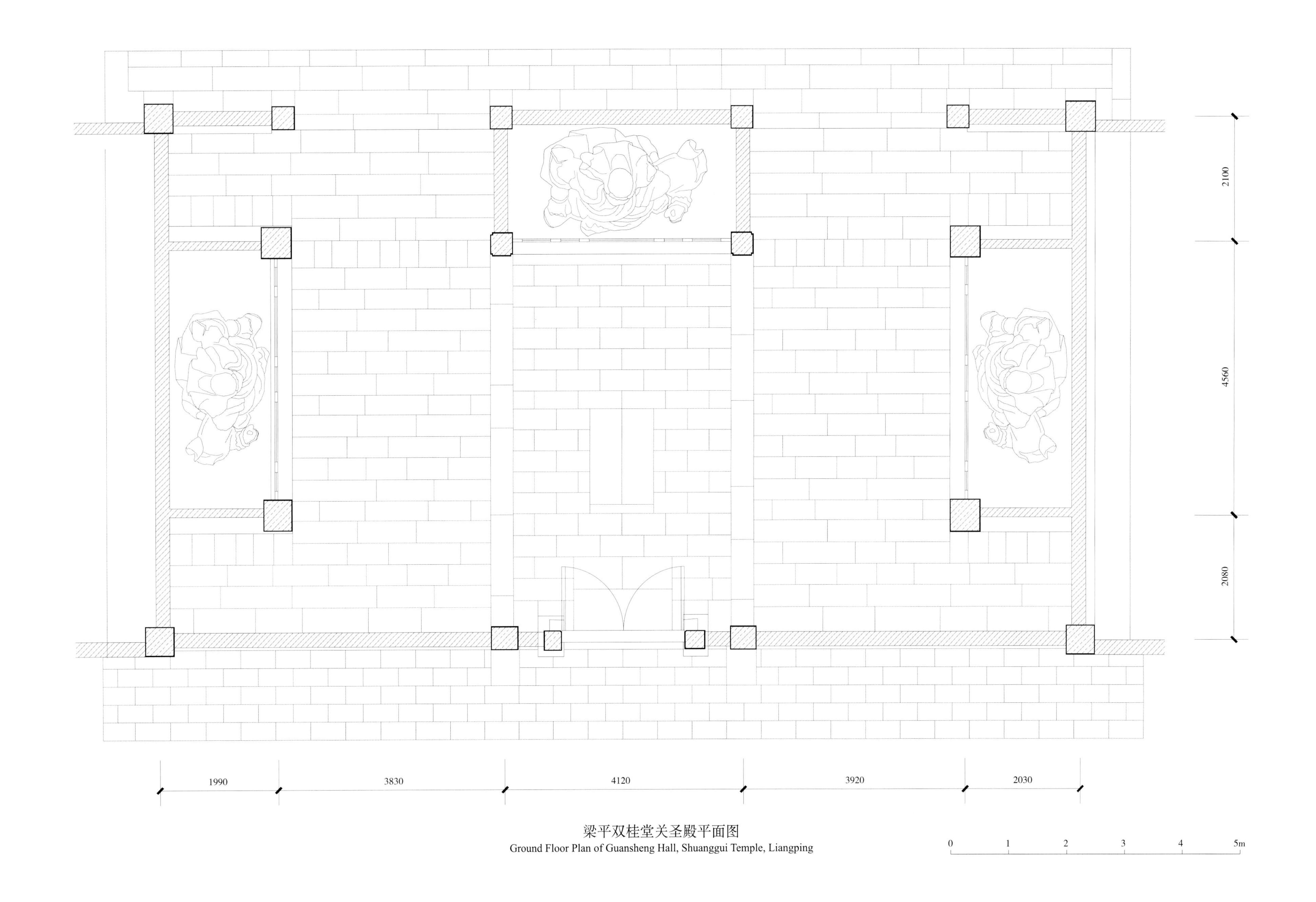

2100
4560
2080
1990
3830
4120
3920
2030
0
1
2
3
4
5m

梁平双桂堂关圣殿平面图
Ground Floor Plan of Guansheng Hall, Shuanggui Temple, Liangping

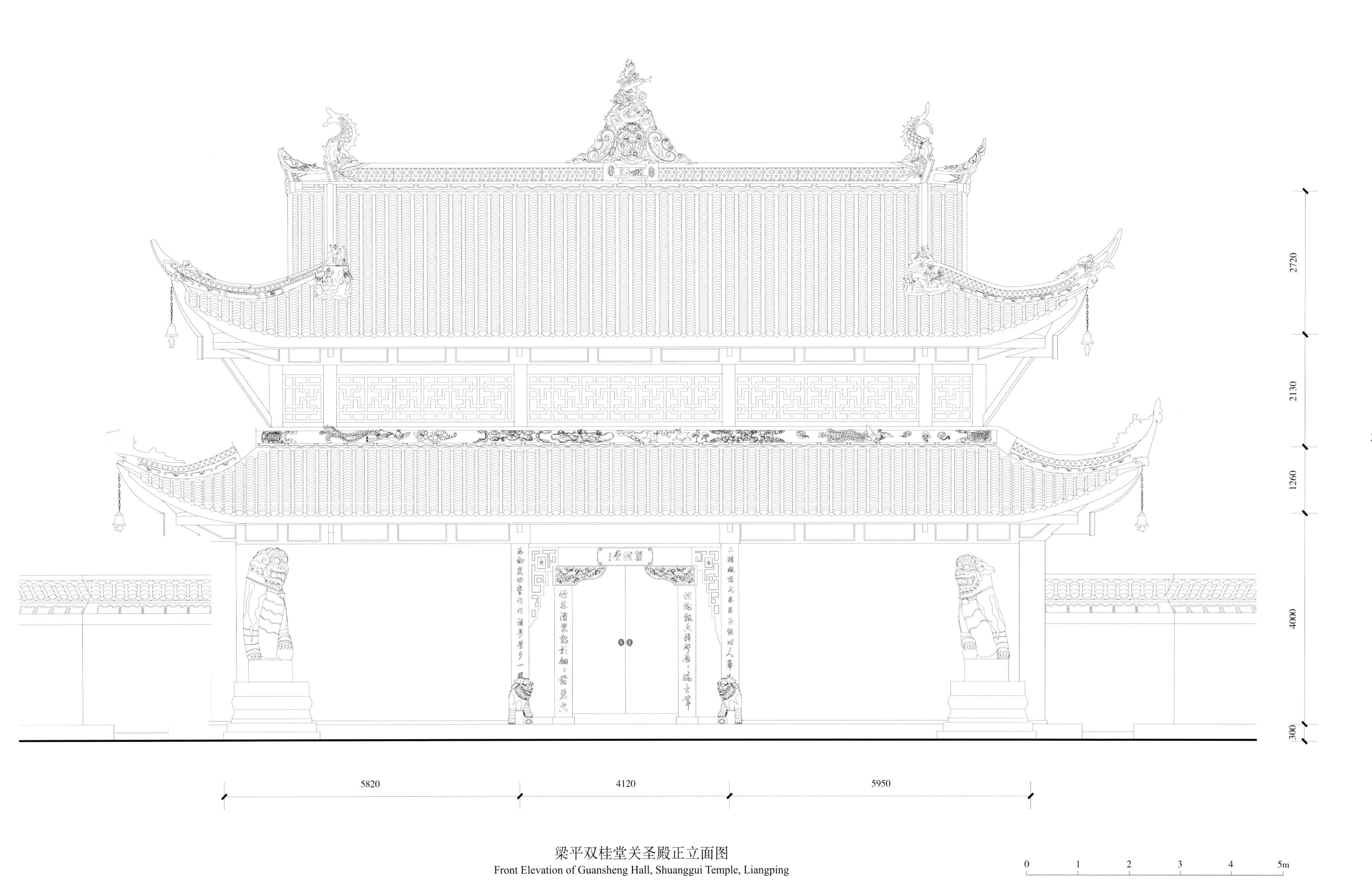
2720
2130
1260
4000
300
5820
4120
5950
0
1
2
3
4
5m

梁平双桂堂关圣殿正立面图
Front Elevation of Guansheng Hall, Shuanggui Temple, Liangping

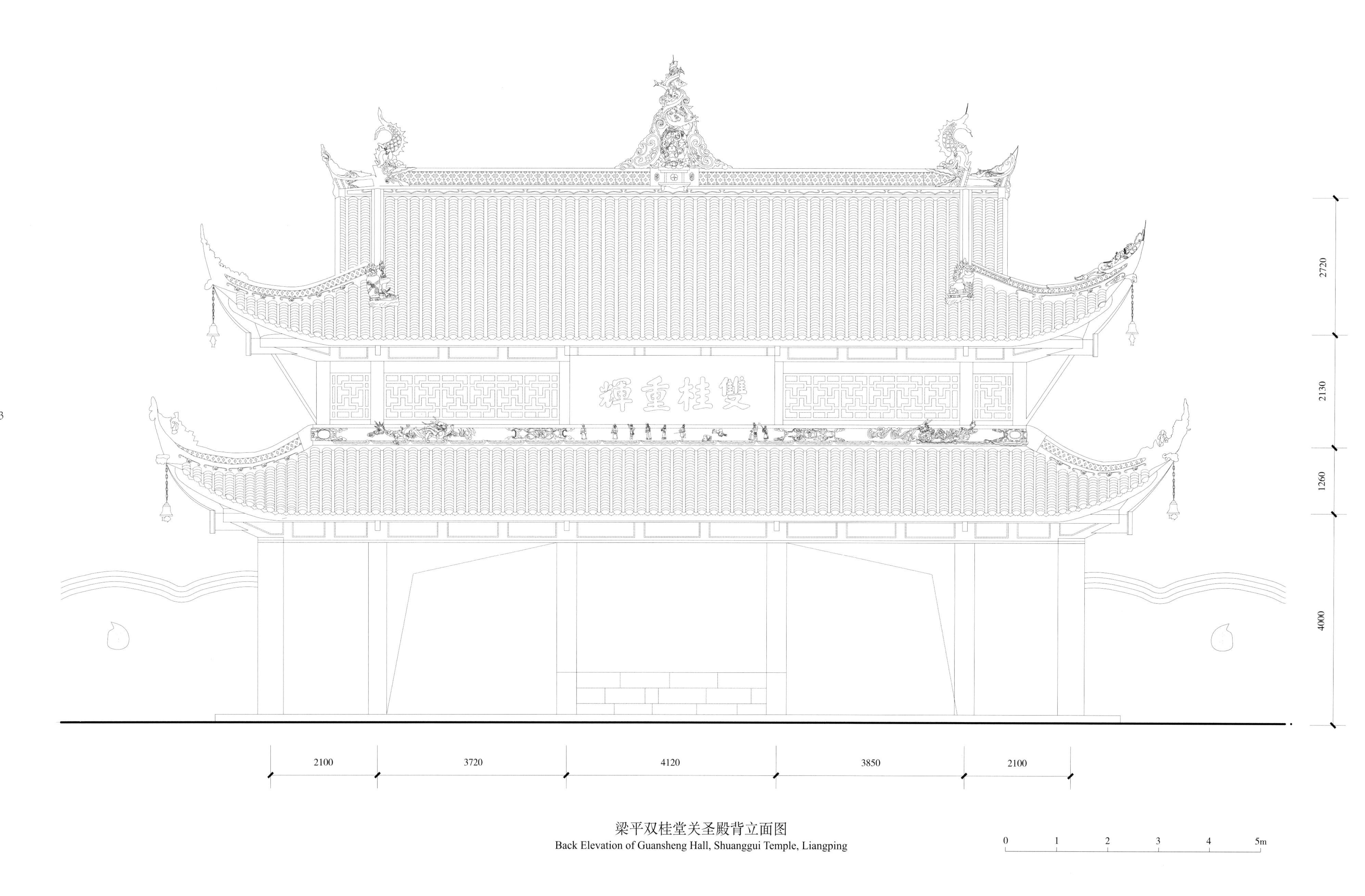
雙桂重輝
2720
2130
1260
4000
2100
3720
4120
3850
2100
0
1
2
3
4
5m

梁平双桂堂关圣殿背立面图
Back Elevation of Guansheng Hall, Shuanggui Temple, Liangping

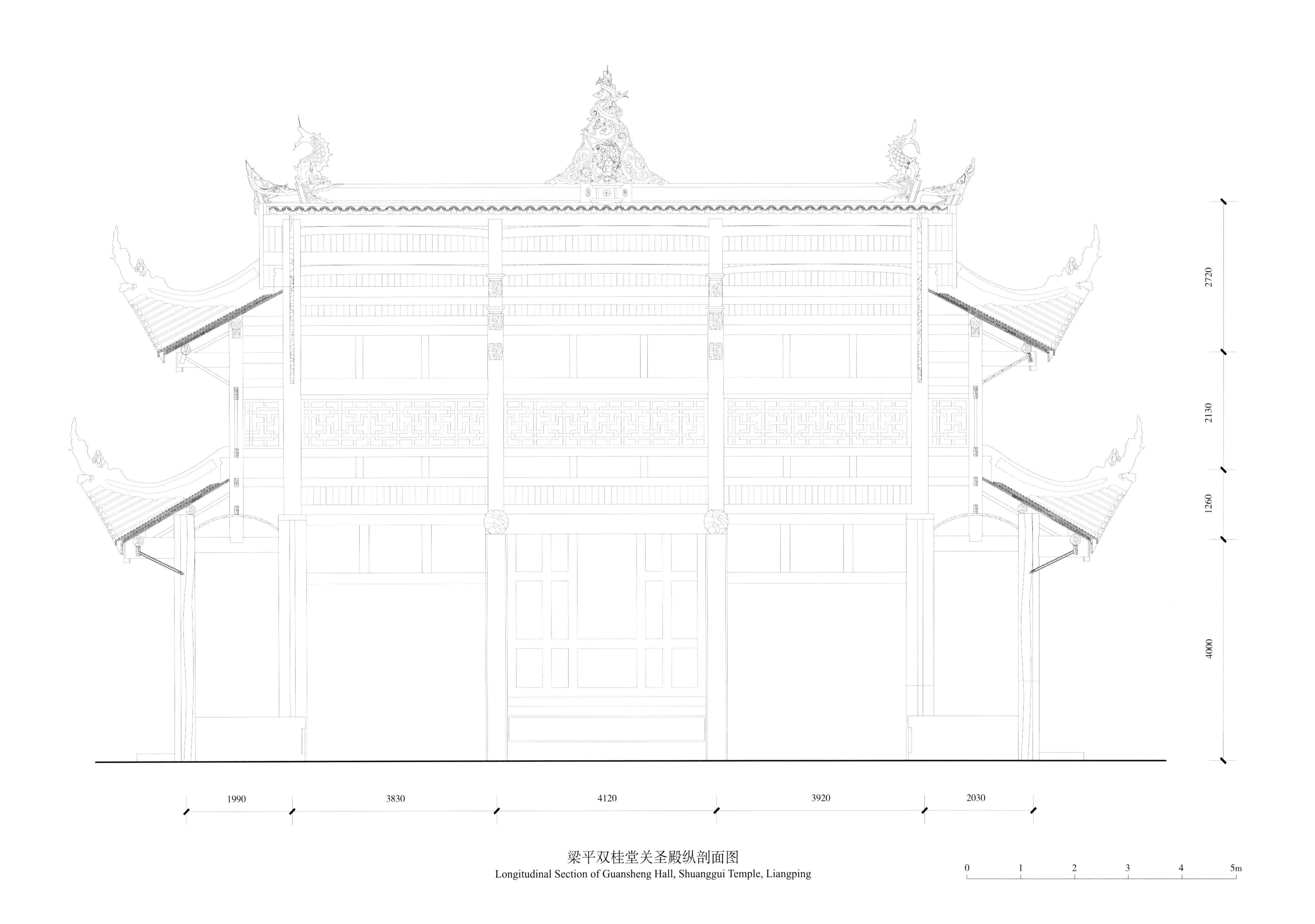
2720
2130
1260
4000
1990
3830
4120
3920
2030
0
1
2
3
4
5m

梁平双桂堂关圣殿纵剖面图
Longitudinal Section of Guansheng Hall, Shuanggui Temple, Liangping

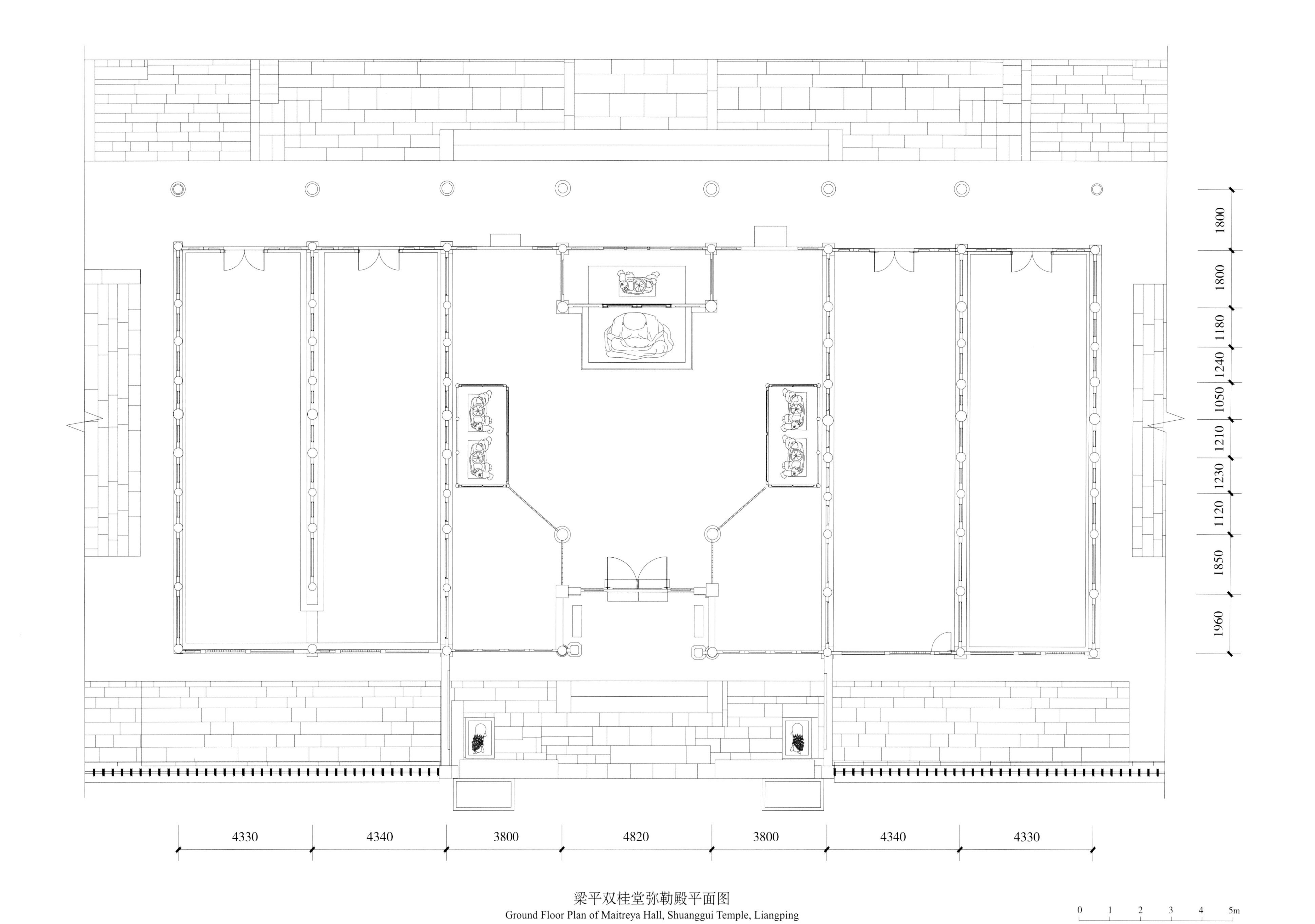
1800
1800
1180
1240
1050
1210
1230
1120
1850
1960
4330
4340
3800
4820
3800
4340
4330
0 1 2 3 4 5m

梁平双桂堂弥勒殿平面图
Ground Floor Plan of Maitreya Hall, Shuanggui Temple, Liangping

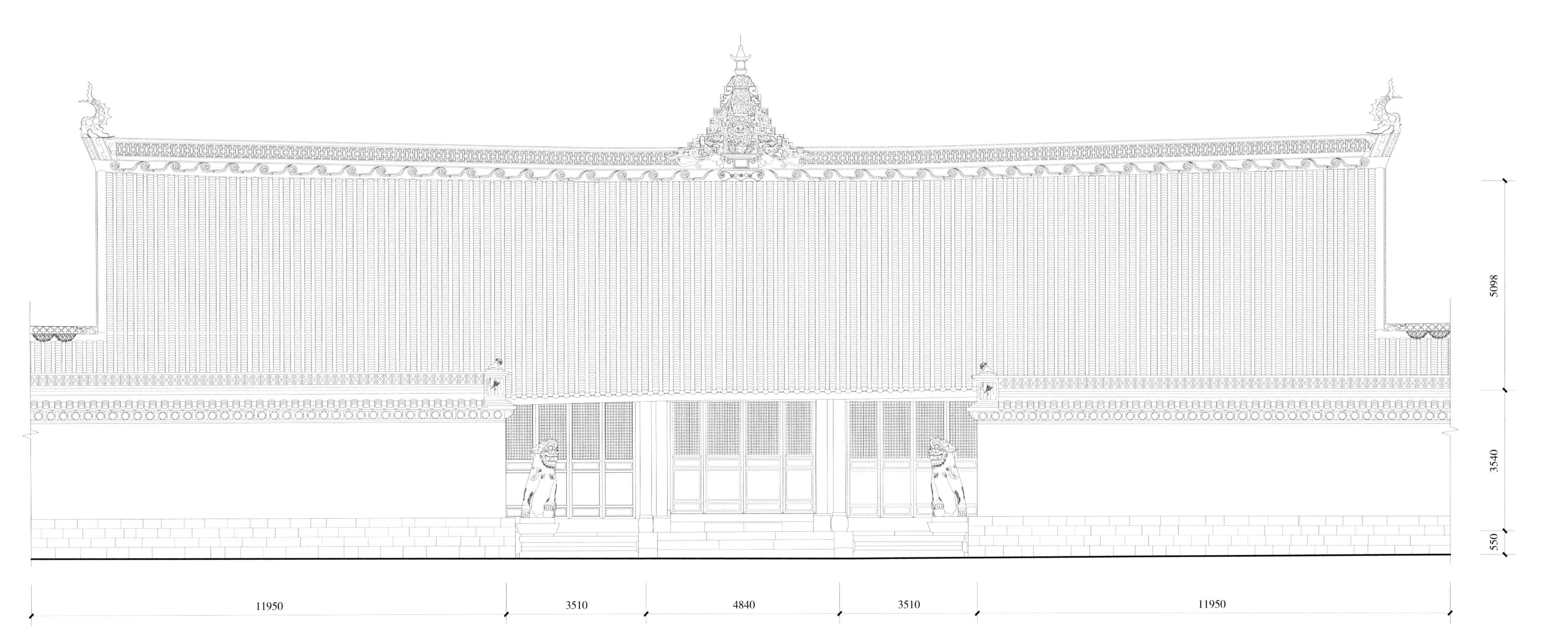
5098
3540
550
11950
3510
4840
3510
11950

0 1 2 3 4 5m

梁平双桂堂弥勒殿正立面图

Front Elevation of Maitreya Hall, Shuanggui Temple, Liangping

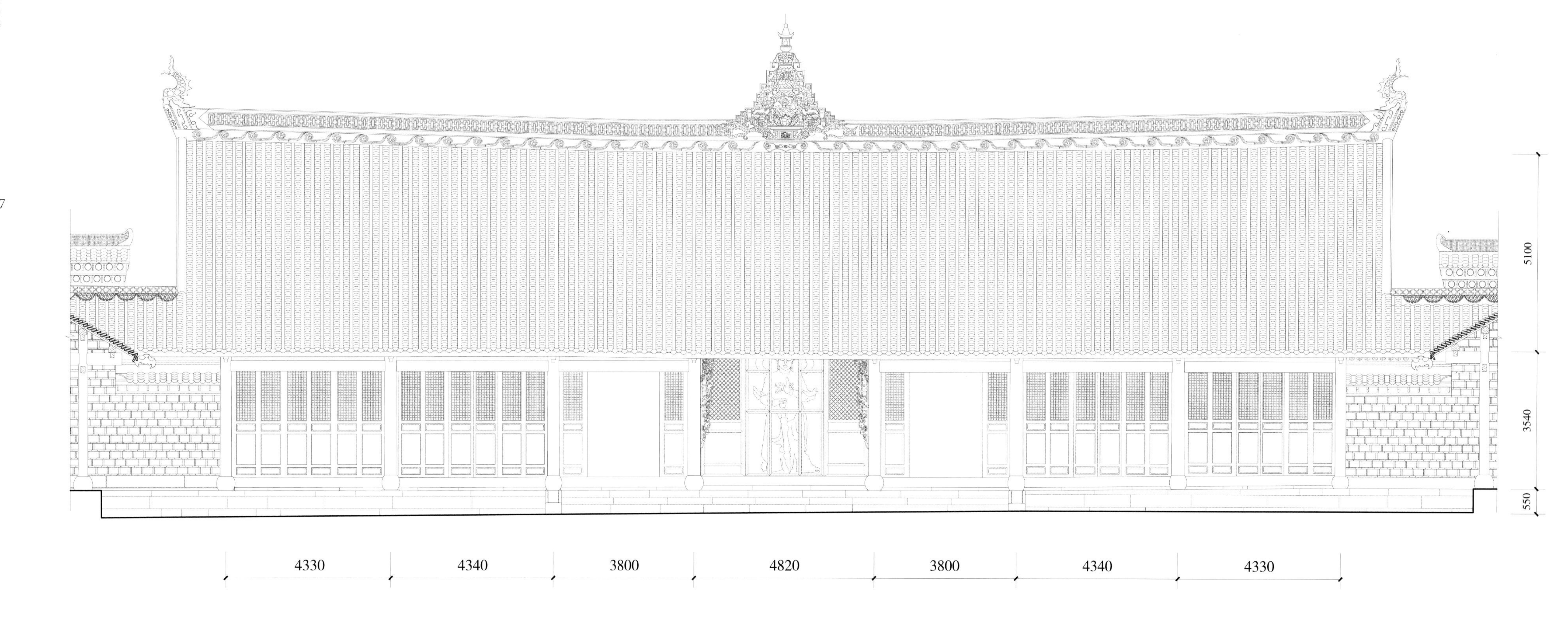

5100
3540
550
4330
4340
3800
4820
3800
4340
4330
0
1
2
3
4
5m

梁平双桂堂弥勒殿背立面图

Back Elevation of Maitreya Hall, Shuanggui Temple, Liangping

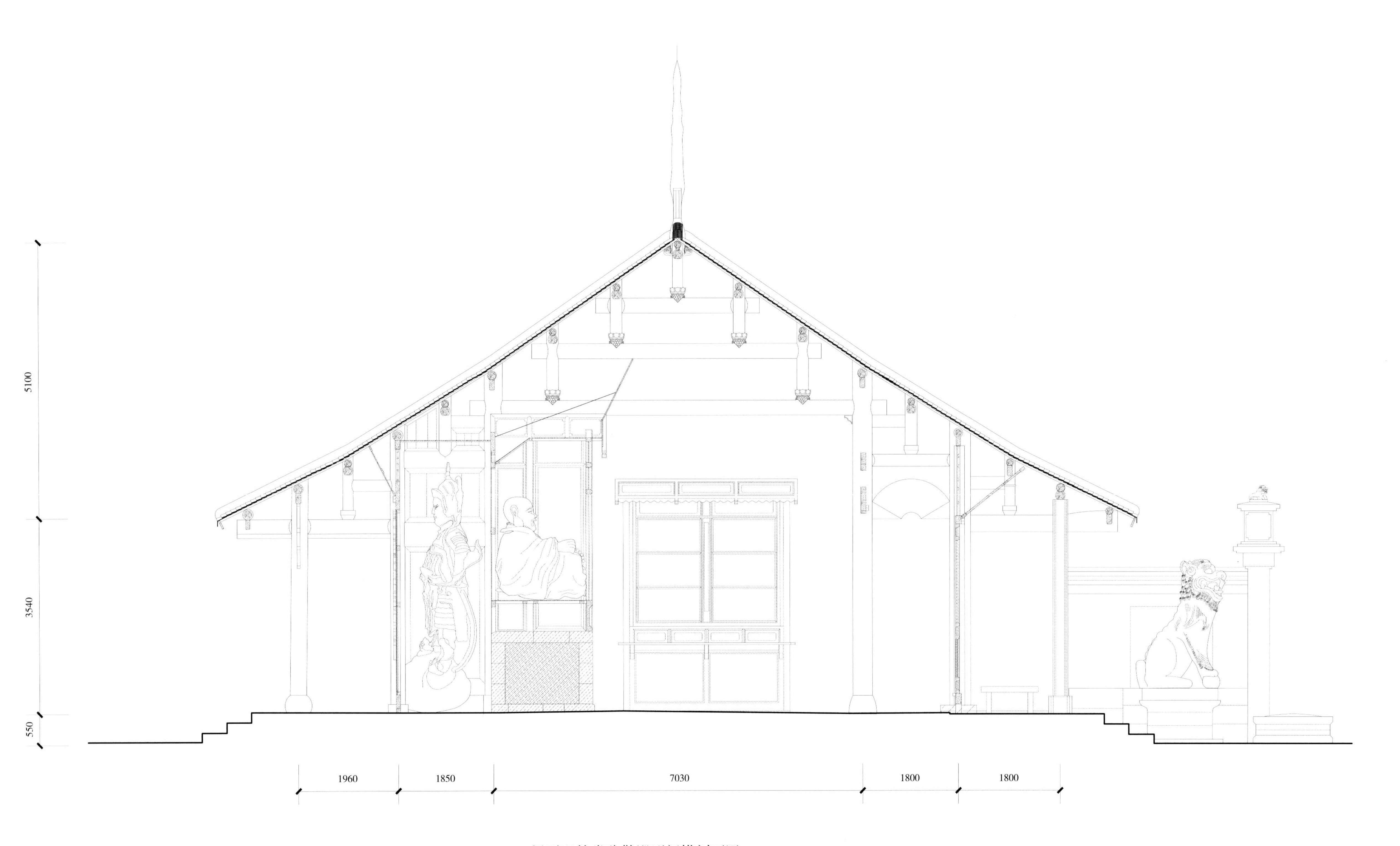
5100
3540
550
1960
1850
7030
1800
1800
0
1
2
3m

梁平双桂堂弥勒殿明间横剖面图
Cross Section of Central Bay (Mingjian), Maitreya Hall, Shuanggui Temple, Liangping

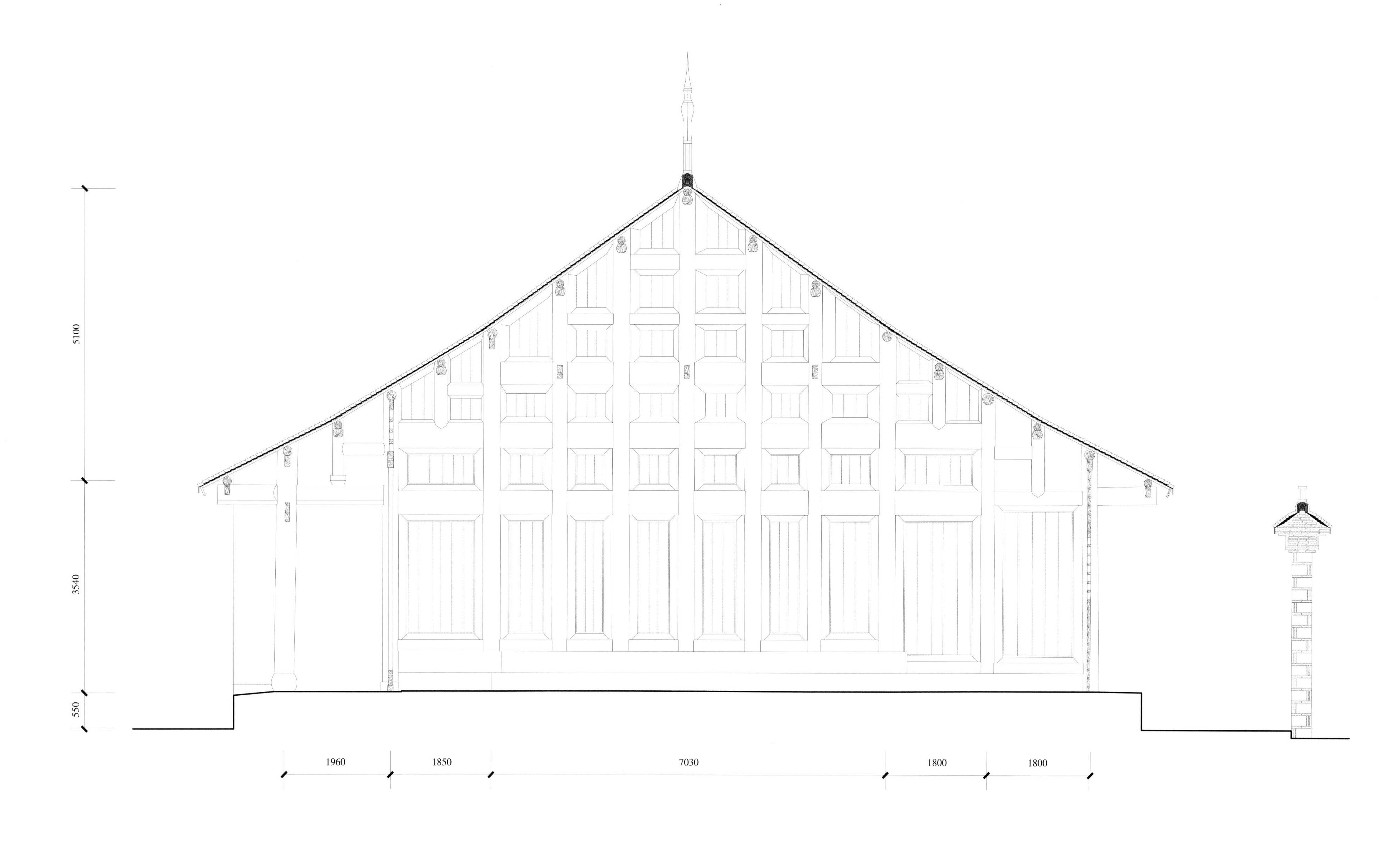
5100
3540
550
1960
1850
7030
1800
1800
0
1
2
3m

梁平双桂堂弥勒殿尽间横剖面图

Cross Section of Ending Bay (Jinjian) Maitreya Hall, Shuanggui Temple, Liangping

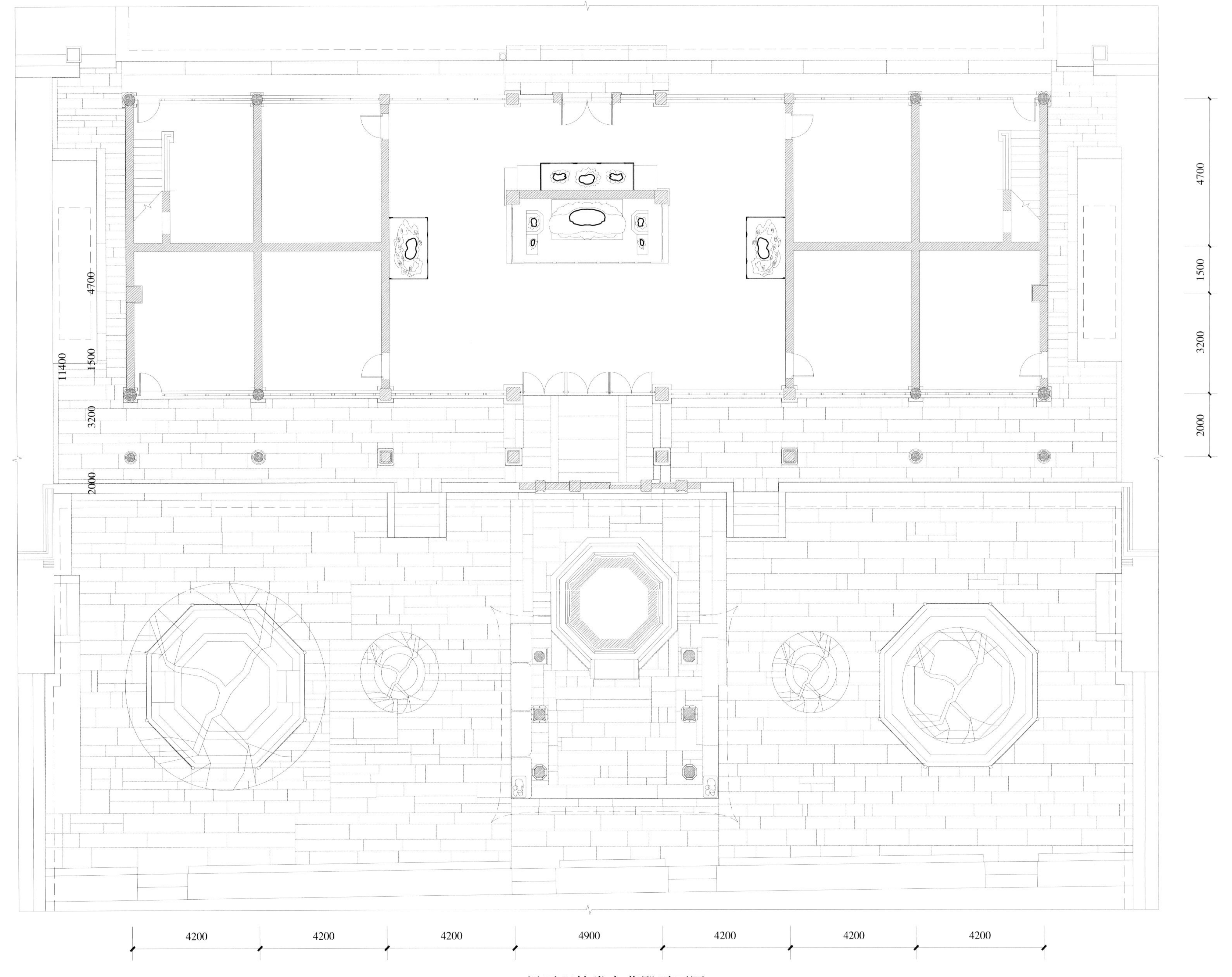
4700
1500
3200
2000
11400
4200
4200
4200
4900
4200
4200
4200
0 1 2 3 4 5m

梁平双桂堂大悲殿平面图
Ground Floor Plan of Dabei Hall, Shuanggui Temple, Liangping

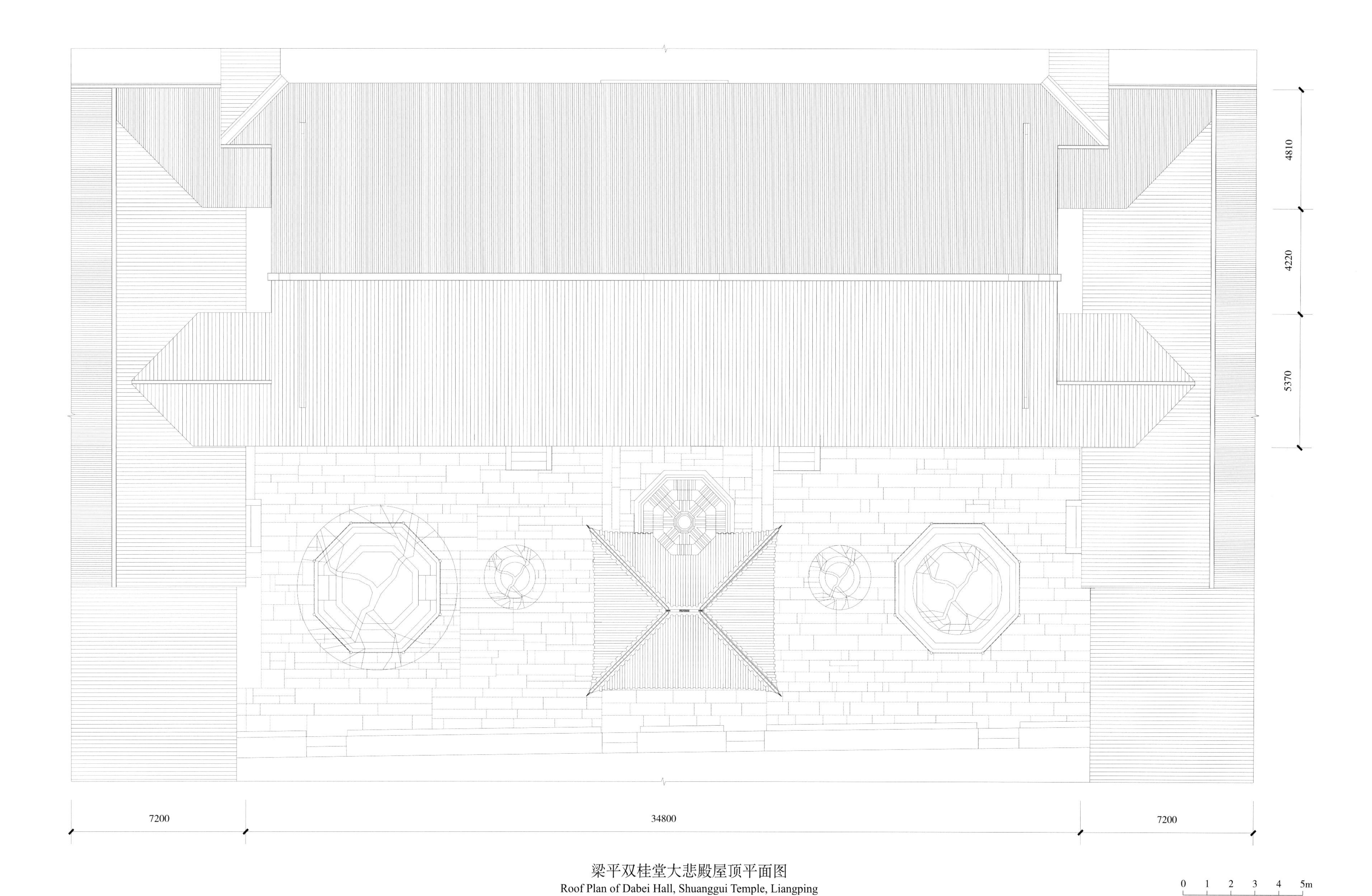
4810
4220
5370
7200
34800
7200
0 1 2 3 4 5m

梁平双桂堂大悲殿屋顶平面图
Roof Plan of Dabei Hall, Shuanggui Temple, Liangping

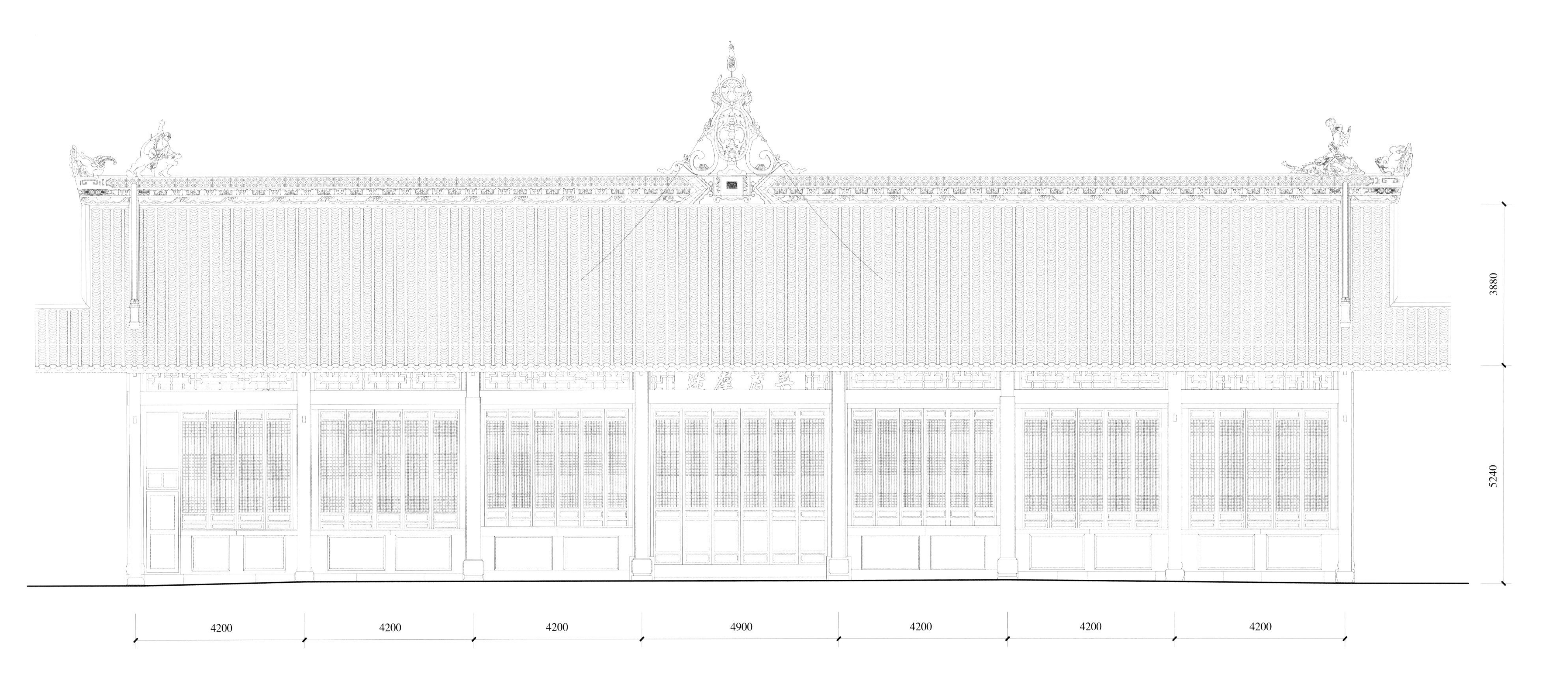
3880
5240
4200
4200
4200
4900
4200
4200
4200
0 1 2 3 4 5m

梁平双桂堂大悲殿正立面图
Front Elevation of Dabei Hall, Shuanggui Temple, Liangping

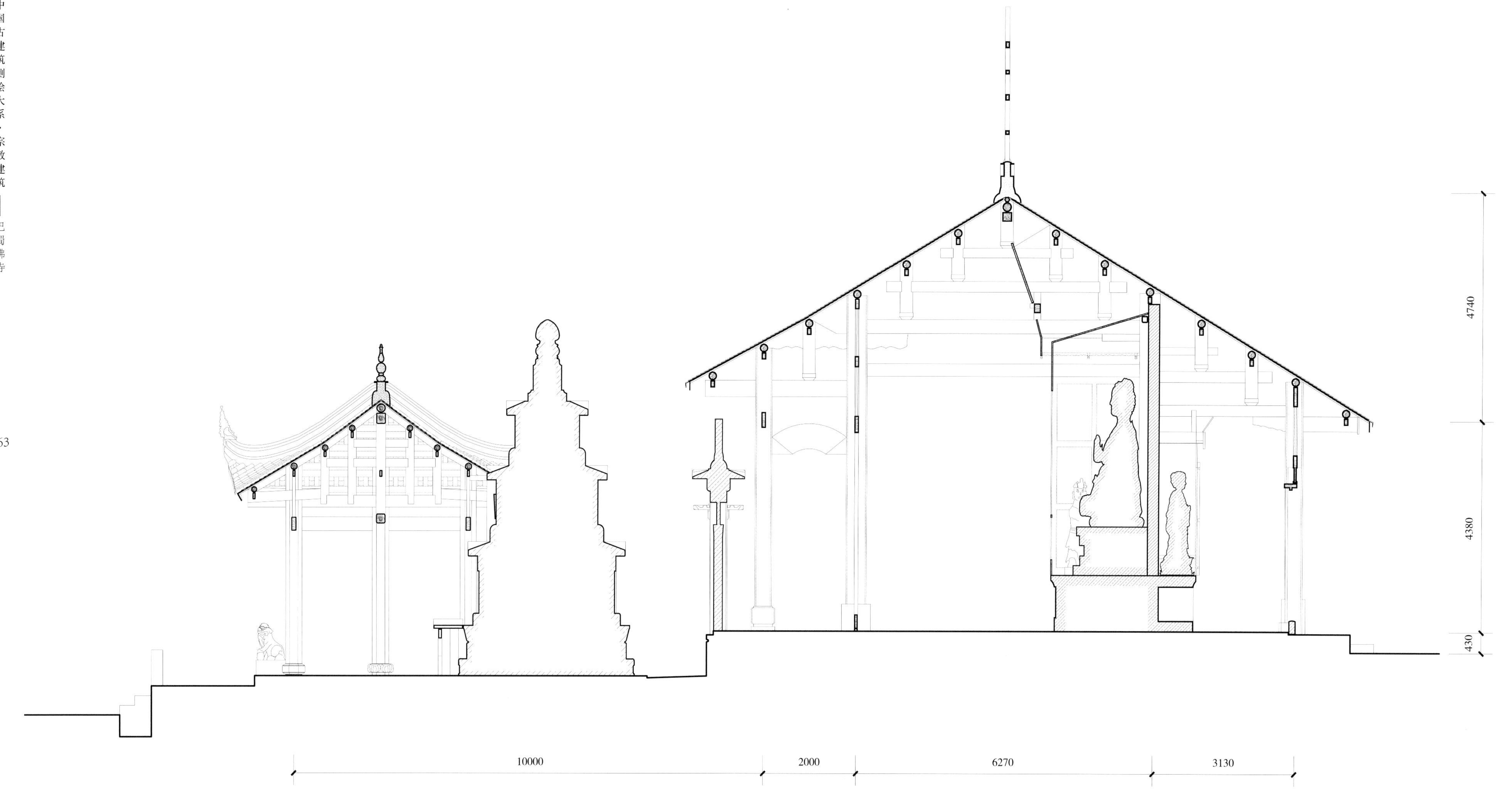
4740
4380
430
10000
2000
6270
3130
0
1
2
3
4
5m

梁平双桂堂大悲殿明间横剖面图

Cross Section of Central Bay (Mingjian), Dabei Hall, Shuanggui Temple, Liangping

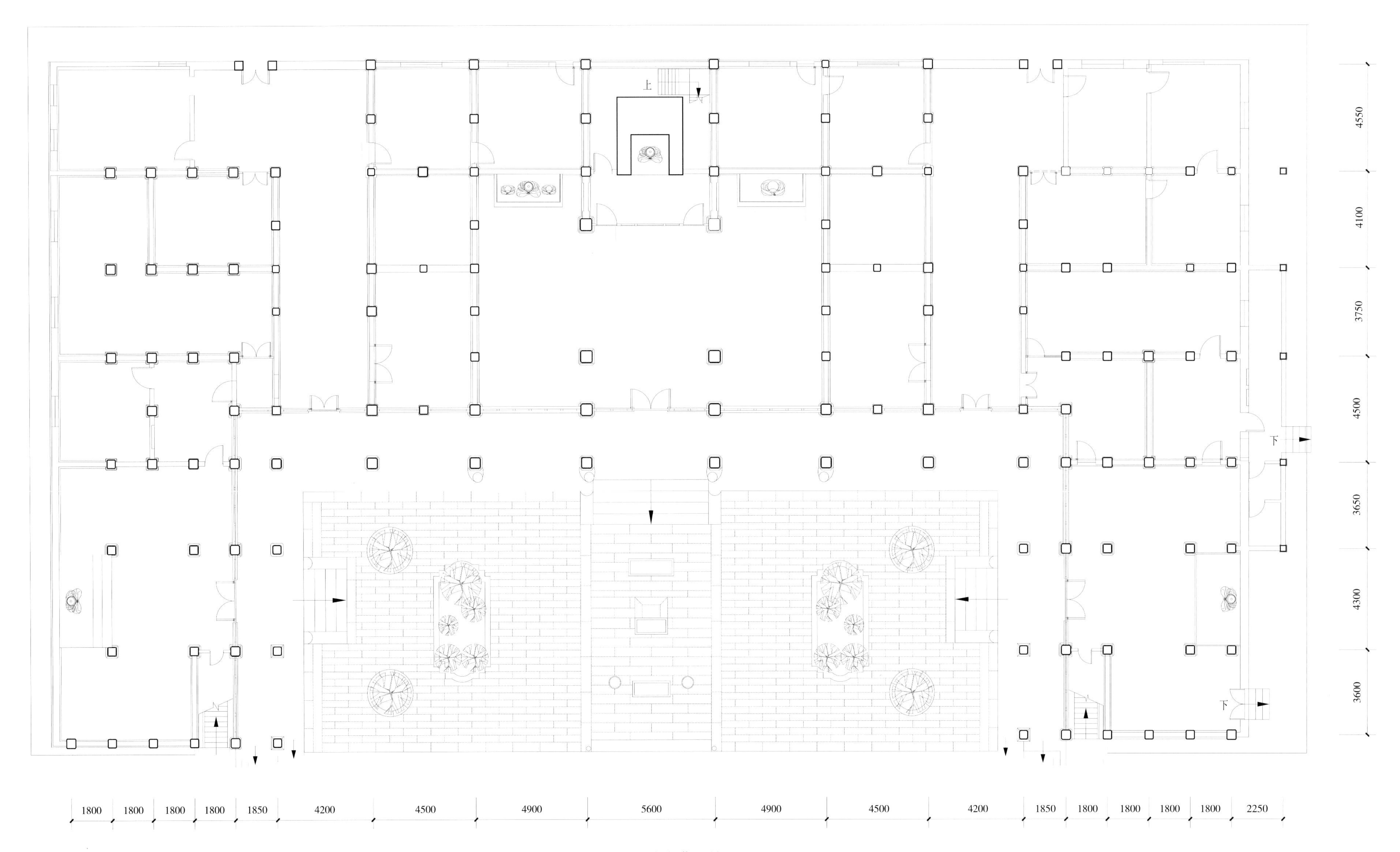

上
下
下
4550
4100
3750
4500
3650
4300
3600
1800
1800
1800
1800
1850
4200
4500
4900
5600
4900
4500
4200
1850
1800
1800
1800
1800
2250
0 1 2 3 4 5m

梁平双桂堂藏经楼一层平面图

Ground Floor Plan of Sutrua Library, Shuanggui Temple, Liangping

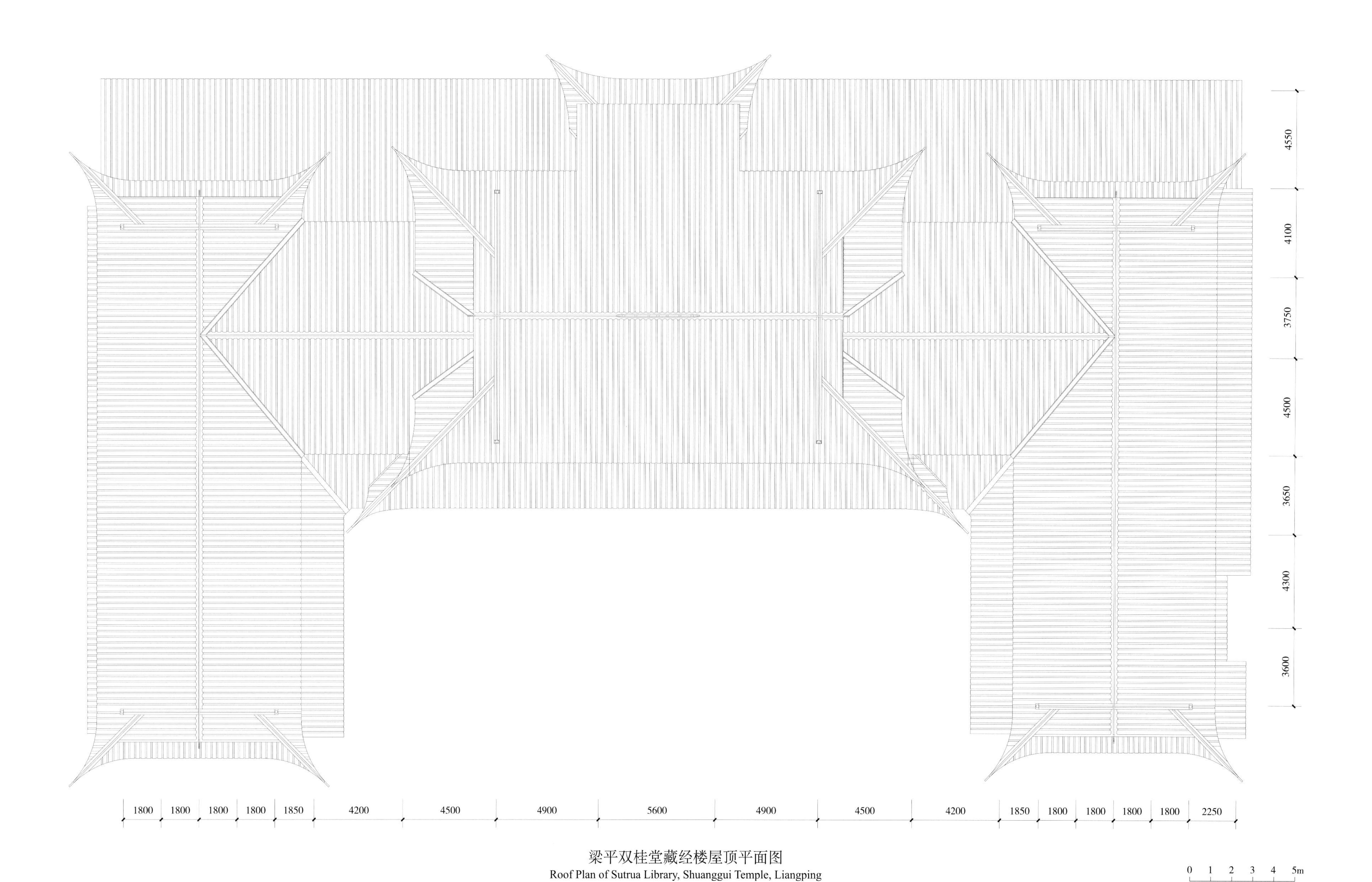
1800
1800
1800
1800
1850
4200
4500
4900
5600
4900
4500
4200
1850
1800
1800
1800
1800
2250
4550
4100
3750
4500
3650
4300
3600
0 1 2 3 4 5m

梁平双桂堂藏经楼屋顶平面图

Roof Plan of Sutrua Library, Shuanggui Temple, Liangping

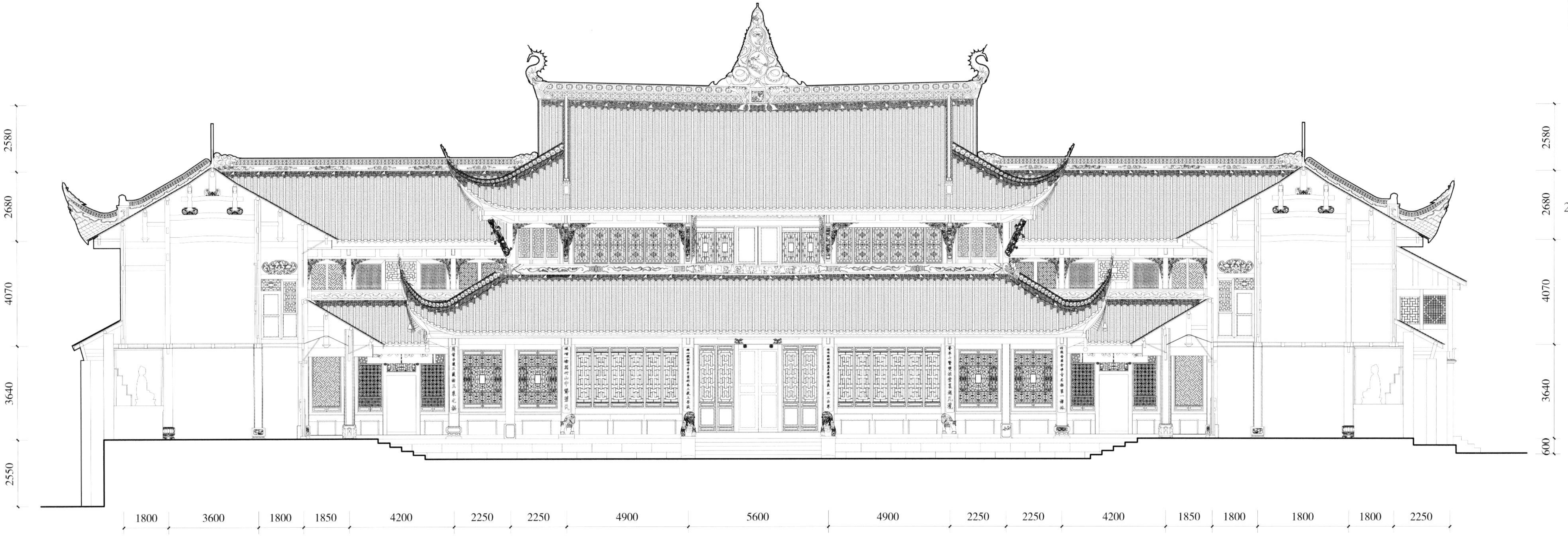

梁平双桂堂藏经楼正立面图

Front Elevation of Sutrua Library, Shuanggui Temple, Liangping

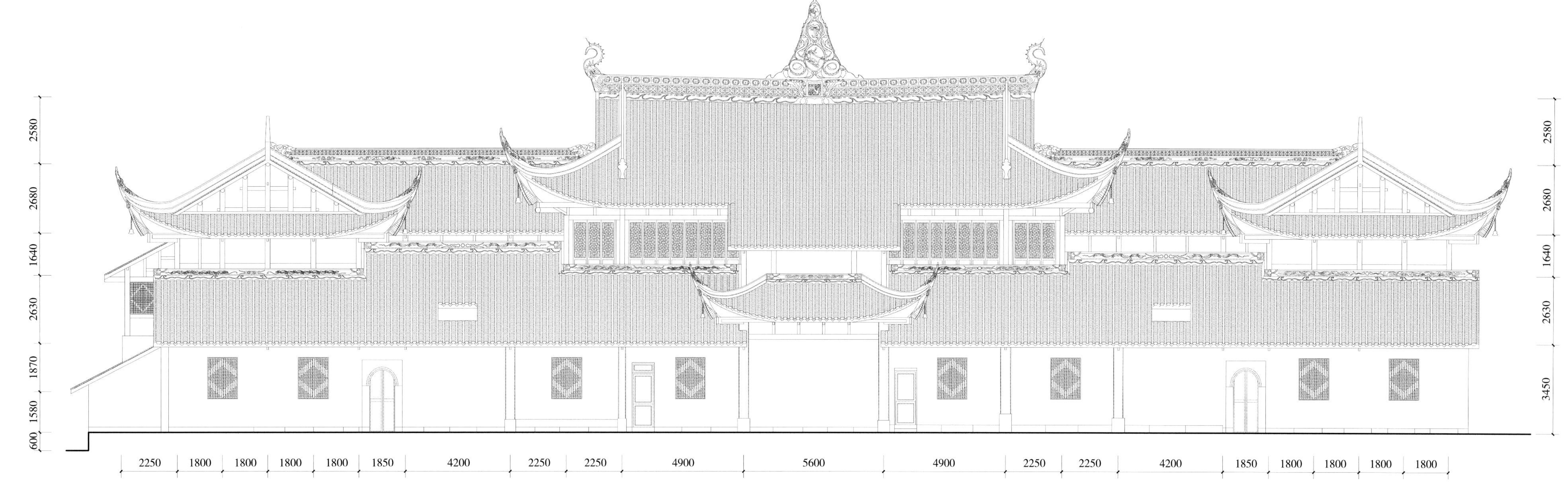
2580
2680
1640
2630
1870
1580
600
2580
2680
1640
2630
3450
2250
1800
1800
1800
1800
1850
4200
2250
2250
4900
5600
4900
2250
2250
4200
1850
1800
1800
1800
1800

梁平双桂堂藏经楼背立面图

Back Elevation of Sutrua Library, Shuanggui Temple, Liangping

0 1 2 3 4 5m

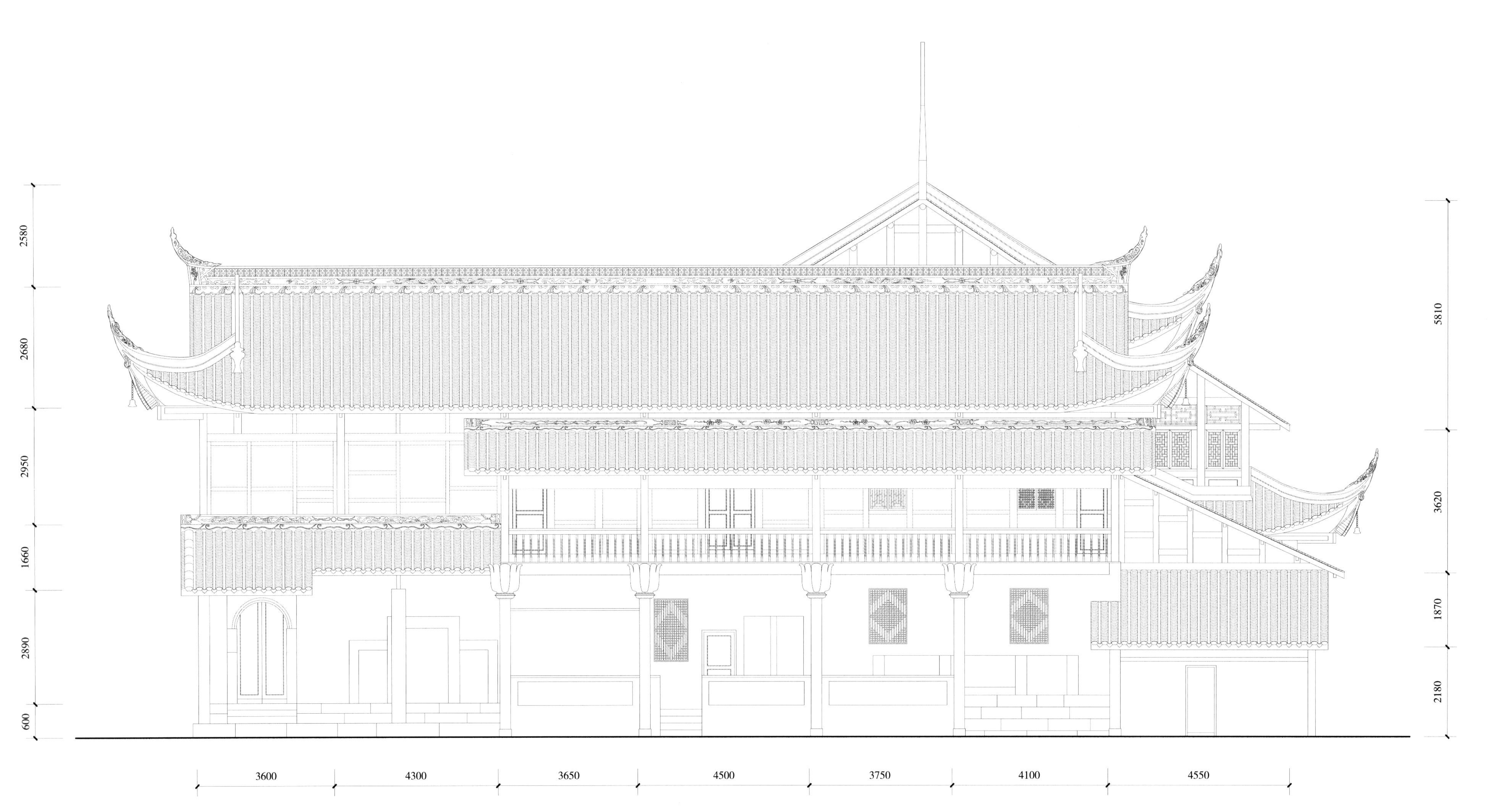
2580
2680
2950
1660
2890
600
5810
3620
1870
2180
3600
4300
3650
4500
3750
4100
4550

梁平双桂堂藏经楼侧立面图

Side Elevation of Sutrua Library, Shuanggui Temple, Liangping

0 1 2 3 4 5m

5610
2850
4520
4025
4870
4090
600
4550
2250
5600
2250
2250
0
1
2
3
4
5m

梁平双桂堂藏经楼明间横剖面图

Cross Section of Central Bay (Mingjian), Sutrua Library, Shuanggui Temple, Liangping

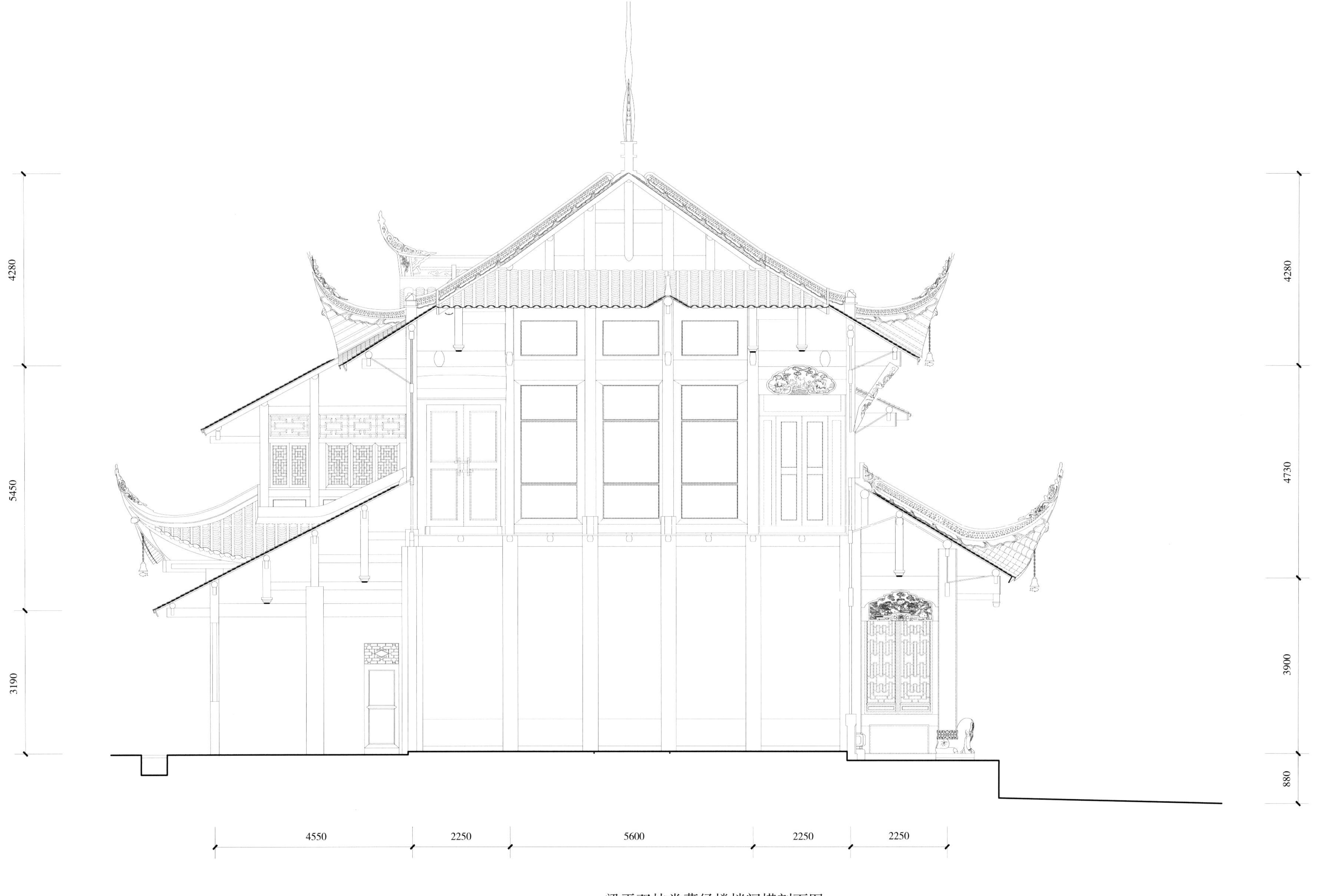
4280
5450
3190
4280
4730
3900
880
4550
2250
5600
2250
2250
0
1
2
3
4
5m

梁平双桂堂藏经楼梢间横剖面图

Cross Section of In-between Bay (Shaojian), Sutrua Library, Shuanggui Temple, Liangping

2580
2680
1910
6010
600
2580
2680
4370
3510
600
2250
1800
1800
1800
1800
1850
4200
2250
2250
4900
5600
4900
2250
2250
4200
1850
3600
3600

梁平双桂堂藏经楼纵剖面图

Longitudinal Section of Sutrua Library, Shuanggui Temple, Liangping

0 1 2 3 4 5m

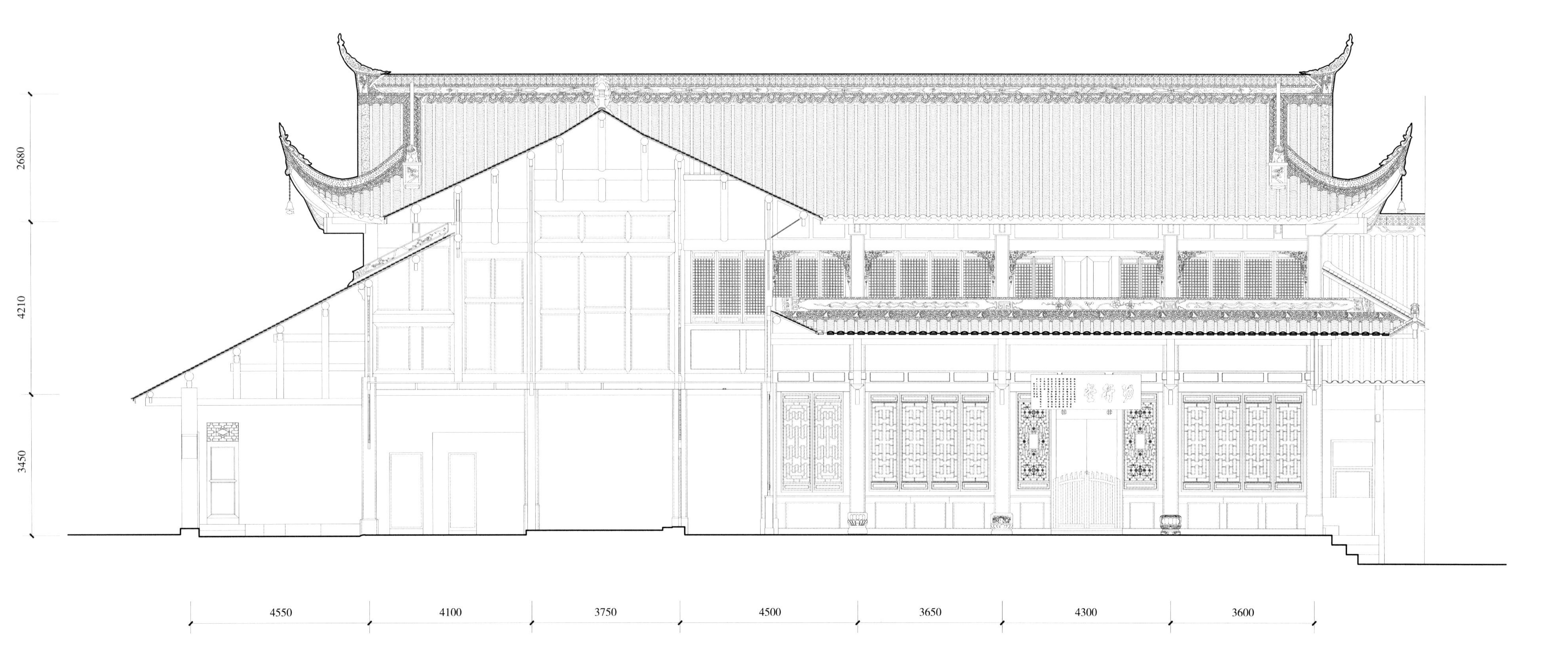
2680
4210
3450
4550
4100
3750
4500
3650
4300
3600

梁平双桂堂藏经楼厢房剖面图
Cross Section of Wings, Sutrua Library, Shuanggui Temple, Liangping

0 1 2 3 4 5m

3200
40
250
50
300
25
320

1600
500

梁平双桂堂藏经楼正脊式样图

Detail, Main Ridge Decoration, Sutrua Library, Shuanggui Temple, Liangping

0 20 40 60cm

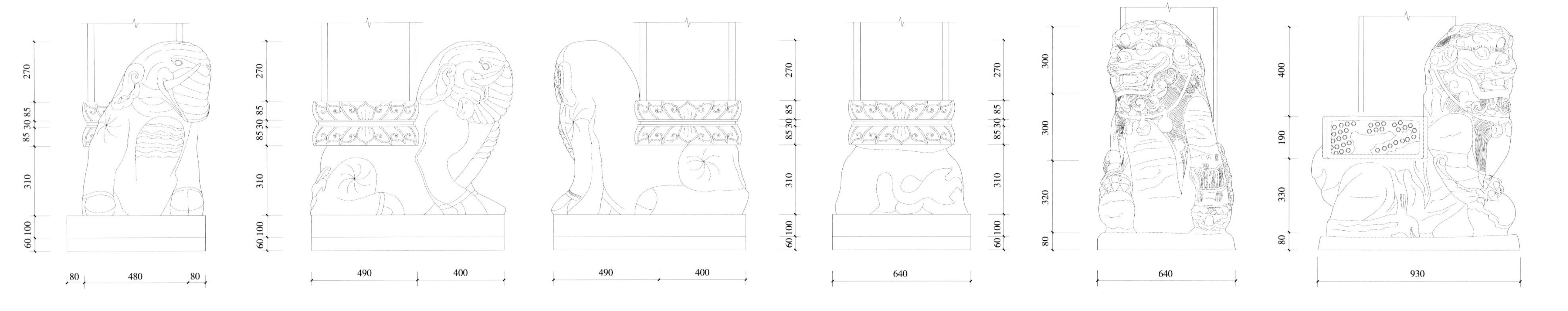
270
85
30
85
310
100
60
80
480
80
490
400
490
400
640
300
300
320
80
640
400
190
330
80
930

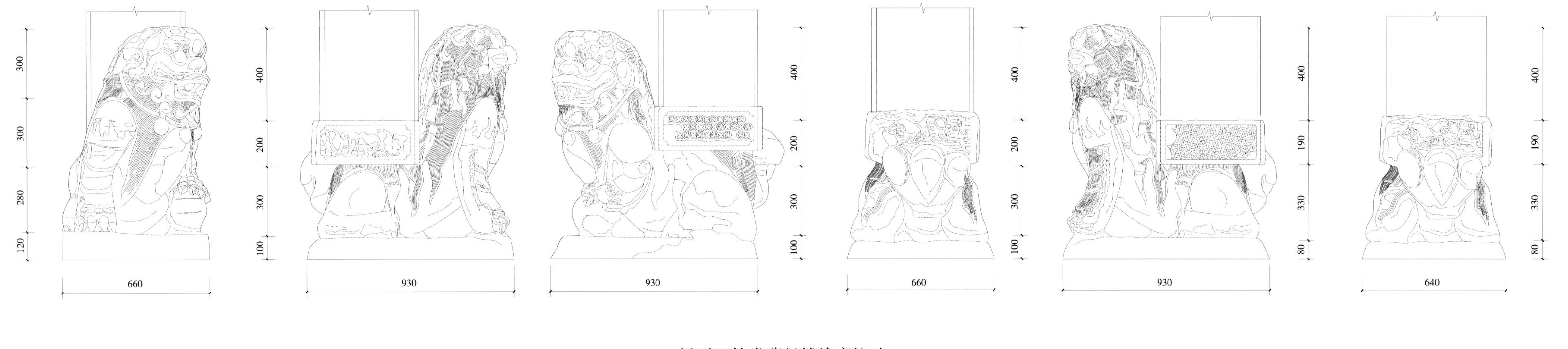
300
300
280
120
660
400
200
300
100
930
930
660
400
190
330
80
930
640

梁平双桂堂藏经楼檐廊柱础

Veranda Column Base, Sutrua Library, Shuanggui Temple, Liangping

0 20 40 60cm

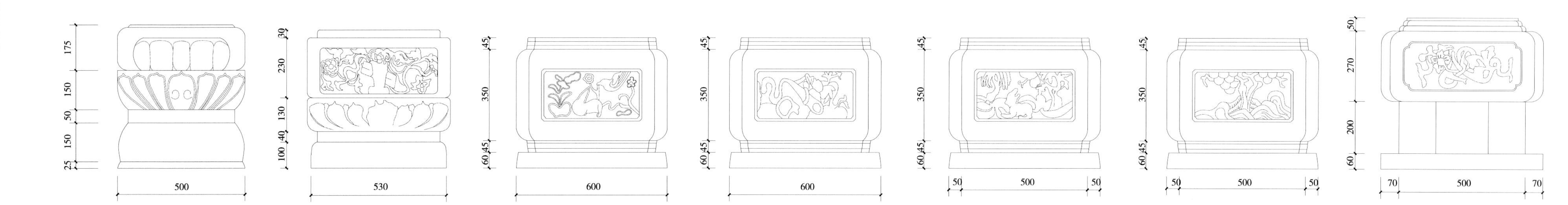
175
150
50
150
25
500
30
230
130
40
100
530
45
350
45
60
600
45
350
45
60
600
45
350
45
60
50
500
50
45
350
45
60
50
500
50
50
270
200
60
70
500
70

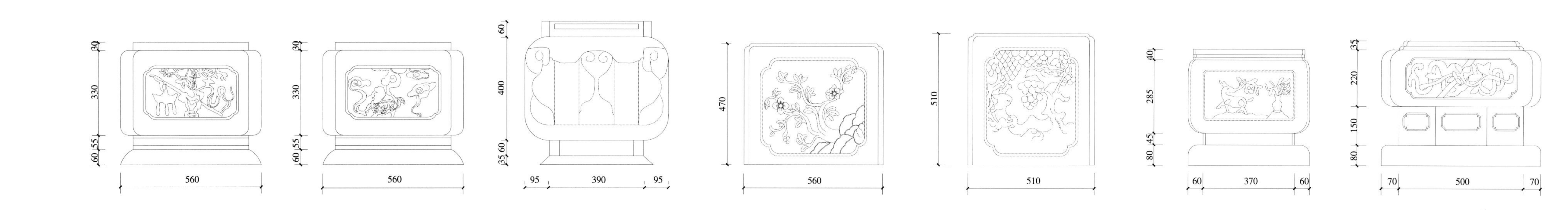
30
330
55
60
560
30
330
55
60
560
60
400
60
35
95
390
95
470
560
510
510
40
285
45
80
60
370
60
35
220
150
80
70
500
70

梁平双桂堂藏经楼柱础

Column Base of Sutrua Library, Shuanggui Temple, Liangping

0 20 40 60cm

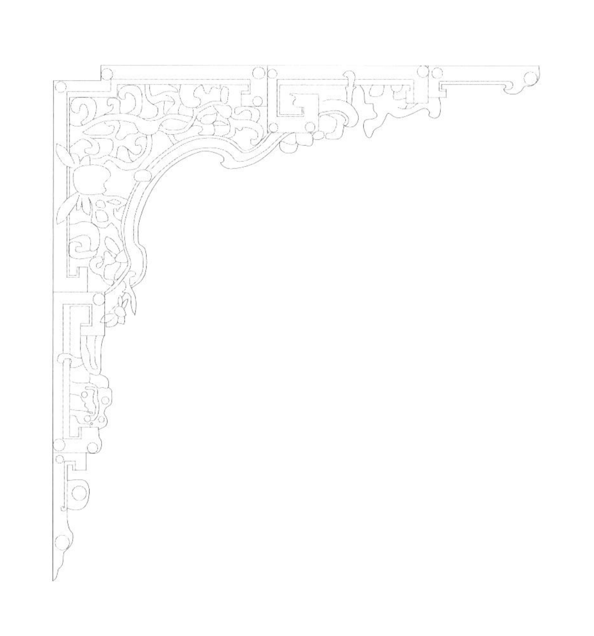

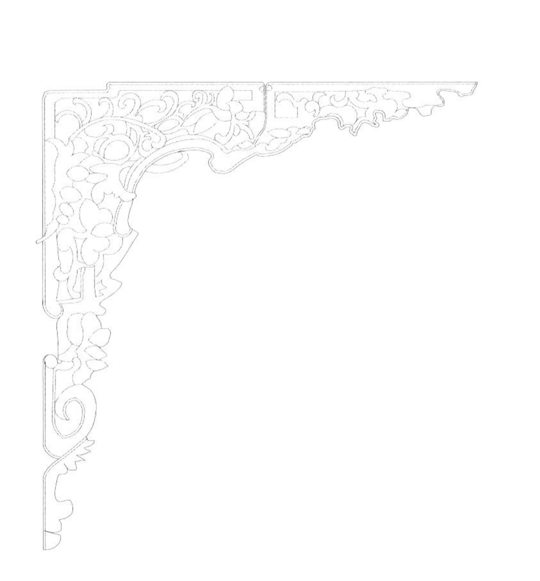

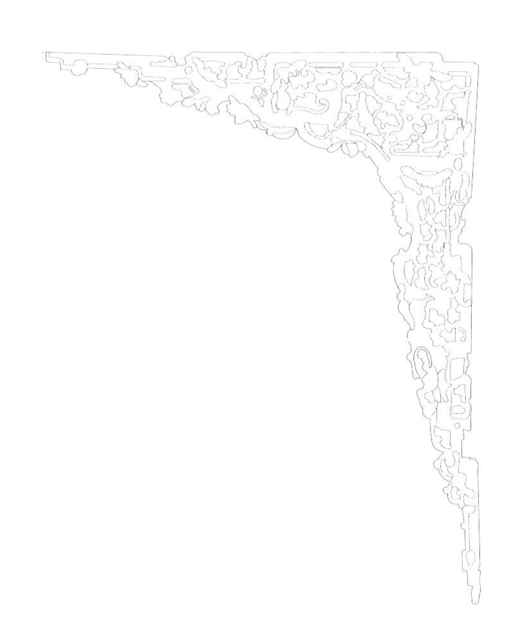

梁平双桂堂藏经楼檐廊花牙子式样图

Detail, Huayazi, Veranda of Sutrua Library, Shuanggui Temple, Liangping

0 20 40 60 80 100cm

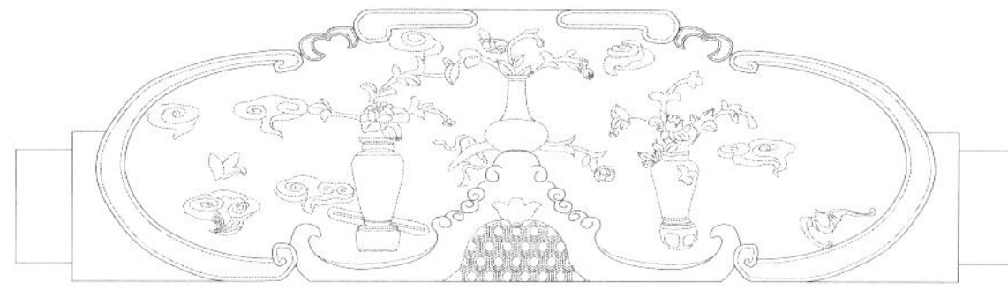

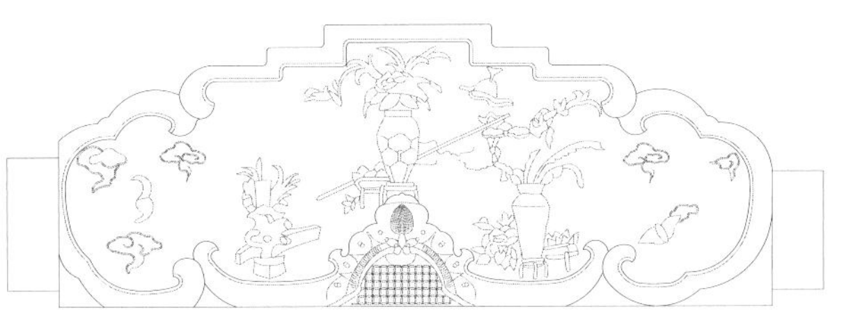
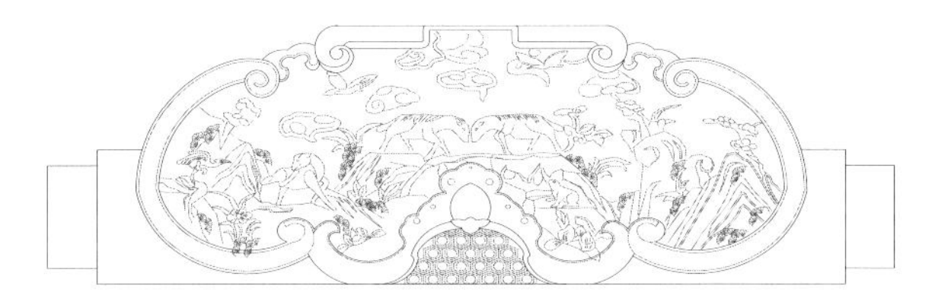

梁平双桂堂藏经楼扇面枋式样图
Detail, Shanmian Fang, Sutrua Library, Shuanggui Temple, Liangping

0 20 40 60 80 100cm

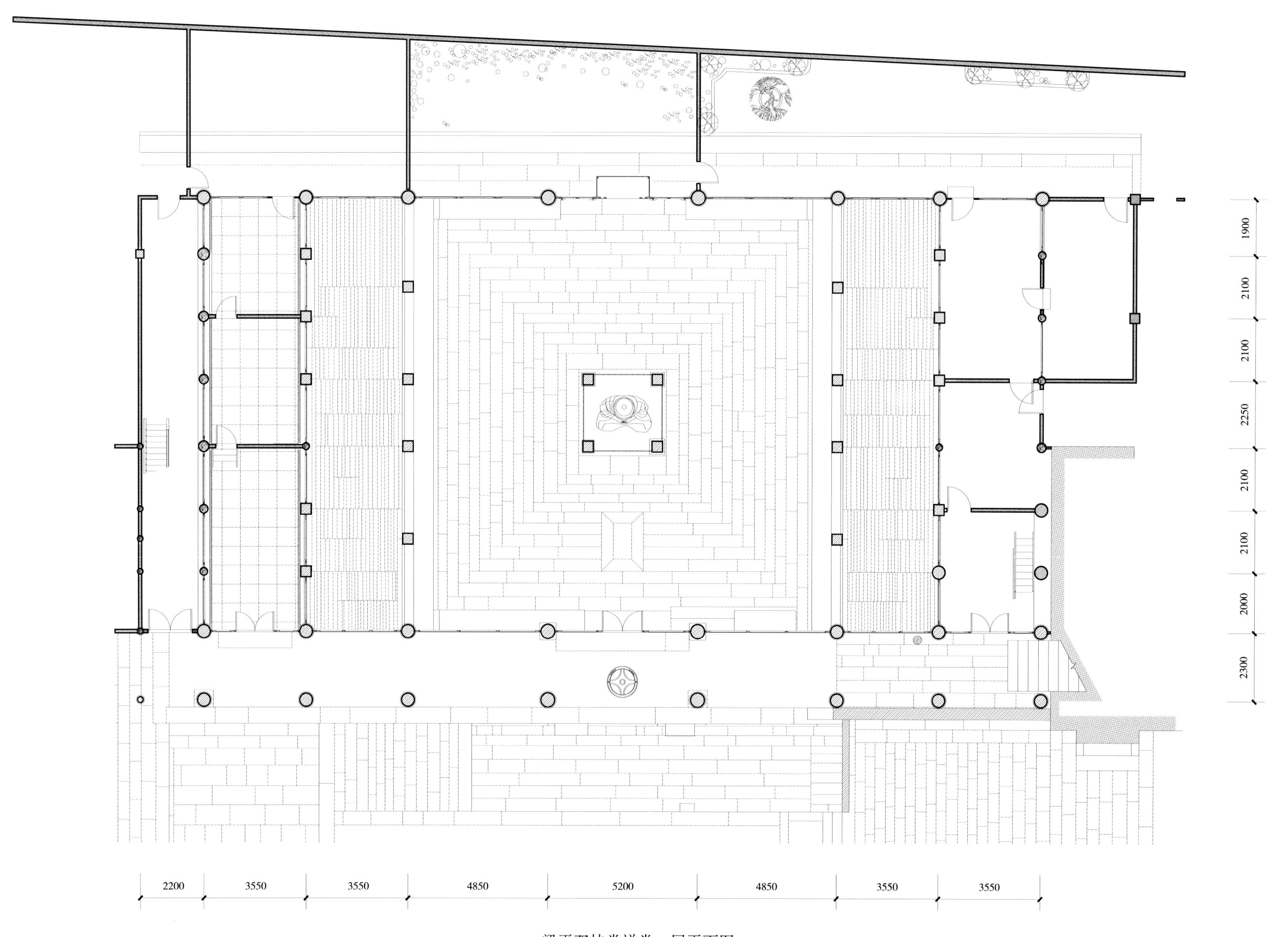
1900
2100
2100
2250
2100
2100
2000
2300
2200
3550
3550
4850
5200
4850
3550
3550
0 1 2 3 4 5m

梁平双桂堂禅堂一层平面图

Ground Floor Plan of Meditation Hall, Shuanggui Temple, Liangping

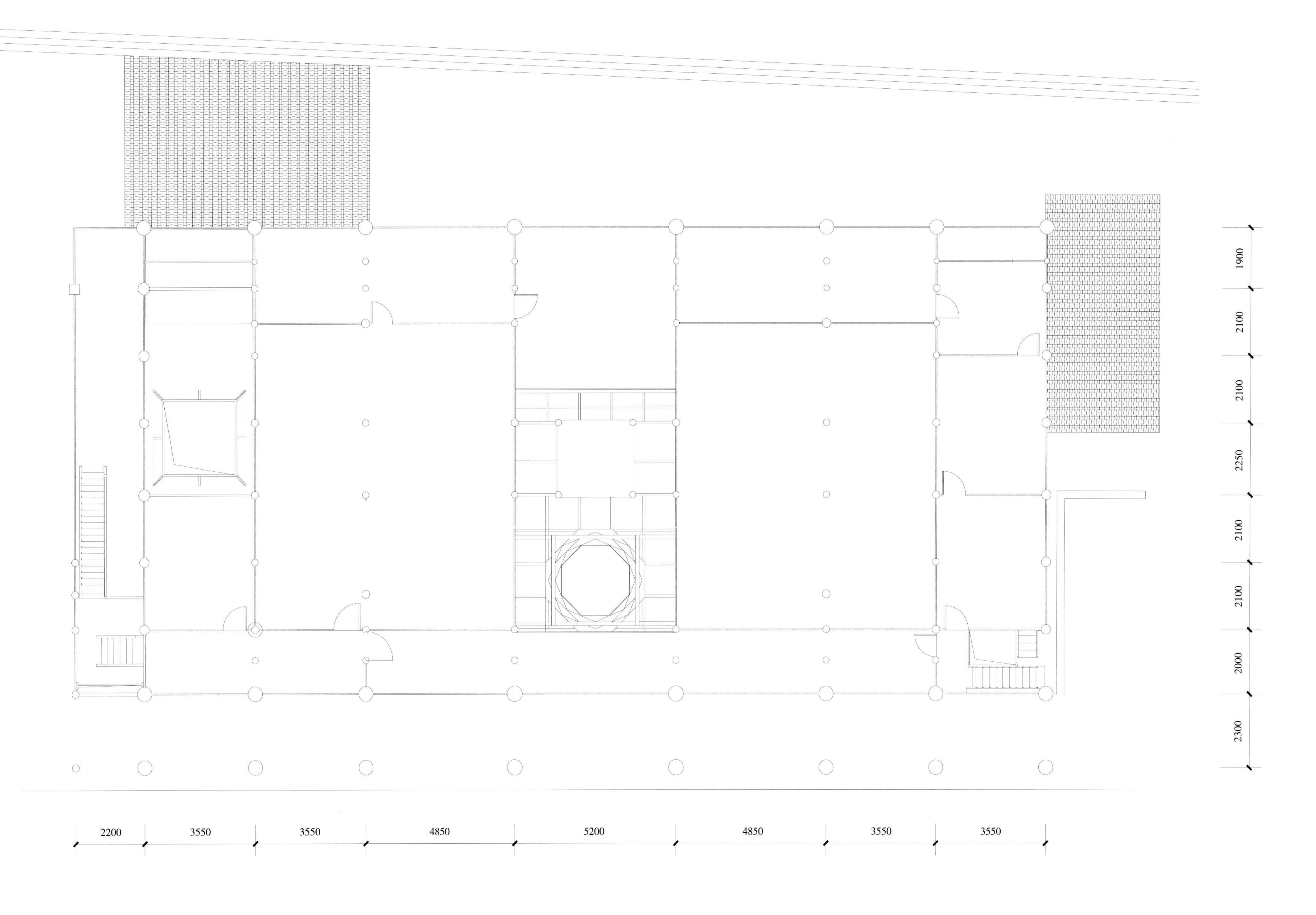
1900
2100
2100
2250
2100
2100
2000
2300
2200
3550
3550
4850
5200
4850
3550
3550
0 1 2 3 4 5m

梁平双桂堂禅堂二层平面图
First Floor Plan of Meditation Hall, Shuanggui Temple, Liangping

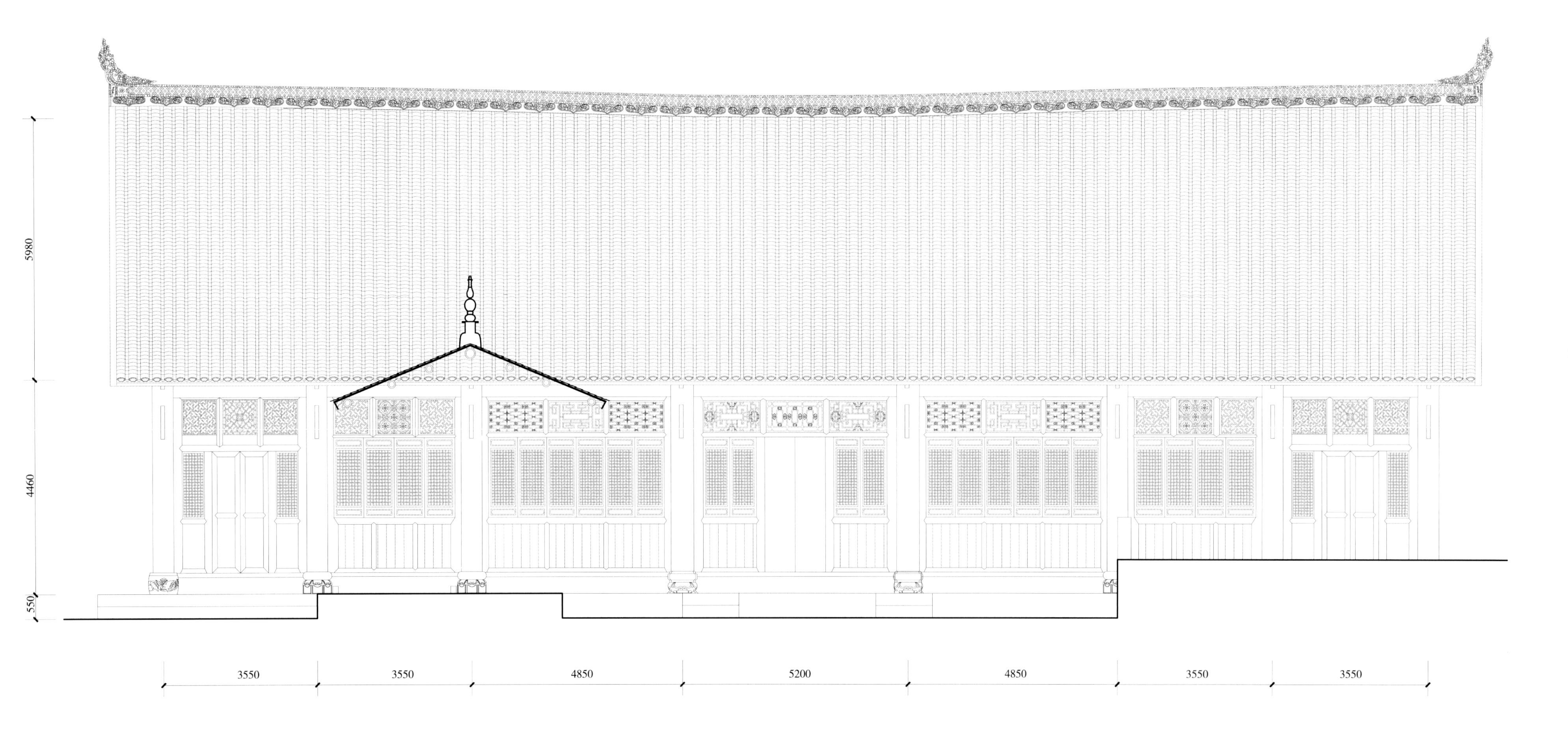
5980
4460
550
3550
3550
4850
5200
4850
3550
3550

梁平双桂堂禅堂正立面图
Front Elevation of Meditation Hall, Shuanggui Temple, Liangping

0 1 2 3 4 5m

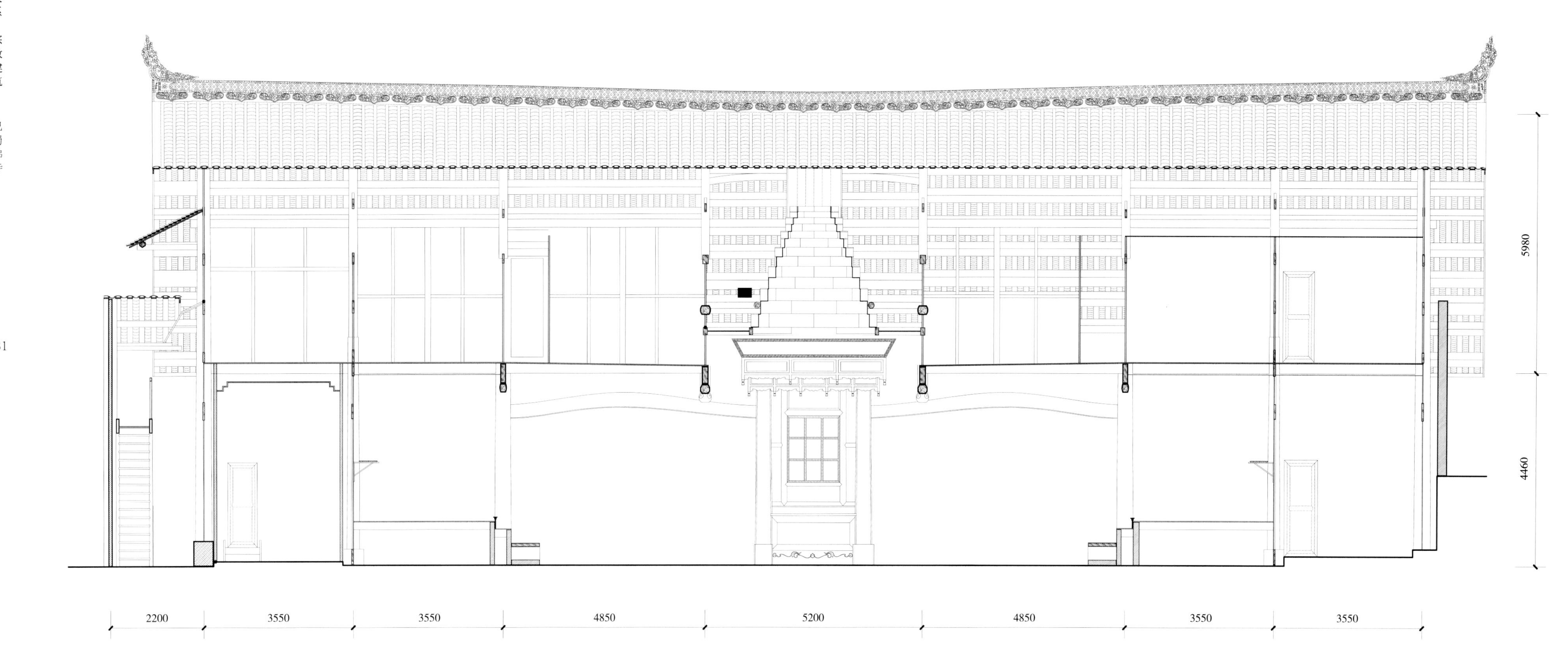
5980
4460
2200
3550
3550
4850
5200
4850
3550
3550

梁平双桂堂禅堂纵剖面图
Longitudinal Section of Meditation Hall, Shuanggui Temple, Liangping

0 1 2 3 4 5m

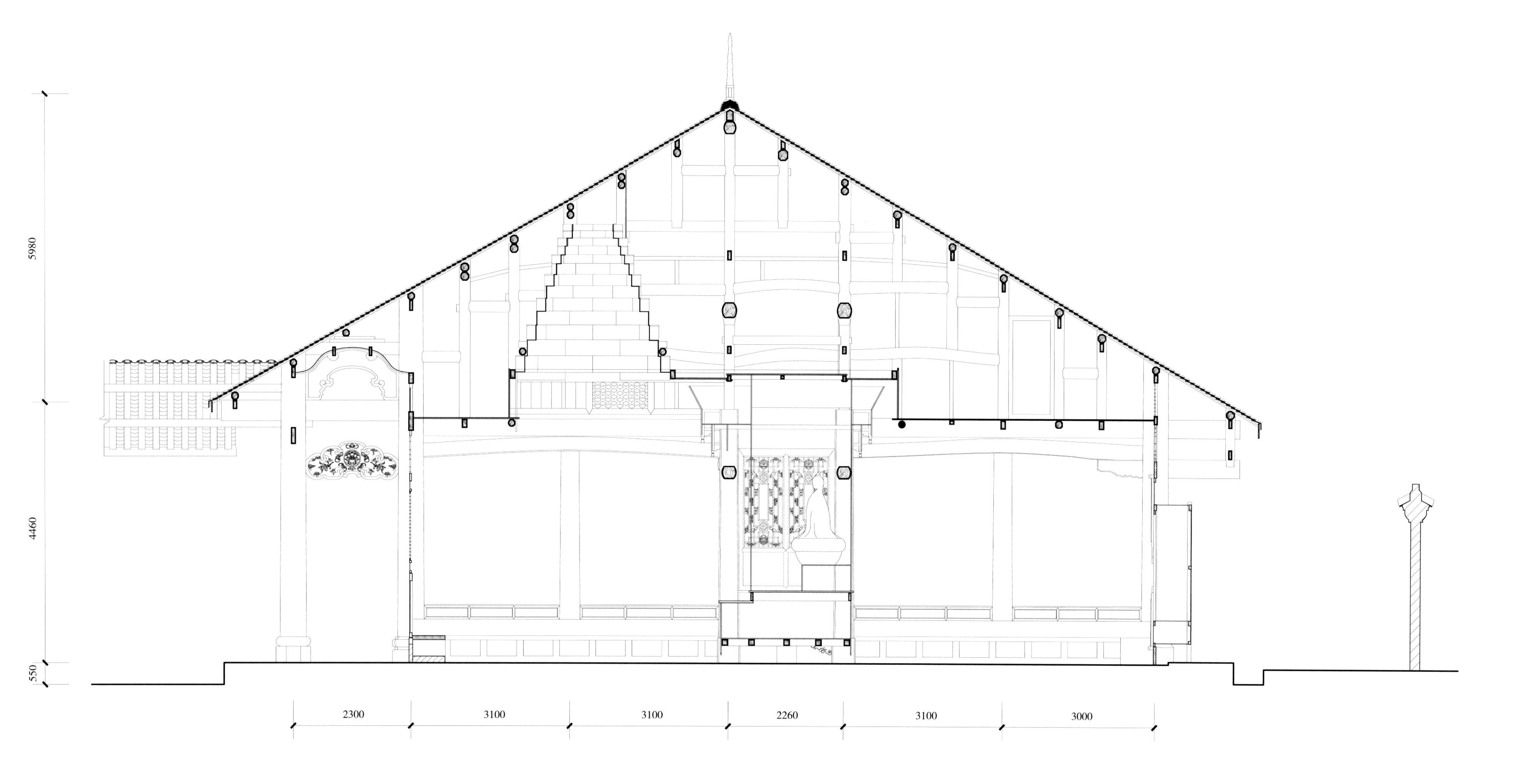
5980
4460
550
2300
3100
3100
2260
3100
3000
0
1
2
3
4
5m

梁平双桂堂禅堂明间横剖面图
Cross Section of Central Bay (Mingjian), Meditation Hall, Shuanggui Temple, Liangping

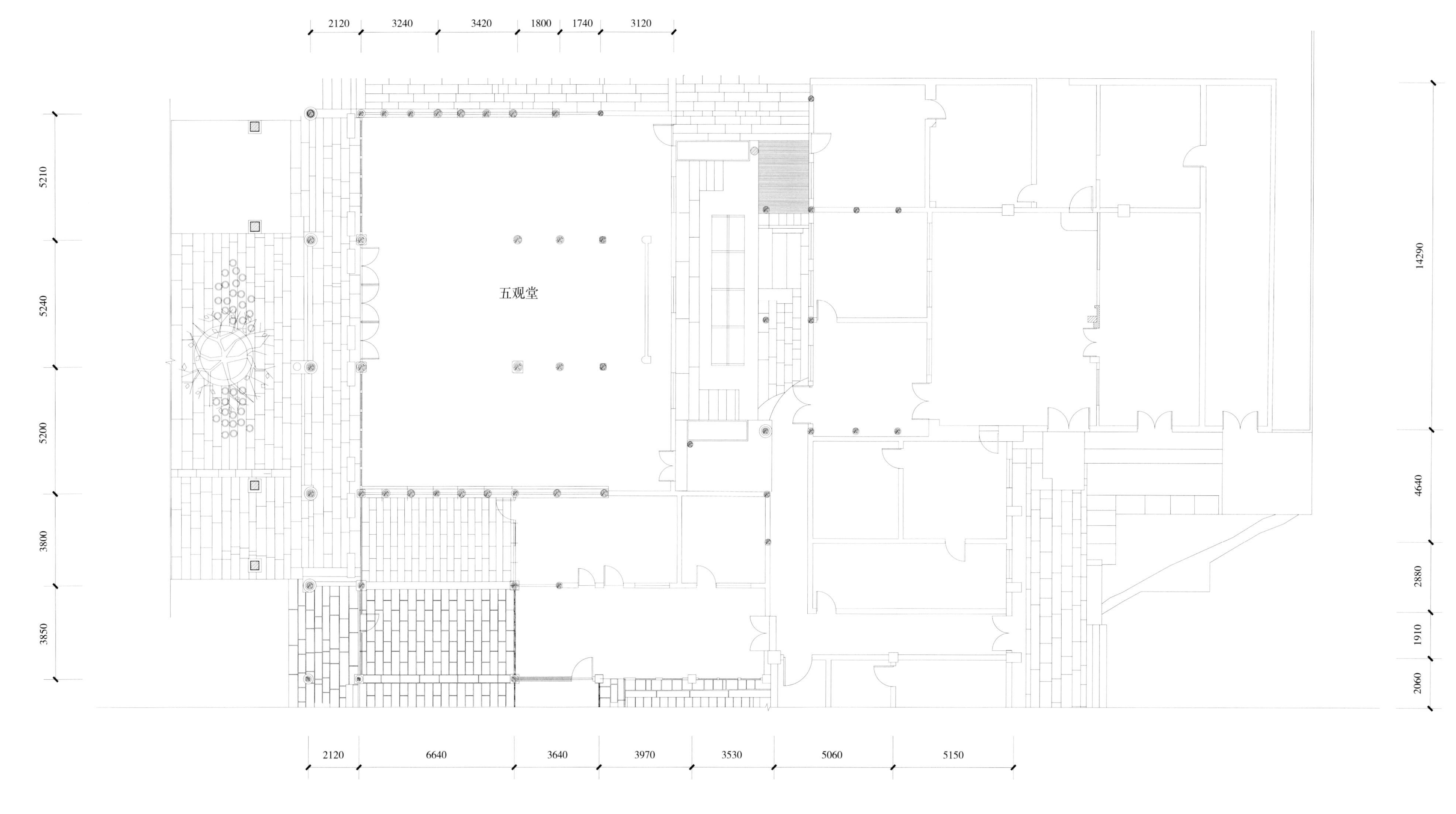
2120
3240
3420
1800
1740
3120
5210
5240
5200
3800
3850
五观堂
14290
4640
2880
1910
2060
2120
6640
3640
3970
3530
5060
5150
0 1 2 3 4 5m

梁平双桂堂五观堂平面图

Ground Floor Plan of Wuguan Tang, Shuanggui Temple, Liangping

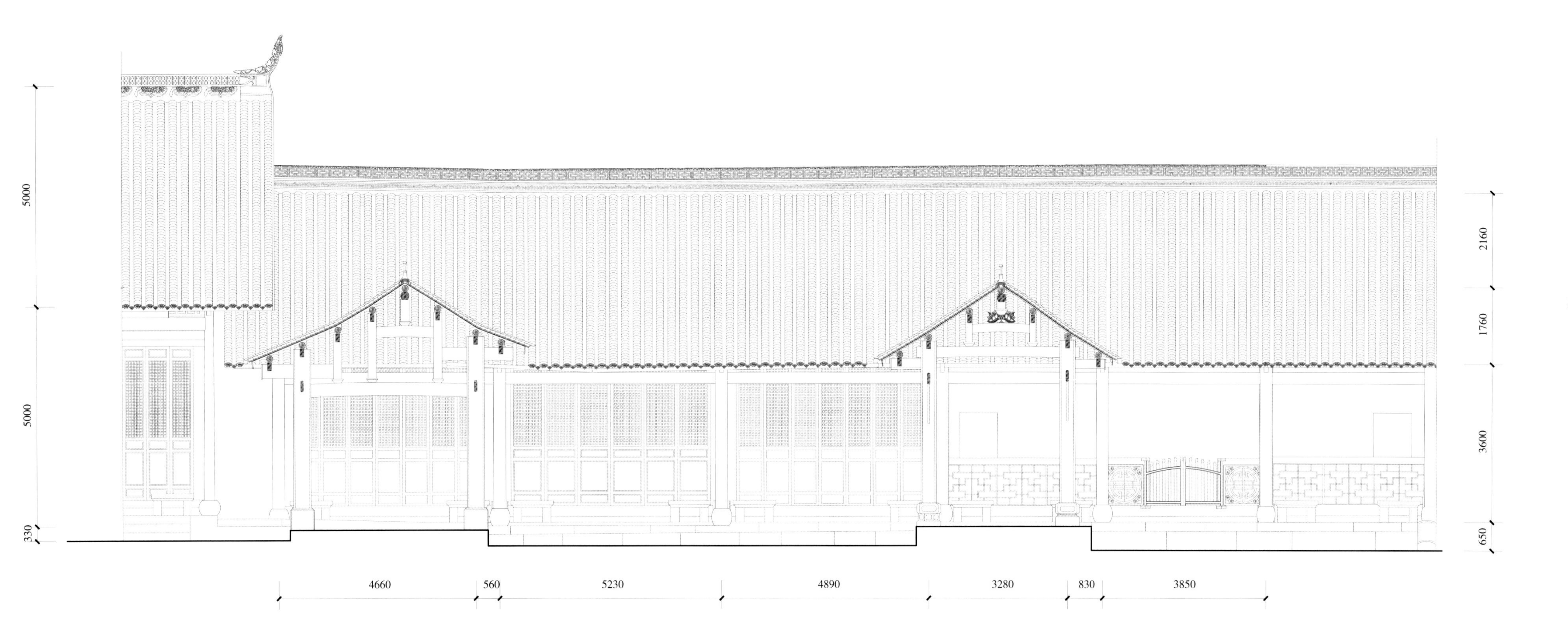
5000
5000
330
2160
1760
3600
650
4660
560
5230
4890
3280
830
3850

梁平双桂堂五观堂正立面图
Front Elevation of Wuguan Tang, Shuanggui Temple, Liangping

0 1 2 3 4 5m

2120
3240
3420
1800
11000
5300
11240
2880

五观堂

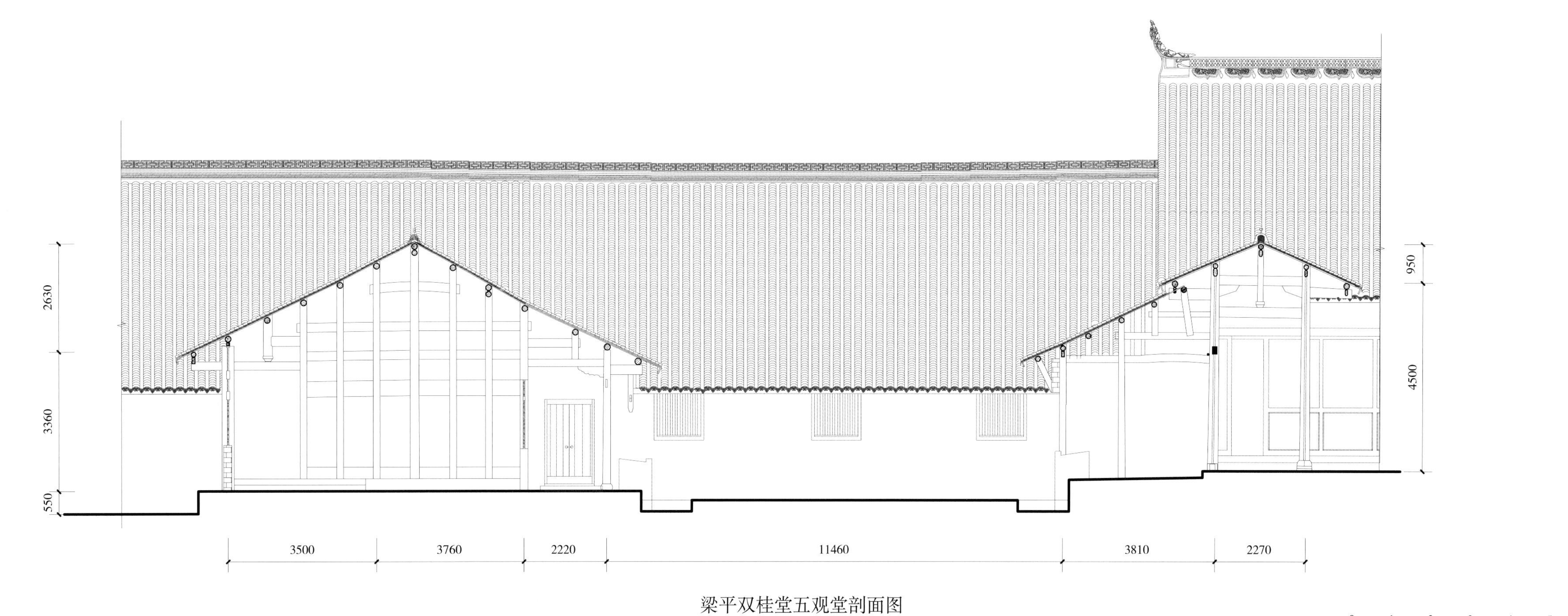
2630
3360
550
950
4500
3500
3760
2220
11460
3810
2270
0 1 2 3 4 5m

梁平双桂堂五观堂剖面图

Cross Section of Wuguan Tang, Shuanggui Temple, Liangping

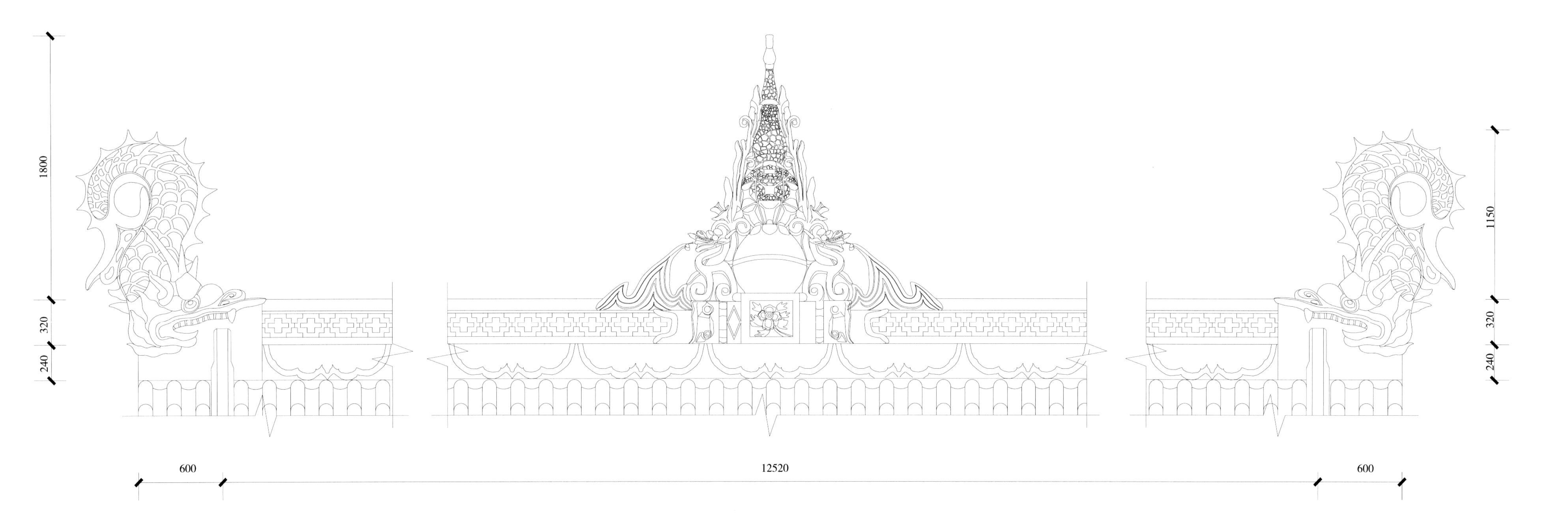
1800
320
240
1150
320
240
600
12520
600

梁平双桂堂南山门正脊式样图

Detail, Main Ridge Decoration, South Main Gate (Shan Men), Shuanggui Temple, Liangping

0
1m

3250
650
1300
30100
1300

梁平双桂堂大悲殿正脊式样图
Detail, Main Ridge Decoration, Dabei Hall, Shuanggui Temple, Liangping

0 1m

梁平双桂堂放生池拱桥望柱石雕图案
Detail, Engraved Pattern, Arched Stone Bridge over Free Life Pond, Shuanggui Temple, Liangping

0 10 20 30cm

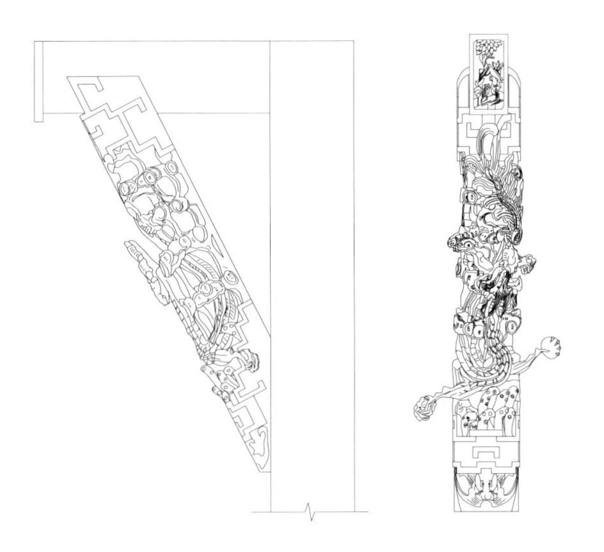
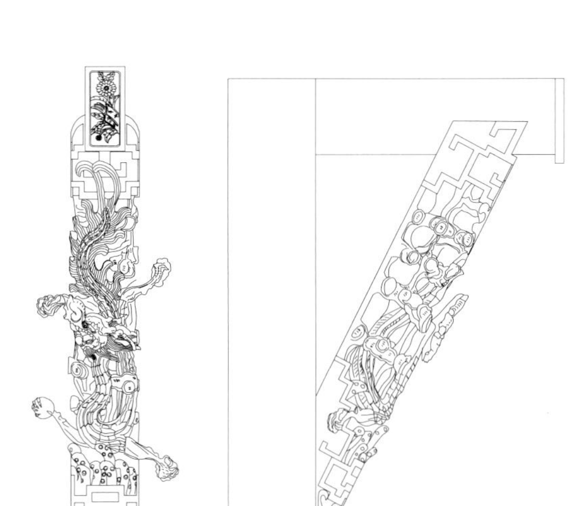
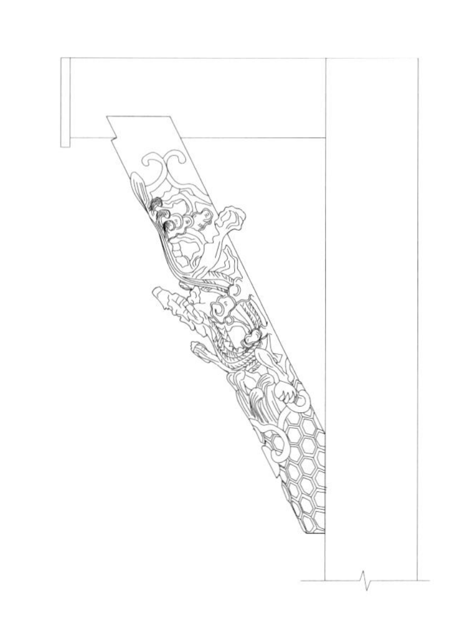
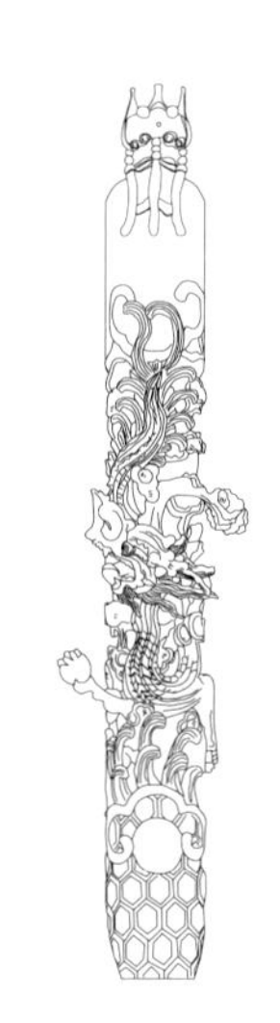
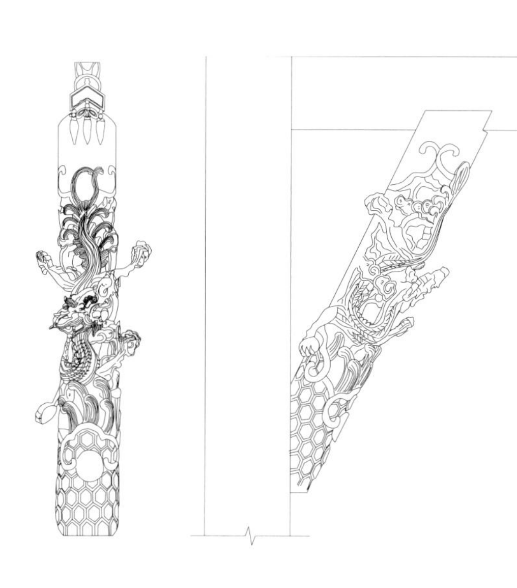

梁平双桂堂大雄宝殿撑栱式样图
Detail, Supporting Gong, Mahavira Hall, Shuanggui Temple, Liangping

0 1m

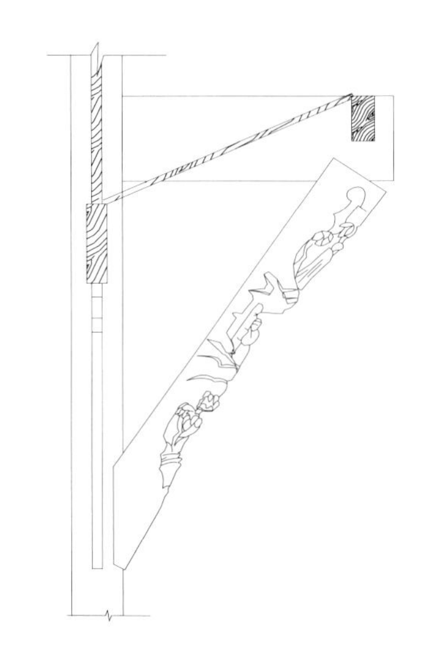

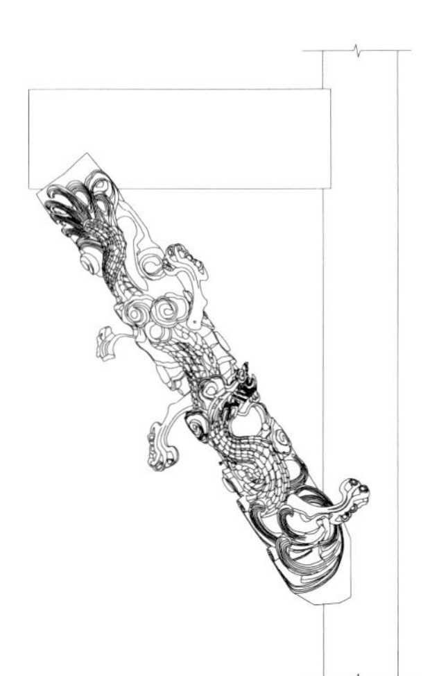
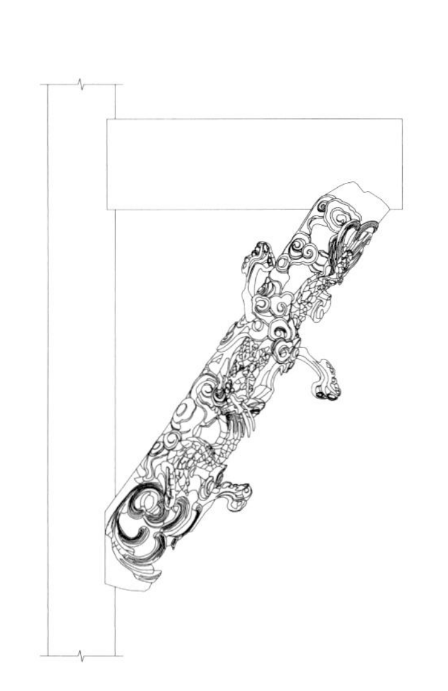
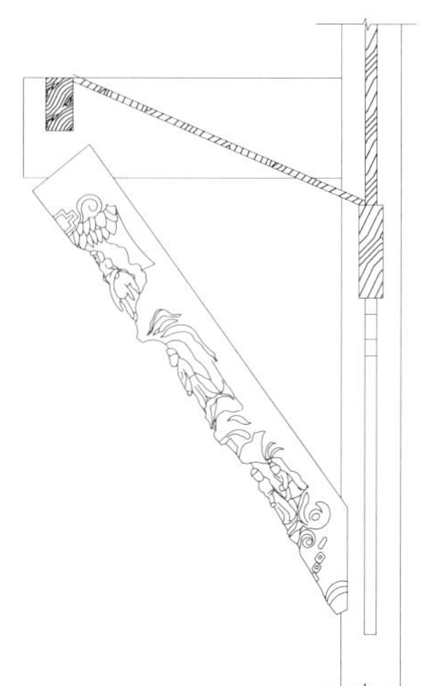

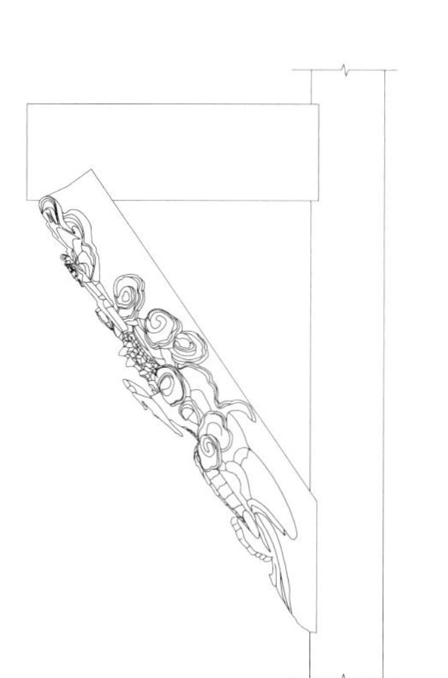
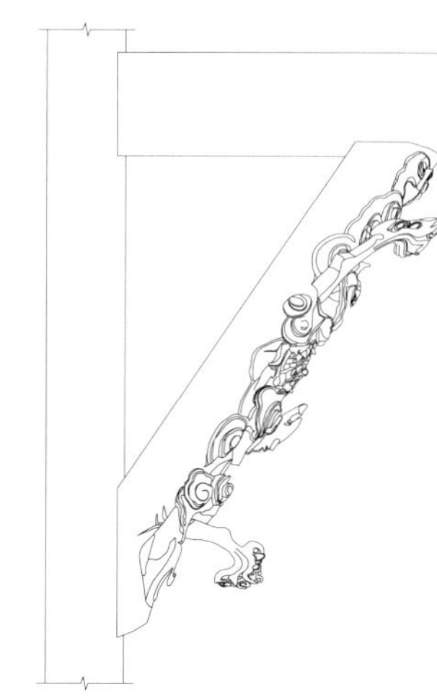

梁平双桂堂藏经楼撑栱式样图
Detail, Supporting Gong, Sutrua Library, Shuanggui Temple, Liangping

0 1m

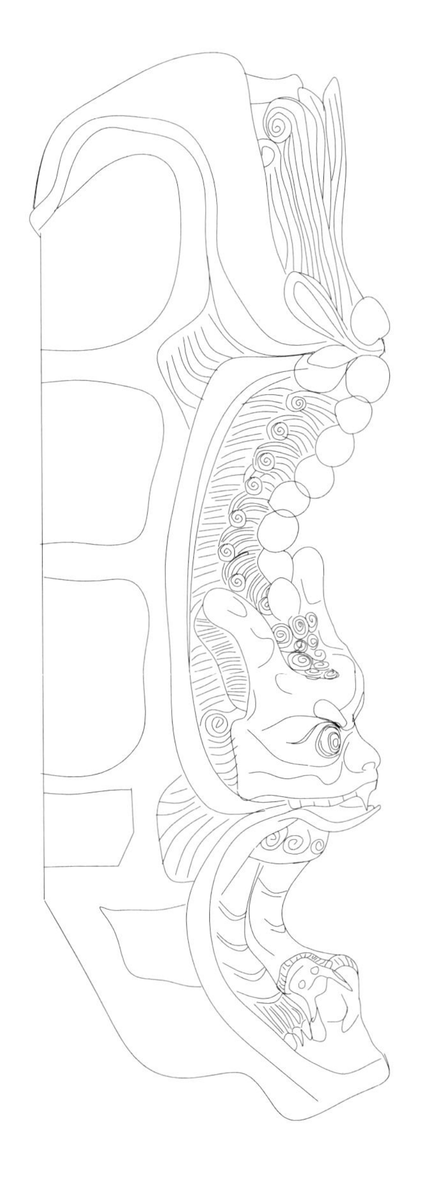
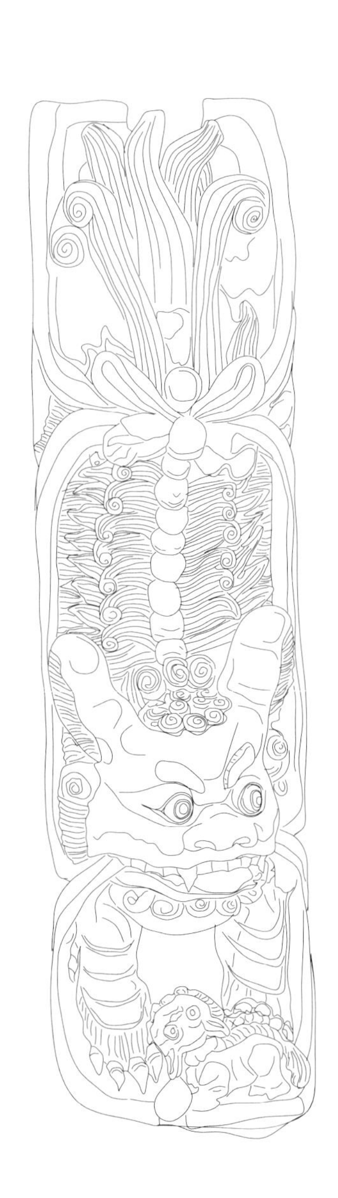
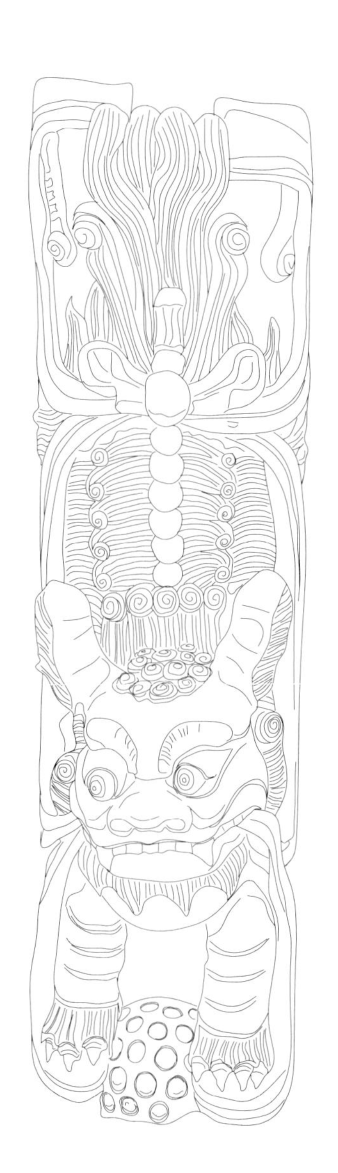
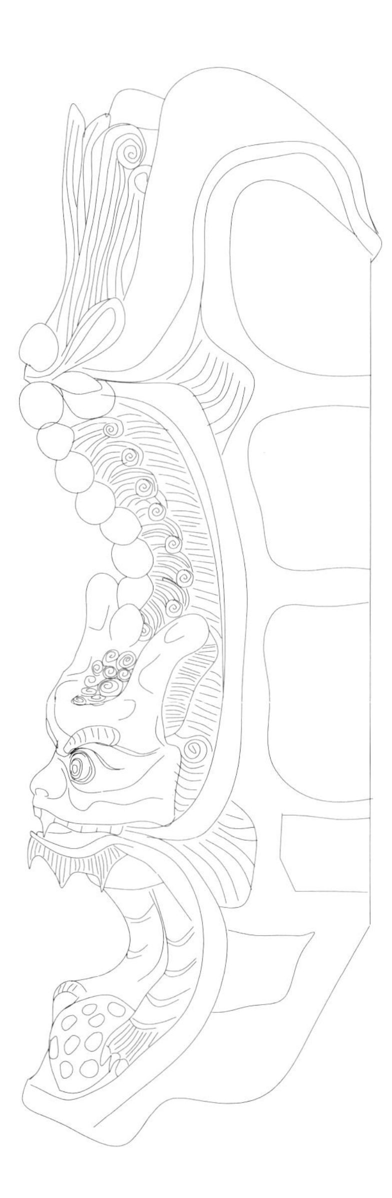

梁平双桂堂大雄宝殿撑栱式样图
Detail, Supporting Gong, Mahavira Hall, Shuanggui Temple, Liangping

0 1m

0 1m

梁平双桂堂南山门扇面枋式样图
Detail, Shanmian Fang, South Main Gate (Shan Men), Shuanggui Temple, Liangping

梁平双桂堂禅堂扇面枋式样图
Detail, Shanmian Fang, Meditation Hall, Shuanggui Temple, Liangping

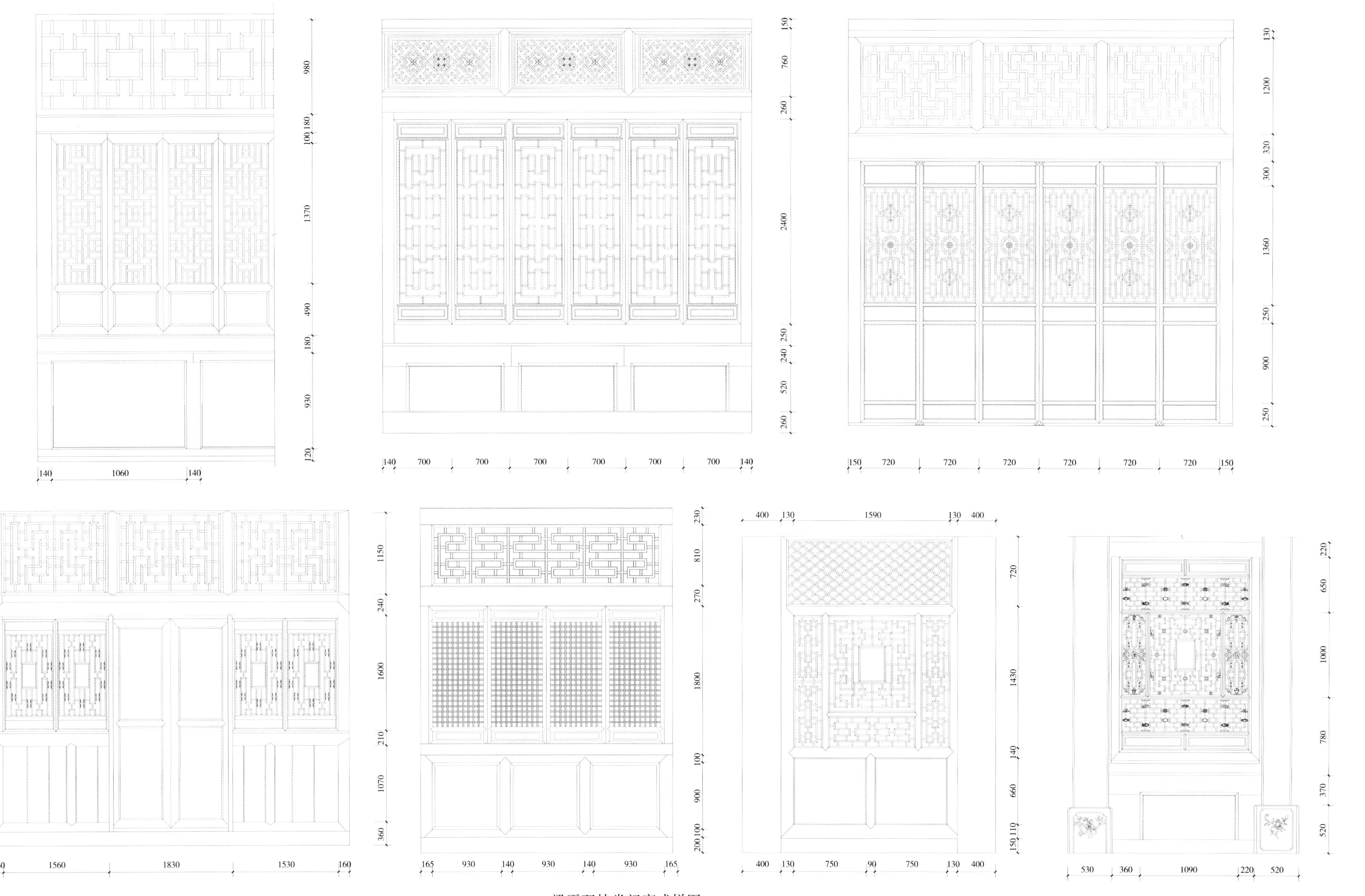
980
100 180
1370
490
180
930
120
140
1060
140
150
760
260
2400
250
240
520
260
140
700
700
700
700
700
700
140
130
1200
320
300
1360
250
900
250
150
720
720
720
720
720
720
150
1150
240
1600
210
1070
360
160
1560
1830
1530
160
230
810
270
1800
100
900
200 100
165
930
140
930
140
930
165
400
130
1590
130
400
720
1430
140
660
150 110
400
130
750
90
750
130
400
220
650
1000
780
370
520
530
360
1090
220
520

梁平双桂堂门窗式样图

Detail, Doors and Windows, Shuanggui Temple, Liangping

潼南大佛寺

大佛寺位于重庆潼南城西的定明山，亦名南禅院，为唐咸通年间所建，因裸露的大佛摩崖造像，南宋建炎元年间（1127年）曾修楼阁覆之，得名大佛阁。南禅院已毁，现存大佛阁、观音阁、玉皇殿等均有摩崖佛像，是巴蜀地区有特色的摩崖建筑。大佛阁内释迦牟尼造像18.43米，唐宋两代雕凿而成，有唐头宋身之说。殿阁总高33米，外部造型是七重檐式楼阁造型，内部空间实际为上下两层，下层约高20米，以满足大佛造像的空间需求，上层置于崖壁顶部，背面形如重檐歇山殿堂，与山顶已毁的南禅院融为一体。外部七重檐楼阁显得雄伟壮观，内部殿堂高畅明朗。现存建筑为晚清建筑风格，主体结构采用穿斗式构架，层层出挑悬空，构架一端嵌固在崖壁上，创造了观赏大佛造像的良好视觉空间环境。

Great Buddha Temple at Tongnan

The Great Buddha Temple is found on Dingming Mountain, to the west of Tongnan District, Chongqing. Also known as Nanchan Monastery, it was built during the Xiantong years in Tang Dynasty. When the great cliff Buddha statue in the open air was sheltered by a building in the first year of Jianyan period in Southern Song Dynasty (AD1127), it was called the Great Buddha Building. The Nanchan Monastery has been destroyed, and the existing Great Buddha Building, the Avalokiteshvara Building, and the Jade Emperor Hall, all of them characteristic Ba-Shu cliff buildings, contain cliff Buddha statues. The Sakyamuni statue in the Great Buddha Building, 18.43m high, was continually carved in Tang and Song dynasties, which led to the popular saying of “a Tang head on a Song body”. The total height of the hall is 33m, with an external appearance of a seven-story building, whereas the internal space has two stories. The lower story is about 20m high, to accommodate the space for the great Buddha statue. The upper story sits on top of the cliff. Seen from the rear, it is like a double-eave gabled hall, integrated with the Nanchan Monastery. The external seven-story building has a magnificent look, and the internal hall is high and bright. The existing building belongs to the architectural style of late Qing Dynasty, and for the main structure tying braces are used, with multiple overhangs and frames that are fixed into the cliff. All of these have created a pleasant visual spatial environment for viewing the great Buddha statue.

北
①大佛殿
②观音殿
③玉皇殿
④鉴亭
⑤文物保护石碑
⑥牌坊
0 5 10 15m

潼南大佛寺总平面图
Site Plan of Great Buddha Temple, Tongnan

图一 潼南大佛寺建筑群鸟瞰图

图一 潼南大佛寺建筑群鸟瞰图

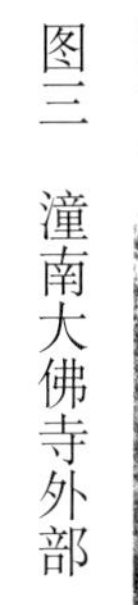

图三 潼南大佛寺外部

图四 潼南大佛寺大佛殿内部

Fig.1 Bird's Eye View of Great Buddha Temple, Tongnan
Fig.2 Bird's Eye View of Great Buddha Temple, Tongnan
Fig.3 Exterior View of Great Buddha Temple, Tongnan
Fig.4 Interior View of Great Buddha Hall, Great Buddha Temple, Tongnan

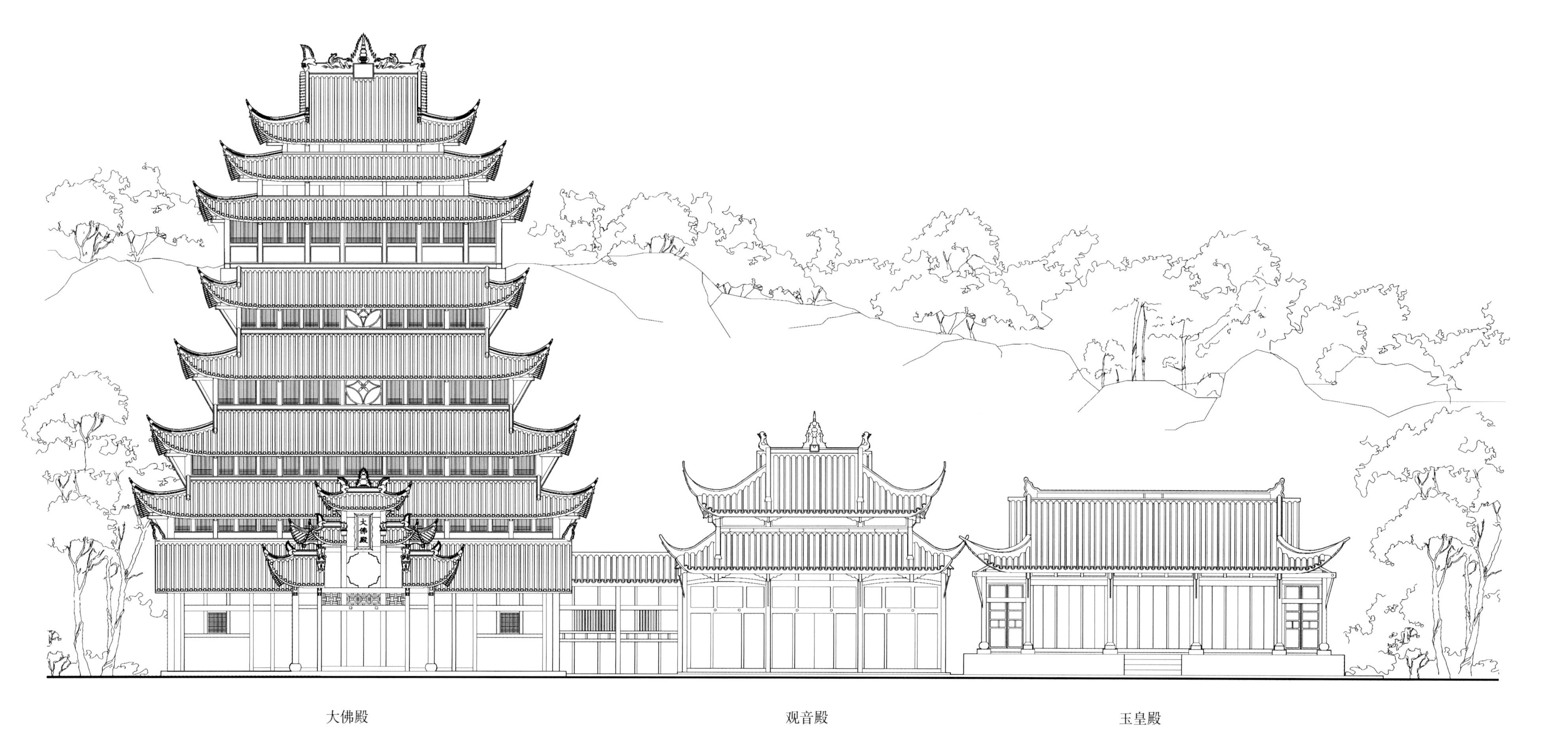
大佛殿
大佛殿
观音殿
玉皇殿
0
5
10m

潼南大佛寺总立面图
Elevation of Entire Compound, Great Buddha Temple, Tongnan

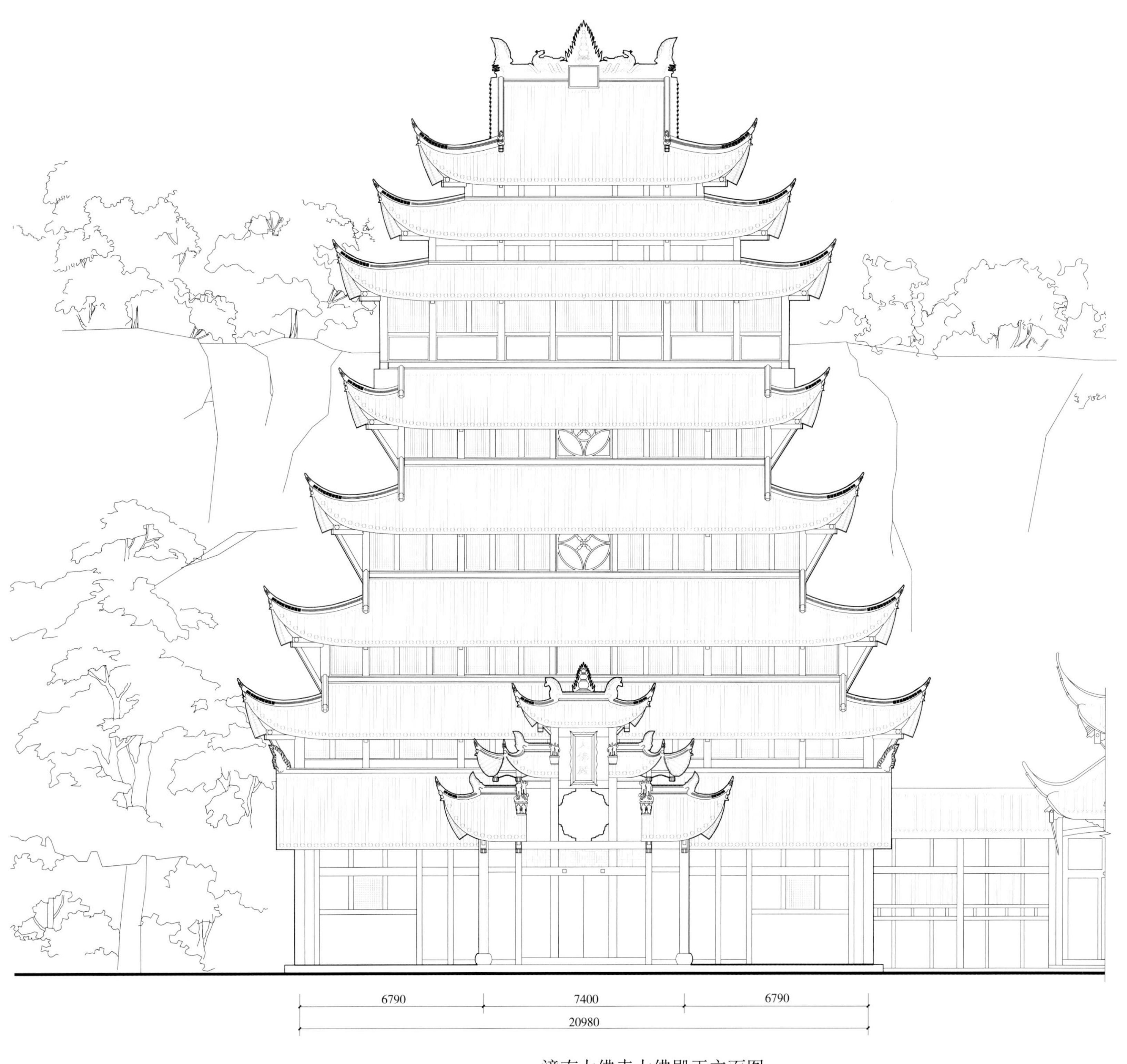
33.560
32.050
30.060
27.810
25.800
23.750
21.500
19.500
17.810
15.500
13.700
11.500
11.060
7.150
4.100
± 0.000
-0.250
1510
1990
2250
2010
2050
2250
2000
1690
2310
1800
2200
440
3910
3050
4100
250
33810
6790
7400
6790
20980

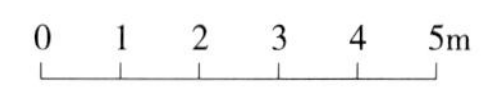
0 1 2 3 4 5m

潼南大佛寺大佛殿正立面图
Front Elevation of Great Buddha Hall, Great Buddha Temple, Tongnan

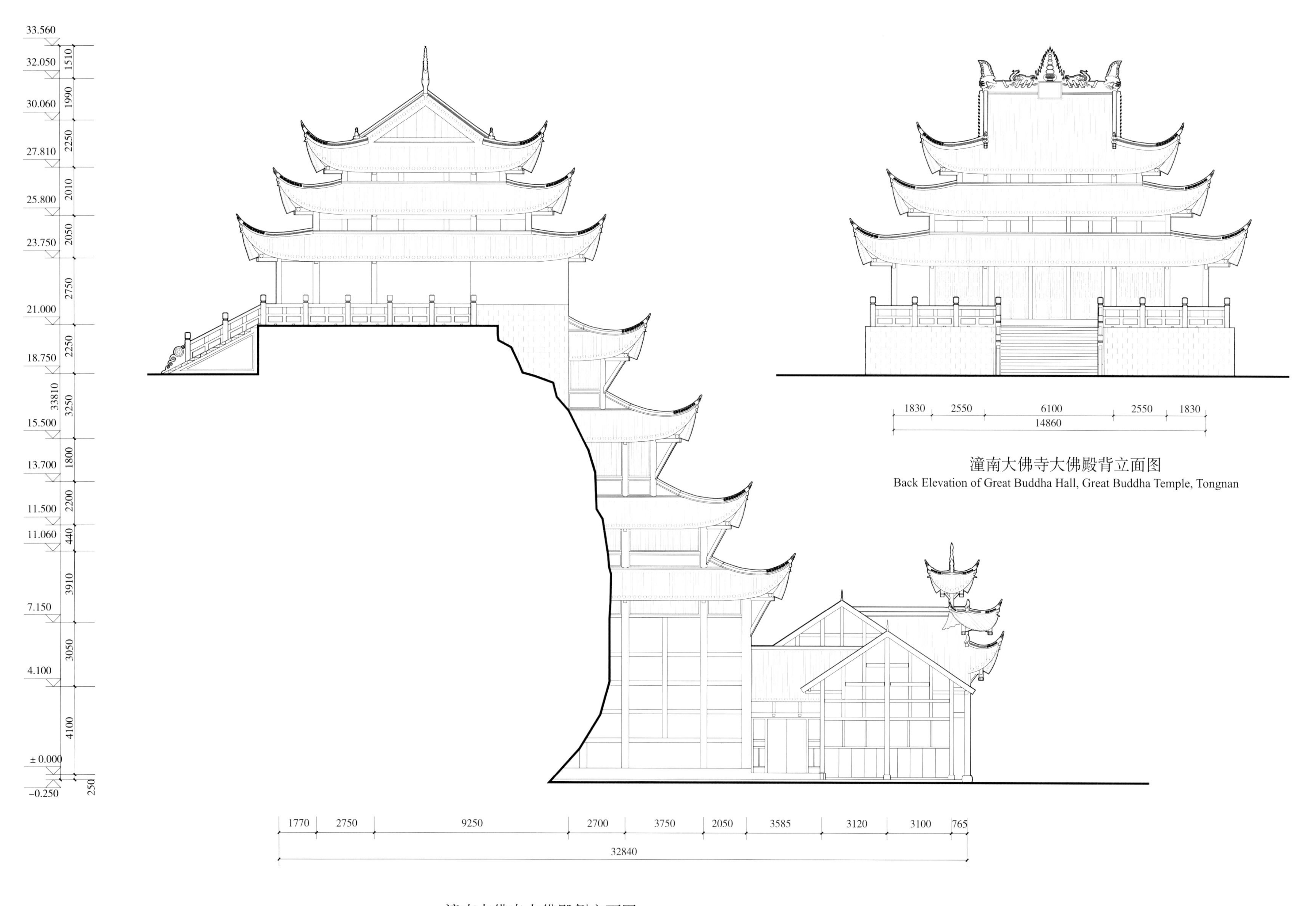
33.560
32.050
30.060
27.810
25.800
23.750
21.000
18.750
15.500
13.700
11.500
11.060
7.150
4.100
± 0.000
−0.250
1510
1990
2250
2010
2050
2750
2250
3250
33810
1800
2200
440
3910
3050
4100
250
1770
2750
9250
2700
3750
2050
3585
3120
3100
765
32840
33.560
32.050
30.060
27.810
25.700
23.650
20.900
18.650
1510
1990
2250
2110
14910
2050
2750
2250
1830
2550
6100
2550
1830
14860
0 1 2 3 4 5m

潼南大佛寺大佛殿背立面图
Back Elevation of Great Buddha Hall, Great Buddha Temple, Tongnan

潼南大佛寺大佛殿侧立面图
Side Elevation of Great Buddha Hall, Great Buddha Temple, Tongnan

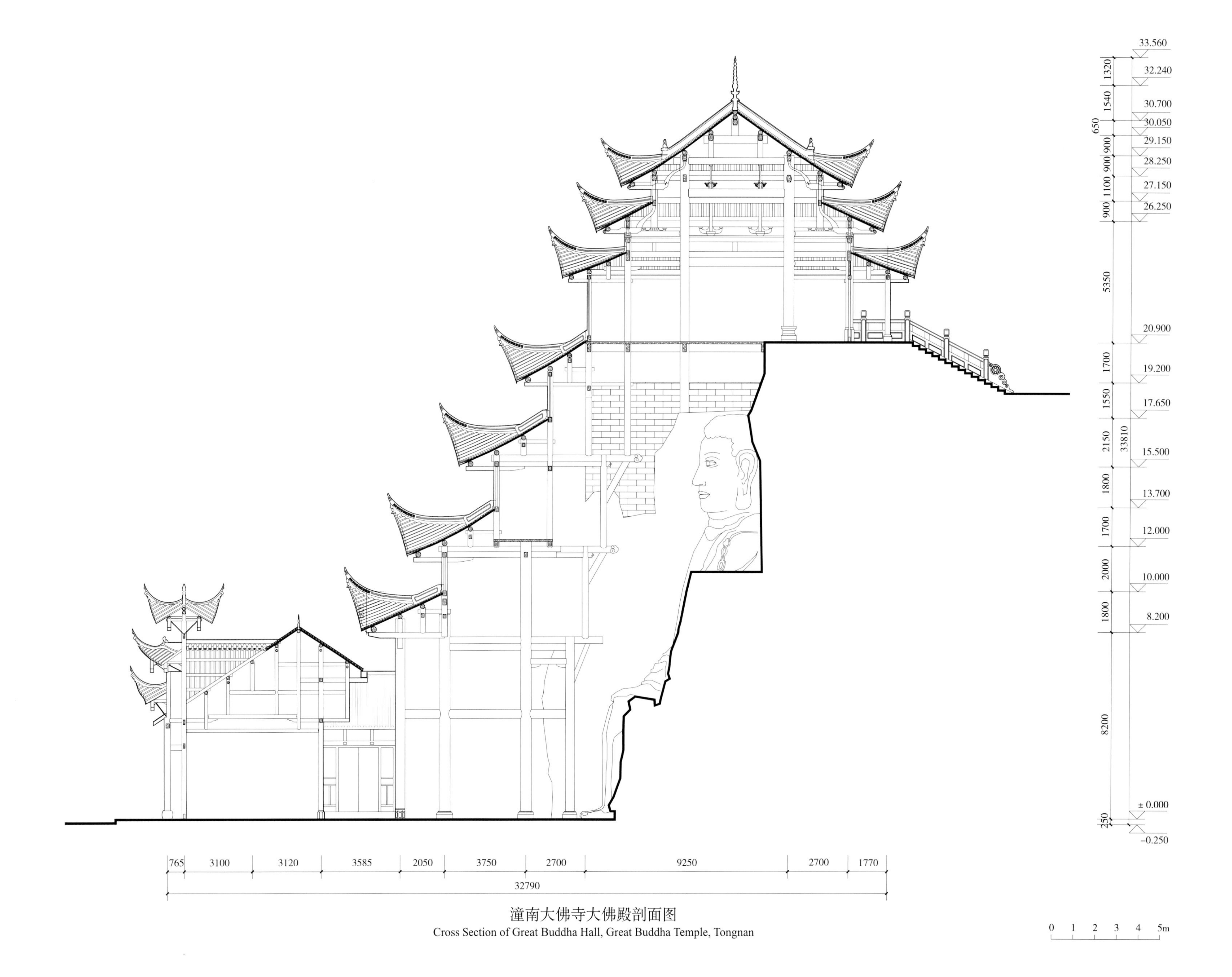
33.560
32.240
30.700
30.050
29.150
28.250
27.150
26.250
20.900
19.200
17.650
15.500
13.700
12.000
10.000
8.200
± 0.000
−0.250
1320
1540
650
900
900
1100
900
5350
1700
1550
2150
33810
1800
1700
2000
1800
8200
250
765
3100
3120
3585
2050
3750
2700
9250
2700
1770
32790
0 1 2 3 4 5m

潼南大佛寺大佛殿剖面图

Cross Section of Great Buddha Hall, Great Buddha Temple, Tongnan

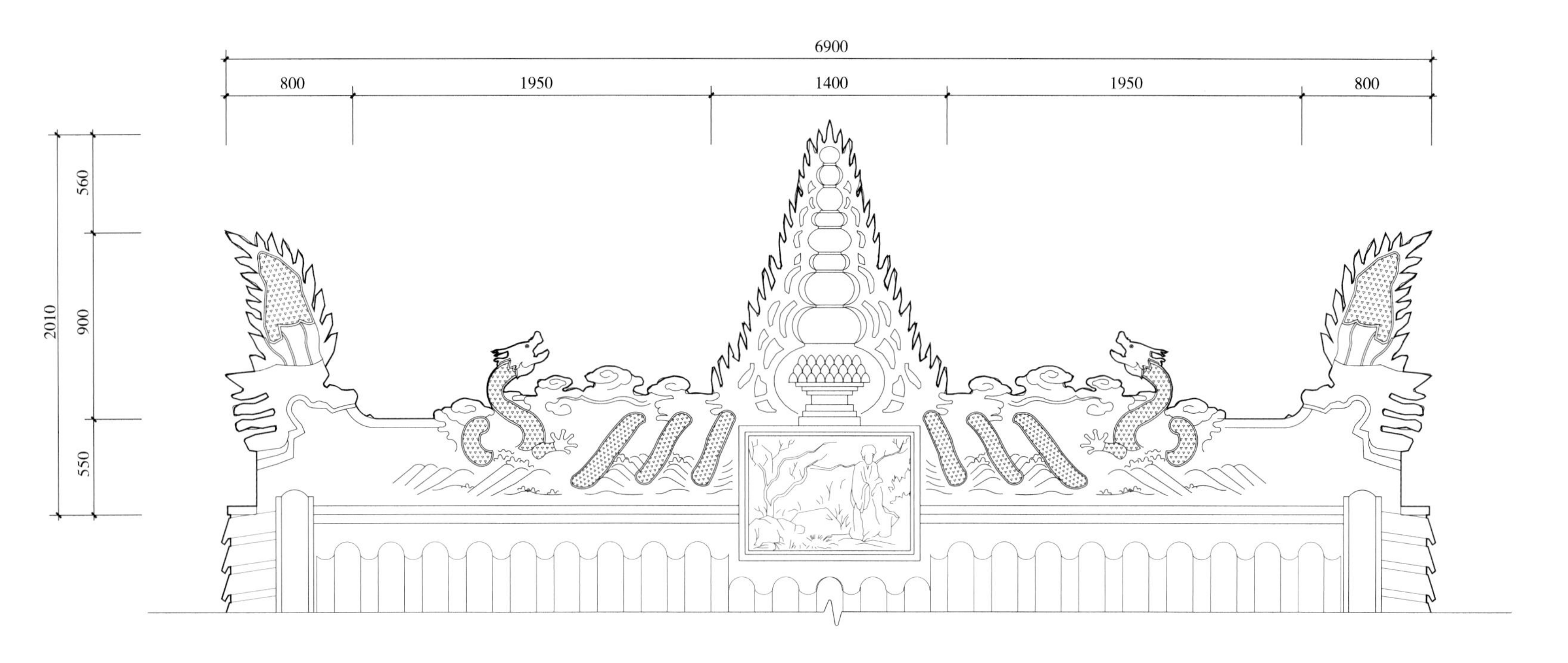
6900
800
1950
1400
1950
800
2010
560
900
550

潼南大佛寺大佛殿脊饰正立面图
Detail, Front Elevation of Main Ridge Decoration, Great Buddha Hall, Great Buddha Temple, Tongnan

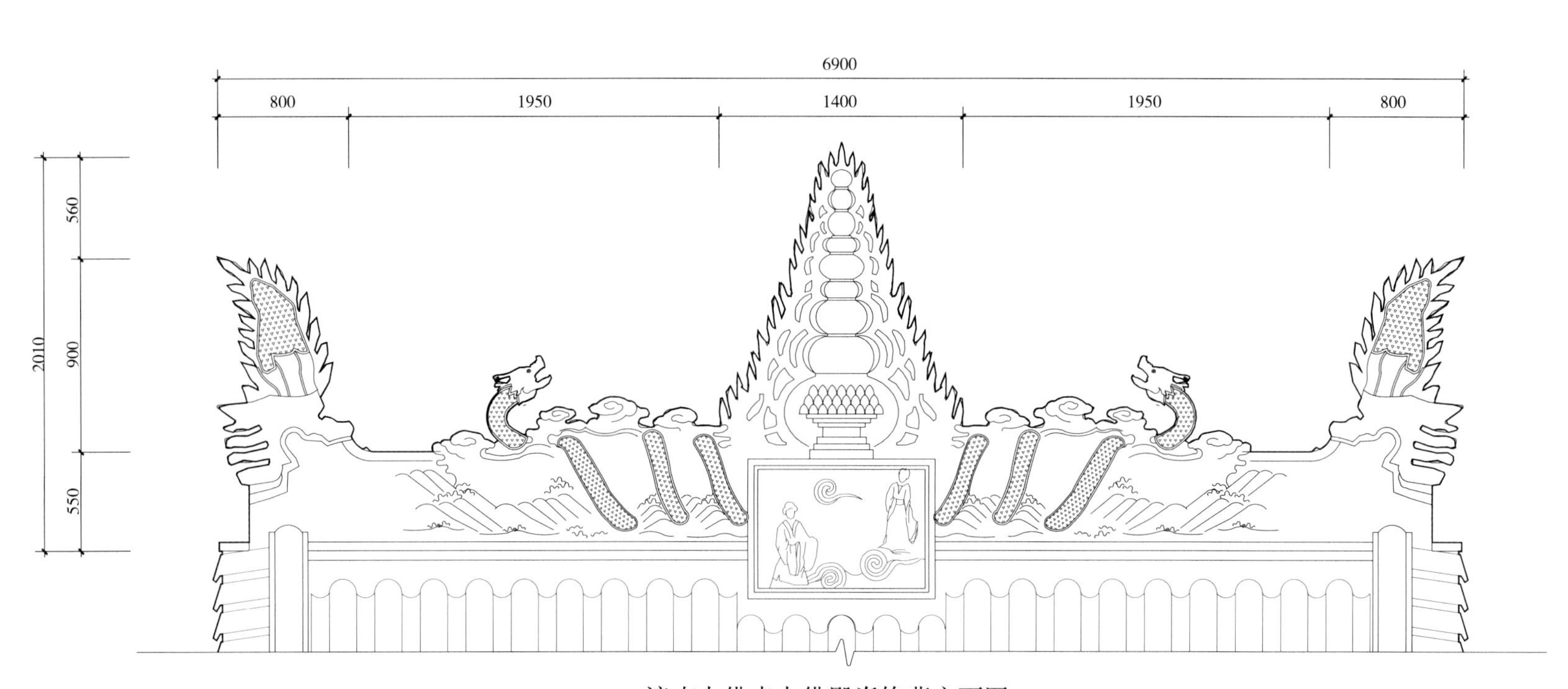
6900
800
1950
1400
1950
800
2010
560
900
550

潼南大佛寺大佛殿脊饰背立面图
Detail, Back Elevation of Main Ridge Decoration, Great Buddha Hall, Great Buddha Temple, Tongnan

潼南大佛寺大佛殿垂脊脊饰式样图
Detail, Hip (Chuiji) Decoration, Great Buddha Hall, Great Buddha Temple, Tongnan

参与测绘及相关工作的人员名单

峨眉山佛寺建筑群测绘

指导教师

邵俊逸　杨嵩林　白佐民　李先逵

本科生

1981级：邓一灵　夏　彦　唐　露　袁晓霞　张　林　艾春梅　张　放　蒋魏川　赵元超　陈　刚
刘　旭　卢春晖　徐　力　邓　凯　沈益明　林晓东　苏晓河　李　纯　郭邦毅　徐　锋
石健和　谭邵江　彭高峰　张　枬　唐　健　谢海春　闫　峰　叶林春　刘成栓　李玉梅
胡可时

1982级：于　佳　张　劲　饶　劼　赵　磊　陈　榕　李　纯　杜春兰　郭　艳　周　波　侯百镇
余保华　邹建华　林瑜初　黄　捷　闫　岩　朱永纯　何浩东　林　嘉　熊建华　田　鸥
曹　群　段晓丹　龚　建　郭　松　胡元月　魏原江　华　强　谭　靖　陈　硕　章兴泉
高力峰　高亮全　周　丹　周西蒙　陈　红　泰　丽　石　维　黄　欢　王　艳　王东方
伍　波　朱　琦　唐　竹　齐　炜　江培格　郑承劼　吕正中　陈　武　王建成　陈建平
罗　兵　陈　进　冀建中　杨晓敏　马列岗　王　培　高小周　沈方涛　樊　勇　余静赣
丁　宏　张克风　刘　涛

后期补充测绘及数据化资料整理

指导教师

张兴国　冯　棣　汪智洋

研究生

莫唯书　刘天琪　晋兆东　查红叶　郭钰晶　黄　鹤　邓小萌　李　翠　文　艺　赵鉴一
赵浏阳　王　盼　俸　瑜　马昕苗　赵　爽　玮　宝　周梦涵

本科生

2014级：葛　臻　何子懿　伍　洲　袁沁心　高秀干　籍梦玥　覃　丹　曾智静　赵晨西　崔若文
周冰鸿　向　兵　罗梦杰　刘世昂　胡曦钰　刘　豪　王于楠　庄惟仁　张政远　沈曼竹
吴卓亚　李枭幸　曾韵璇　刘国阳　付冰昂　于　沐　梁思齐　吕成栋　夏　天　王俊懿
钟　航　杨明语

List of Participants Involved in Surveying and Related Works

Buddhist Temple Complex of Mount Emei Surveying and Mapping

Instructor: SHAO Junyi, YANG Songlin, BAI Zuomin, LI Xiankui

Bachelor Student

1981 Year: DENG Yiling, XIA Yan, TANG Lu, YUAN Xiaoxia, ZHANG Lin, AI Chunmei, HANG Fang, JIANG Weichuan, ZHAO Yuanchao, CHEN Gang, LIU Xu, LU Chunhui, XU Li, DENG Kai, SHEN Yiming , LIN Xiaodong, SU Xiaohe, LI Chun, GUO Bangyi, XU Feng, SHI Jianhe, TAN Shaojiang, PENG Gaofeng, ZHANG Zhan, TANG Jian, XIE Haichun, YAN Feng, YE Linchun, LIU Chengshuan, LI Yumei, HU Keshi

1982 Year: YU Jia, ZHANG Jin, RAO Jie, ZHAO Lei, CHEN Rong, LI Chun, DU Chunlan, GUO Yan, ZHOU Bo, HOU Baizhen, YU Baohua, ZOU Jianhua, LIN Yuchu, HUANG Jie, YAN Yan, ZHU Yongchun, HE Haodong, LIN Jia, XIONG Jianhua, TIAN Ou, CAO Qun, DUAN Xiaodan, GONG Jian, GUO Song, HU Yuanyue, WEI Yuanjiang, HUA Qiang, TAN Jing, CHEN Shuo, ZHANG Xingquan, GAO Lifeng, GAO Liangquan, ZHOU Dan, ZHOU Ximeng, CHEN Hong,
TAI Li, SHI Wei, HUANG Huan, WANG Yan, WANG Dongfang, WU Bo, ZHU Qi, TANG Zhu, QI Wei, JIANG Peige, ZHENG Chengjie, LV Zhengzhong, CHEN Wu, WANG Jiancheng, CHEN Jianping, LUO Bing, CHEN Jin, JI Jianzhong, YANG Xiaomin, MA Liegang, WANG Pei, GAO Xiaozhou, SHEN Fangtao, FAN Yong, YU Jinggan, DING Hong, ZHANG Kefeng, LIU Tao

Later Supplementary Surveying and Data Processing

Instructor: ZHANG Xingguo, FENG Di, WANG Zhiyang

Master Student

MO Weishu, LIU Tianqi, JIN Zhaodong, CHA Hongye, GUO Yujing, HUANG He, DENG Xiaomeng, LI Cui, WEN Yi, ZHAO Jianyi, ZHAO Liuyang, WANG Pan, FENG Yu, MA Xinzhuo, ZHAO Shuang, WEI Bao, ZHOU Menghan

Bachelor Student

2014 Year: GE Zhen, HE Ziyi, WU Zhou, YUAN Qinxin, GAO Xiugan, JI Mengyue, TAN Dan, CENG Zhijing, ZHAO Chenxi, CUI Ruowen, ZHOU Binghong, XIANG Bing, LUO Mengjie, LIU Shiang, HU Xiyu, LIU Hao, WANG Yunan, ZHUANG Weiren, ZHANG Zhengyuan, SHEN Manzhu, WU Zhuoya, LI Xiaoxing, CENG Yunxuan, LIU Guoyang, FU Bingang, YU Mu, LIANG Siqi, LV Chengdong, XIA Tian, WANG Junyi, ZHONG Hang, YANG Mingyu

Bao'en Temple in Pingwu County Surveying and Mapping

Instructor: YE Qishen, BAI Zuomin, LI Xiankui, SHAO Junyi

Master Student

ZHANG Shuming, PENG Wenzheng, CUI Yanyu, ZHANG Yujia, ZHANG Zhuling, YIN Zixiang, WU Yilin, ZHANG Xiaolei

Bachelor Student

1977 Year: ZOU Zhengyu, WANG Dingshi, SHI Weilin, XU Yi, LI Yaqin, TAN Ying, YANG Wenhua, LI Xiaojing, MAO Yongning, ZHAO Hongyu, CAO Guanghui, HU Yi, HUANG Ping, LIU Zhengrong, WANG Qiushi, XU Xingchuan, ZHONG Hua, LIU Jiakun, PAN An, HUA Lin, WANG Haoming, PAN Guozhu, HU Xiaobin, DAI Zhizhong, YANG Wenyan, FU Qin, PAN Xin, GUO Deqiao, LU Qi, FU Haicong, CENG Ziwen, SUN Ping, FAN Jinhua, YANG Ying, YIN Yuanliang, LI Qi, CHENG Wende

Later Supplementary Surveying and Data Processing

Instructor: GUO Xuan, DAI Qiusi

Bachelor Student

2006 Year: ZHANG Pu, FAN Juanjuan, WANG Yingnan, JIN Shengxian, DI Yibo, YU Dao, WU Xiaofan, YAN Xinyi, GAO Shuai, CHEN Si, XU Teng, LUO Mi, YAO Yuan, SHUAI Yannan, ZHU Guangxu, WANG Yuqi, GUO Yu, XU Qi, YANG

平武报恩寺测绘

指导教师

叶启燊 白佐民 李先逵 邵俊义等

本科生

1977级：邹正瑜 王丁士 施维琳 徐 艺 李亚琴 谭 英 杨文华 李小静 毛永宁 赵洪宇
曹光辉 胡 翌 黄 平 刘正荣 王求是 徐行川 钟 华 刘家琨 潘 安 华 林
王浩明 潘国柱 胡小滨 戴志中 杨文炎 付 勤 潘 欣 郭德桥 陆 琦 付海聪
曾子文 孙 平 范晋华 杨 鹰 尹元良 李 奇 程文德

后期补充测绘及数据化资料整理

指导教师

郭 璇 戴秋思

研究生

张书铭 彭文峥 崔燕宇 张羽佳 张著灵 尹子祥 吴奕霖 张潇镭

本科生

2006级：张 溥 范娟娟 王英楠 金圣现 翟逸波 余 岛 吴晓帆 晏心奕 高 帅 陈 斯
徐 腾 罗 米 姚 远 帅彦男 朱光旭 王玉琦 郭 彧 徐 绮 杨翔宇 何鹏程
刘生兵 杨成垒 刘 菁 陈力然 王 聪 舒晨箫 江玉林 蔡尚君 张 寒 于 杨
张文青 张 亮 刘 锦 樊俊苏 陈劲帆 何 磊 张欢欢 潘 皓 付 逸

大足圣寿寺测绘

指导老师

李先逵 邱书杰 谢吾同

本科生

1989级：龙 亮 刘晓燕 孙 智 薛昆陵 王 锴 郭春胜 程 涛 龚六六 王庆广 范晓东
赖俭醒 李 忠 米 军 唐桂斌 高 东 徐 恒 郭 伟 张春生 张健飞 王 蕾
黄 茵 张 梅 胡江淳 张 黎 徐林华 银 华 金 虹 王雪然 赵 亮 李 韬
唐 华

后期补充测绘及数据化资料整理

指导教师

郭 璇 戴秋思 冷 婕

研究生

崔燕宇 张羽佳 张著灵 尹子祥 吴奕霖 杨浩祥 张潇镭

本科生

2006级：张 溥 范娟娟 王英楠 金圣现 翟逸波 余 岛 吴晓帆 晏心奕 高 帅 陈 斯
徐 腾 罗 米 姚 远 帅彦男 朱光旭 王玉琦 郭 彧 徐 绮 杨翔宇 何鹏程
刘生兵 杨成垒 刘 菁 陈力然 王 聪 舒晨箫 江玉林 蔡尚君 张 寒 于 杨
张文青 张 亮 刘 锦 樊俊苏 陈劲帆 何 磊 张欢欢 潘 皓 付 逸

Xiangyu, HE Pengcheng, LIU Shengbing, YANG Chenglei, LIU Jing, CHEN Liran, WANG Cong, SHU Chenxiao, JIANG Yulin, CAI Shangjun, ZHANG Han, YU Yang, ZHANG Wenqing, ZHANG Liang, LIU Jin, FAN Junsu, CHEN Jinfan, HE Lei, ZHANG Huanhuan, PAN Hao, FU Yi

Shengshou Temple at Dazu Surveying and Mapping

Instructor: LI Xiankui, QIU Shujie, XIE Wutong

Bachelor Student

1989 Year: LONG Liang, LIU Xiaoyan, SUN Zhi, XUE Kunling, WANG Kai, GUO Chunsheng, CHENG Tao, GONG Liuliu, WANG Qingguang, FAN Xiaodong, LAI Jianxing, LI Zhong, MI Jun, TANG Guibin, GAO Dong, XU Heng, GUO Wei, ZHANG Chunsheng, ZHANG Jianfei, WANG Lei, HUANG Yin, ZHANG Mei, HU Jiangchun, ZHANG Li, XU Linhua, YIN Hua, JIN Hong, WANG Xueran, ZHAO Liang, LI Tao, TANG Hua

Later Supplementary Surveying and Data Processing

Instructor: GUO Xuan, DAI Qiusi, LENG Jie

Master Student

CUI Yanyu, ZHANG Yujia, ZHANG Zhuling, YIN Zixiang, WU Yilin, YANG Haoxiang, ZHANG Xiaolei

Bachelor Student

2006 Year: ZHANG Pu, FAN Juanjuan, WANG Yingnan, JIN Shengxian, DI Yibo, YU Dao, WU Xiaofan, YAN Xinyi, GAO Shuai, CHEN Si, XU Teng, LUO Mi, YAO Yuan, SHUAI Yannan, ZHU Guangxu, WANG Yuqi, GUO Yu, XU Qi, YANG Xiangyu, HE Pengcheng, LIU Shengbing, YANG Chenglei, LIU Jing, CHEN Liran, WANG Cong, SHU Chenxiao, JIANG Yulin, CAI Shangjun, ZHANG Han, YU Yang, ZHANG Wenqing, ZHANG Liang, LIU Jin, FAN Junsu, CHEN Jinfan, HE Lei, ZHANG Huanhuan, PAN Hao, FU Yi

Great Buddha Temple at Tongnan Surveying and Mapping

Instructor: ZHANG Xingguo, LU Rui

Bachelor Student

1989 Year: YAO Bo, HUANG Hongzhou, ZHU Hongbo, DU Yang, SUN Nanfei, WENG Ji, WU Jianhui, WU Daoyuan, YUE Tao, SHEN Wentao, CHEN Yuntao, LIANG Guobiao, ZHENG Shengfeng, MO Haijun, JIA Jianghai, LI Libin, CAO Jin, SONG Lifeng, CHI Xiaobin, HU Jianhua, WU Yili, PANG Lifeng, WANG Mei, XIANG Hua, YUAN Man, WANG Zhiling, GONG Xia, TAN Yinhong, YU Chunhui, CHEN Wei, LI Xiuping, HU Yumei, ZHANG Liming

Later Drawing Arrangement

Bachelor Student

2013 Year: LI Zimo, SUN Kun

Shuanggui Temple at Liangping Surveying and Mapping

Instructor: ZHANG Xingguo, CHEN Wei, GUO Xuan, DAI Qiusi, FENG Di, JIANG Jialong, LIAO Yudi, HU Bin, LENG Jie, WANG Zhiyang, TANG Miao, LIU Zhiyong, LUO Qiang, LI Zhenzhe, XIONG Hailong

Master Student

LIN Chenyang, XIAO Guanlan, ZENG Yu, XU Jiongjiong, WANG Wenjing, CHEN Lin, XIA Min, YANG Chuan, QIN Hao

Bachelor Student

2008Year: ZHANG Han, HE Wenqu, CHEN Ying, GAO Pengfei, CENG Xianming, SHEN Qi, SUN Han, GAO Yun, YANG Li, TANG Kun, CHEN Qiao, WANG Zhe, TANG Mingjie, ZHENG Shizhong, WANG Yang, WANG Geyao, KONG Weimao, ZHANG Xiaolong, PAN Shuang, XU Xiaoyu, ZHOU Jing, MA Xin, LIU Lu, LIANG Yan, ZHOU Lin, XU Miao, MA Xiulian, CHAO Yang, LIU Yao, CHEN Hongwei, SUN Yanchen, WANG Lingyun, YAO Chong, GUO Weiling, ZOU Luying, FAN Yindian,

潼南大佛寺测绘

指导教师

张兴国　鲁　锐

本科生

1989级：姚　波　黄弘州　诸洪波　杜　洋　孙南飞　翁　季　吴健辉　吴道远　岳　涛　沈文涛
陈云涛　梁国标　郑圣峰　莫海军　贾江海　李励斌　曹　进　宋立峰　池小斌　胡建华
吴轶力　庞立峰　王　梅　向　华　袁　满　王志玲　宫　霞　谭音洪　于春辉　陈　蔚
李秀萍　胡育梅　张黎明

后期图纸整理

本科生

2013级：李姿默　孙　锟

梁平双桂堂测绘

指导教师

张兴国　陈　蔚　郭　璇　戴秋思　冯　棣　蒋佳龙　廖屿荻　胡　斌　冷　捷　汪智洋
唐　淼　刘志勇　罗　强　李臻哲　熊海龙

研究生

林晨阳　肖冠兰　曾　宇　徐炯炯　王文婧　陈　琳　夏　敏　杨　钏　秦　浩

本科生

2008级：张　晗　何汶蕖　陈　颖　高鹏飞　曾宪明　沈　奇　孙　涵　高　云　杨　力　唐　坤
陈　桥　王　喆　唐铭杰　郑世中　王　洋　王戈瑶　孔维懋　张小龙　潘　爽　许潇予
周　靓　马　欣　刘　璐　梁　艳　周　琳　徐　苗　马秀莲　晁　阳　刘　瑶　陈泓蔚
孙艳晨　王凌云　姚　翀　郭蔚玲　邹露滢　范银典　张子涵　陈　茗　刘壬可　张令泽
田　丰　黄　川　宋河舟　黄建伟　赵富强　邹蕴波　柴　鑫　范润平　谭　诚　韩艺文
陈　渝　徐徕严　李鑫月　许　晨　杨龚华　王琢著　杜云桥　史文滔　黄霜翼　白苏日吐
李长军　王迪超　黄言言　游　畅　刘亚之　蒲　澍　张小龙　刘　勇　刘志坚　郭绍波
邓　捷　罗延力　蒋　帅　王亦平　刘德清　徐海军　何松育　冯天铭　张　睿　王艺儒
刘　韶　余嘉琦　罗　斌　罗玉熊　李　卓　蔡坤妤　鞠娜娜　朱海华　罗瑾琰　王启慧
袁　烨　秦　朗　陈　果　赵玉立　邓静静　何芸荻　陈　鹏　张　恒　刘　博　李　恒
王　益　刘　逍　陈　程　涂　钦　肖凌骁　郭宇翔　杨　波　许　建　胡　秧　郑伟强
廖礼才　杨　阳　郭伟男　宁小庚　胡卓霖　农　睿　吴星辰　张浥尘　王振文　冉　乐
苏枭枭　秦　岭　马梦迎　周晓宇　赵筱丹　刘　佩　王宇靓　贾慧泉　于晓原　徐　霞
和丹丹　杨　帆

后期补充测绘及数据化资料整理

高鹏飞　齐一聪　张　霁　范银典　陈　果　刘　璐　李鹏飞　滕文皓　欧亚美　何曼琪

ZHANG Zihan, CHEN Ming, LIU Renke, ZHANG Lingze, TIAN Feng, HUANG Chuan, SONG Hezhou, HUANG Jianwei, ZHAO Fuqiang, ZOU Yunbo, CHAI Xin, FAN Runping, TAN Cheng, HAN Yiwen, CHEN Yu, XU Laiyan, LI Xinyue, XU Chen, YANG Gonghua, WANG Zuozhu, DU Yunqiao, SHI Wentao, HUANG Shuangyi, BAI Suritu, LI Zhangjun, WANG Dichao, HUANG Yanyan, YOU Chang, LIU Yazhi, PU Shu, ZHANG Xiaolong, LIU Yong, LIU Zhijian, GUO Shaobo, DENG Jie, LUO Yanli, JIANG Shuai, WANG Yiping, LIU Deqing, XU Haijun, HE Songyu, FENG Tianming, ZHANG Rui, WANG Yiru, LIU Shao, YU Jiaqi, LUO Bin, LUO Yuxiong, LI Zhuo, CAI Kunyu, JU Nana, ZHU Haihua, LUO Jinyan, WANG Qihui, YUAN Ye, QIN Lang, CHEN Guo, ZHAO Yuli, DENG Jingjing, HE Yundi, CHEN Peng, ZHANG Heng, LIU Bo, LI Heng, WANG Yi, LIU Xiao, CHEN Cheng, TU Qin, XIAO Lingxiao, GUO Yuxiang, YANG Bo, XU Jian, HU Yang, ZHENG Weiqiang, LIAO Licai, YANG Yang, GUO Weinan, NING Xiaogeng, HU Zhuolin, NONG Rui, WU Xingchen, ZHANG Yichen, WANG Zhenwen, RAN Le, SU Xiaoxiao, QIN Ling, MA Mengying, ZHOU Xiaoyu, ZHAO Xiaodan, LIU Pei, WANG Yujing, JIA Huiquan, YU Xiaoyuan, XU Xia, HE Dandan, YANG Fan

Later Supplementary Surveying and Data Processing

GAO Pengfei, QI Yicong, ZHANG Ji, FAN Yindian, CHEN Guo, LIU Lu, LI Pengfei, TENG Wenhao, OU Yamei, HE Manqi

图书在版编目（CIP）数据

巴蜀佛寺 = BA-SHU BUDDHIST ARCHITECTURE：汉英对照 / 张兴国等主编；重庆大学建筑城规学院编写. —北京：中国建筑工业出版社，2019.12
（中国古建筑测绘大系. 宗教建筑）
ISBN 978-7-112-24547-5

Ⅰ.①巴… Ⅱ.①张…②重… Ⅲ.①寺庙－宗教建筑－建筑艺术－四川－图集 Ⅳ.①TU-885

中国版本图书馆CIP数据核字（2019）第284623号

丛书策划 / 王莉慧
责任编辑 / 李 鸽 刘 川
英文翻译 / 尚 晋 金 田
书籍设计 / 付金红
责任校对 / 王 烨

中国古建筑测绘大系 · 宗教建筑
巴蜀佛寺
BA-SHU BUDDHIST ARCHITECTURE
重庆大学建筑城规学院 编写
张兴国 冯 棣 郭 璇 汪智洋 主编

Traditional Chinese Architecture Surveying and Mapping Series: Religious Architecture
BA-SHU BUDDHIST ARCHITECTURE
Compiled by School of Architecture and Urban Planning, Chongqing University
Edited by ZHANG Xingguo, FENG Di, GUO Xuan, WANG Zhiyang

*
中国建筑工业出版社出版、发行（北京海淀三里河路9号）
各地新华书店、建筑书店经销
北京雅盈中佳图文设计公司制版
北京雅昌艺术印刷有限公司印刷
*
开本：787毫米×1092毫米 横1/8 印张：39½ 字数：1057千字
2022年12月第一版 2022年12月第一次印刷
定价：**298.00**元
ISBN 978-7-112-24547-5
（35219）